BMW 6 Series
Gold Portfolio
1976-1989

Compiled by
R.M.Clarke

ISBN 1 85520 3618

 **BROOKLANDS BOOKS LTD.**
P.O. BOX 146, COBHAM,
SURREY, KT11 1LG. UK

Brooklands Books

MOTORING

BROOKLANDS ROAD TEST SERIES

Abarth Gold Portfolio 1950-1971
AC Ace & Aceca 1953-1983
Alfa Romeo Giulietta Gold Portfolio 1954-1965
Alfa Romeo Giulia Berlinas 1962-1976
Alfa Romeo Giulia Coupés 1963-1976
Alfa Romeo Giulia Coupés Gold P. 1963-1976
Alfa Romeo Spider 1966-1990
Alfa Romeo Spider Gold Portfolio 1966-1991
Alfa Romeo Alfasud 1972-1984
Alfa Romeo Alfetta Gold Portfolio 1972-1987
Alfa Romeo Alfetta GTV6 1980-1986
Allard Gold Portfolio 1937-1959
Alvis Gold Portfolio 1919-1967
AMX & Javelin Muscle Portfolio 1968-1974
Armstrong Siddeley Gold Portfolio 1945-1960
Aston Martin Gold Portfolio 1948-1971
Aston Martin Gold Portfolio 1972-1985
Aston Martin Gold Portfolio 1985-1995
Audi Quattro Gold Portfolio 1980-1991
Austin A30 & A35 1951-1962
Austin Healey 100 & 100/6 Gold P. 1952-1959
Austin Healey 3000 Gold Portfolio 1959-1967
Austin Healey Sprite 1958-1971
Barracuda Muscle Portfolio 1964-1974
BMW 1600 Collection No.1 1966-1981
BMW 2002 Gold Portfolio 1968-1976
BMW 316, 318, 320 (4 cyl.) Gold P. 1975-1990
BMW 320, 323, 325 (6 cyl.) Gold P. 1977-1990
BMW M Series Performance Portfolio 1976-1993
BMW 5 Series Gold Portfolio 1981-1987
BMW 6 Series Gold Portfolio 1976-1989
Bricklin Gold Portfolio 1974-1975
Bristol Cars Gold Portfolio 1946-1992
Buick Automobiles 1947-1960
Buick Muscle Cars 1965-1970
Cadillac Allanté 1986-1993
Cadillac Automobiles 1949-1959
Cadillac Automobiles 1960-1969
Charger Muscle Portfolio 1966-1974
Checker ☆ Limited Edition
Chevrolet 1955-1957
Impala & SS Muscle Portfolio 1958-1972
Chevrolet Corvair 1959-1969
Chevy II & Nova SS Muscle Portfolio 1962-1974
Chevy El Camino & SS 1959-1987
Chevelle & SS Muscle Portfolio 1964-1972
Chevrolet Muscle Cars 1966-1971
Chevy Blazer 1969-1981
Chevrolet Corvette Gold Portfolio 1953-1962
Chevrolet Corvette Sting Ray Gold P. 1963-1967
Chevrolet Corvette Gold Portfolio 1968-1977
High Performance Corvettes 1983-1989
Camaro Muscle Portfolio 1967-1973
Chevrolet Camaro Z28 & SS 1966-1973
Chevrolet Camaro & Z28 1973-1981
High Performance Camaros 1982-1988
Chrysler 300 Gold Portfolio 1955-1970
Chrysler Valiant 1960-1962
Citroen Traction Avant Gold Portfolio 1934-1957
Citroen 2CV Gold Portfolio 1948-1989
Citroen DS & ID 1955-1975
Citroen DS & ID Gold Portfolio 1955-1975
Citroen SM 1970-1975
Cobras & Replicas 1962-1983
Shelby Cobra Gold Portfolio 1962-1969
Cobras & Cobra Replicas Gold P. 1962-1989
Cunningham Automobiles 1951-1955
Daimler SP250 Sports & V-8 250 Saloon Gold P. 1959-1969
Datsun Roadsters 1962-1971
Datsun 240Z 1970-1973
Datsun 280Z & ZX 1975-1983
DeLorean Gold Portfolio 1977-1995
Dodge Muscle Cars 1967-1970
Dodge Viper on the Road
ERA Gold Portfolio 1934-1994
Excalibur Collection No.1 1952-1981
Facel Vega 1954-1964
Ferrari Dino 1965-1974
Ferrari Dino 308 & Mondial Gold Portfolio1974-1985
Ferrari 328 • 348 • Mondial Gold Portfolio 1986-1994
Fiat 500 Gold Portfolio 1936-1972
Fiat 600 & 850 Gold Portfolio 1955-1972
Fiat Pininfarina 124 & 2000 Spider 1968-1985
Fiat X1/9 Gold Portfolio1973-1989
Fiat Abarth Performance Portfolio 1972-1987
Ford Consul, Zephyr, Zodiac Mk.I & II 1950-1962
Ford Zephyr, Zodiac, Executive, Mk.III & Mk.IV 1962-1971
Ford Cortina 1600E & GT 1967-1970
High Performance Capris Gold Portfolio 1969-1987
Capri Muscle Portfolio 1974-1987
High Performance Fiestas 1979-1991
High Performance Escorts Mk.I 1968-1974
High Performance Escorts Mk.II 1975-1980
High Performance Escorts 1980-1985
High Performance Escorts 1985-1990
High Performance Sierras & Merkurs
Gold Portfolio 1983-1990
Ford Automobiles 1949-1959
Ford Fairlane 1955-1970
Ford Ranchero 1957-1959
Ford Thunderbird 1955-1957
Ford Thunderbird 1958-1963
Ford GT40 Gold Portfolio 1964-1987
Ford Bronco 1966-1977
Ford Bronco 1978-1988
Goggomobil ☆ Limited Edition
Holden 1948-1962
Honda CRX 1983-1987
International Scout Gold Portfolio 1961-1980
Isetta 1953-1964
ISO & Bizzarrini Gold Portfolio 1962-1974
Jaguar and SS Gold Portfolio 1931-1951
Jaguar XK120, 140, 150 Gold P. 1948-1960
Jaguar Mk.VII, VIII, IX, X, 420 Gold P. 1950-1970
Jaguar Mk.1 & Mk.2 Gold Portfolio 1959-1969
Jaguar C-Type & D-Type ☆ Limited Edition
Jaguar E-Type Gold Portfolio 1961-1971
Jaguar E-Type V-12 1971-1975
Jaguar S-Type & 420 ☆ Limited Edition
Jaguar XJ12, XJ5.3, V12 Gold P. 1972-1990
Jaguar XJ6 Series I & II Gold P. 1968-1979
Jaguar XJ6 Series III Perf. Portfolio 1979-1986
Jaguar XJ6 Gold Portfolio 1986-1994
Jaguar XJS Gold Portfolio 1975-1988
Jaguar XJS Gold Portfolio 1988-1995
Jeep CJ5 & CJ6 1960-1976
Jeep CJ5 & CJ7 1976-1986
Jensen Cars 1946-1967
Jensen Cars 1967-1979
Jensen Interceptor Gold Portfolio 1966-1986
Jensen Healey 1972-1976
Lagonda Gold Portfolio 1919-1964
Lamborghini Countach & Urraco 1974-1980
Lamborghini Countach & Jalpa 1980-1985
Lancia Aurelia & Flaminia Gold Portfolio 1950-1970
Lancia Fulvia Gold Portfolio 1963-1976
Lancia Beta Gold Portfolio 1972-1984
Lancia Delta Gold Portfolio 1979-1994
Lancia Stratos 1972-1985
Land Rover Series I 1948-1958
Land Rover Series II & IIa 1958-1971
Land Rover Series III 1971-1985
Land Rover 90 110 Defender Gold Portfolio 1983-1994
Land Rover Discovery 1989-1994
Land Rover Story Part One 1948-1971
Lincoln Gold Portfolio 1949-1960
Lincoln Continental 1961-1969
Lincoln Continental 1969-1976
Lotus Sports Racers Gold Portfolio 1953-1965
Lotus Seven Gold Portfolio 1957-1974
Lotus Caterham Seven Gold Portfolio 1974-1995
Lotus Elan Gold Portfolio 1962-1974
Lotus Elan Collection No. 2 1963-1972
Lotus Elan & SE 1989-1992
Lotus Cortina Gold Portfolio 1963-1970
Lotus Europa Gold Portfolio 1966-1975
Lotus Elite & Eclat 1974-1982
Lotus Turbo Esprit 1980-1986
Marcos Cars 1960-1988
Maserati 1965-1975
Mazda Miata-MX-5 Performance Portfolio 1989-1996
Mazda RX-7 Gold Portfolio 1978-1991
Mercedes 190 & 300 SL 1954-1963
Mercedes G Wagen 1981-1994
Mercedes S & 600 1965-1972
Mercedes S Class 1972-1979
Mercedes 230 • 250 • 280SL Gold Portfolio 1963-1971
Mercedes SLs & SLCs Gold Portfolio 1971-1989
Mercedes SLs Performance Portfolio 1989-1994
Mercury Muscle Cars 1966-1971
Messerschmitt Gold Portfolio 1954-1964
MG Gold Portfolio 1929-1939
MG TA & TC Gold Portfolio 1936-1949
MG TD & TF Gold Portfolio 1949-1955
MGA & Twin Cam Gold Portfolio 1955-1962
MG Midget Gold Portfolio 1961-1979
MGB Roadsters 1962-1980
MGB MGC & V8 Gold Portfolio 1962-1980
MGB GT 1965-1980
MG Y-Type & Magnette ZA/ZB ☆ Limited Edition
Mini Gold Portfolio 1959-1969
Mini Gold Portfolio 1969-1980
High Performance Minis Gold Portfolio 1960-1973
Mini Cooper Gold Portfolio 1961-1971
Mini Moke Gold Portfolio 1964-1994
Mopar Muscle Cars 1964-1967
Morgan Three-Wheeler Gold Portfolio 1910-1952
Morgan Plus 4 & Four 4 Gold P. 1936-1967
Morgan Cars 1960-1970
Morgan Cars Gold Portfolio 1968-1989
Morris Minor Collection No. 1 1948-1980
Shelby Mustang Muscle Portfolio 1965-1970
High Performance Mustang IIs 1974-1978
High Performance Mustangs 1982-1988
Nash-Austin Metropolitan Gold P. 1954-1962
Oldsmobile Automobiles 1955-1963
Oldsmobile Muscle Cars 1964-1971
Oldsmobile Toronado 1966-1978
Opel GT Gold Portfolio 1968-1973
Packard Gold Portfolio 1946-1958
Pantera Gold Portfolio 1970-1989
Panther Gold Portfolio 1972-1990
Plymouth Muscle Cars 1966-1971
Pontiac Tempest & GTO 1961-1965
Pontiac Muscle Cars 1966-1972
Pontiac Firebird & Trans-Am 1973-1981
High Performance Firebirds 1982-1988
Pontiac Fiero 1984-1988
Porsche 356 Gold Portfolio 1953-1965
Porsche 911 1965-1969
Porsche 911 1970-1972
Porsche 911 1973-1977
Porsche 911 Turbo 1975-1984
Porsche 911 SC & Turbo Gold Portfolio 1978-1983
Porsche 911 Carrera & Turbo Gold P. 1984-1989
Porsche 924 Gold Portfolio 1975-1988
Porsche 928 Performance Portfolio 1977-1994
Porsche 944 Gold Portfolio 1981-1991
Range Rover Gold Portfolio 1970-1985
Range Rover Gold Portfolio 1986-1995
Reliant Scimitar 1964-1986
Renault Alpine Gold Portfolio 1958-1994
Riley Gold Portfolio 1924-1939
Riley 1.5 & 2.5 Litre Gold Portfolio 1945-1955
Rolls Royce Silver Cloud & Bentley 'S' Series
Gold Portfolio 1955-1965
Rolls Royce Silver Shadow Gold P. 1965-1980
Rolls Royce & Bentley Gold P. 1980-1989
Rover P4 1949-1959
Rover P4 1955-1964
Rover 3 & 3.5 Litre Gold Portfolio 1958-1973
Rover 2000 & 2200 1963-1977
Rover 3500 1968-1977
Rover 3500 & Vitesse 1976-1986
Saab Sonett Collection No.1 1966-1974
Saab Turbo 1976-1983
Studebaker Gold Portfolio 1947-1966
Studebaker Hawks & Larks 1956-1963
Avanti 1962-1990
Sunbeam Tiger & Alpine Gold P. 1959-1967
Toyota MR2 1984-1988
Toyota Land Cruiser 1956-1984
Triumph Dolomite Sprint ☆ Limited Edition
Triumph TR2 & TR3 Gold Portfolio 1952-1961
Triumph TR4, TR5, TR250 1961-1968
Triumph TR6 Gold Portfolio 1969-1976
Triumph TR7 & TR8 Gold Portfolio 1975-1982
Triumph Herald 1959-1971
Triumph Vitesse 1962-1971
Triumph Spitfire Gold Portfolio 1962-1980
Triumph 2000, 2.5, 2500 1963-1977
Triumph GT6 Gold Portfolio 1966-1974
Triumph Stag Gold Portfolio 1970-1977
TVR Gold Portfolio 1959-1986
TVR Performance Portfolio 1986-1994
VW Beetle Gold Portfolio 1935-1967
VW Beetle Gold Portfolio 1968-1991
VW Beetle Collection No.1 1970-1982
VW Karmann Ghia 1955-1982
VW Bus, Camper, Van 1954-1967
VW Bus, Camper, Van 1968-1979
VW Bus, Camper, Van 1979-1989
VW Scirocco 1974-1981
VW Golf GTI 1976-1986
Volvo PV444 & PV544 1945-1965
Volvo Amazon-120 Gold Portfolio 1956-1970
Volvo 1800 Gold Portfolio 1960-1973
Volvo 140 & 160 Series Gold Portfolio 1966-1975

Forty Years of Selling Volvo

BROOKLANDS ROAD & TRACK SERIES

Road & Track on Alfa Romeo 1964-1970
Road & Track on Alfa Romeo 1971-1976
Road & Track on Aston Martin 1962-1990
R & T on Auburn Cord and Duesenburg 1952-84
Road & Track on Audi & Auto Union 1952-1980
Road & Track on Audi & Auto Union 1980-1986
Road & Track on Austin Healey 1953-1970
Road & Track on BMW Cars 1966-1974
Road & Track on BMW Cars 1975-1978
Road & Track on BMW Cars 1979-1983
R & T on Cobra, Shelby & Ford GT40 1962-1992
Road & Track on Corvette 1953-1967
Road & Track on Corvette 1968-1982
Road & Track on Corvette 1982-1986
Road & Track on Corvette 1986-1990
Road & Track on Ferrari 1975-1981
Road & Track on Ferrari 1981-1984
Road & Track on Ferrari 1984-1988
Road & Track on Fiat Sports Cars 1968-1987
Road & Track on Jaguar 1950-1960
Road & Track on Jaguar 1961-1968
Road & Track on Jaguar 1968-1974
Road & Track on Jaguar 1974-1982
Road & Track on Jaguar 1983-1989
Road & Track on Lamborghini 1964-1985
Road & Track on Lotus 1972-1981
Road & Track on Maserati 1975-1983
R & T on Mazda RX7 & MX5 Miata 1986-1991
Road & Track on Mercedes 1952-1962
Road & Track on Mercedes 1963-1970
Road & Track on Mercedes 1971-1979
Road & Track on Mercedes 1980-1987
Road & Track on MG Sports Cars 1949-1961
Road & Track on MG Sports Cars 1962-1980
Road & Track on Mustang 1964-1977
R & T on Nissan 300-ZX & Turbo 1984-1989
Road & Track on Pontiac 1960-1983
Road & Track on Porsche 1951-1967
Road & Track on Porsche 1968-1971
Road & Track on Porsche 1972-1975
Road & Track on Porsche 1975-1978
Road & Track on Porsche 1979-1982
Road & Track on Porsche 1985-1988
R & T on Rolls Royce & Bentley 1950-1965
R & T on Rolls Royce & Bentley 1966-1984
Road & Track on Saab 1972-1992
R & T on Toyota Sports & GT Cars 1966-1984
R & T on Triumph Sports Cars 1953-1967
R & T on Triumph Sports Cars 1967-1974
R & T on Triumph Sports Cars 1974-1982
Road & Track on Volkswagen 1951-1968
Road & Track on Volkswagen 1968-1978
Road & Track on Volkswagen 1978-1985
Road & Track on Volvo 1957-1974
Road & Track on Volvo 1977-1994
R & T - Henry Manney at Large & Abroad
R & T - Peter Egan's "Side Glances"
R & T - Peter Egan "At Large"

BROOKLANDS CAR AND DRIVER SERIES

Car and Driver on BMW 1955-1977
Car and Driver on BMW 1977-1985
C and D on Cobra, Shelby & Ford GT40 1963-84
Car and Driver on Corvette 1978-1982
Car and Driver on Corvette 1983-1988
C and D on Datsun Z 1600 & 2000 1966-1984
Car and Driver on Ferrari 1955-1962
Car and Driver on Ferrari 1963-1975
Car and Driver on Ferrari 1976-1983
Car and Driver on Mopar 1956-1967
Car and Driver on Mopar 1968-1975
Car and Driver on Mustang 1964-1972
Car and Driver on Pontiac 1961-1975
Car and Driver on Porsche 1955-1962
Car and Driver on Porsche 1963-1970
Car and Driver on Porsche 1970-1976
Car and Driver on Porsche 1977-1981
Car and Driver on Porsche 1982-1986
Car and Driver on Saab 1956-1985
Car and Driver on Volvo 1955-1986

BROOKLANDS PRACTICAL CLASSICS SERIES

PC on Austin A40 Restoration
PC on Land Rover Restoration
PC on Metalworking in Restoration
PC on Midget/Sprite Restoration
PC on MGB Restoration
PC on Sunbeam Rapier Restoration
PC on Triumph Herald/Vitesse
PC on Spitfire Restoration
PC on Beetle Restoration
PC on 1930s Car Restoration

BROOKLANDS HOT ROD 'MUSCLECAR & HI-PO ENGINES' SERIES

Chevy 265 & 283
Chevy 302 & 327
Chevy 348 & 409
Chevy 350 & 400
Chevy 396 & 427
Chevy 454 thru 512
Chrysler Hemi
Chrysler 273, 318, 340 & 360
Chrysler 361, 383, 400, 413, 426, 440
Ford 289, 302, Boss 302 & 351W
Ford 351C & Boss 351
Ford Big Block

BROOKLANDS RESTORATION SERIES

Auto Restoration Tips & Techniques
Basic Bodywork Tips & Techniques
Camaro Restoration Tips & Techniques
Chevrolet High Performance Tips & Techniques
Chevy Engine Swapping Tips & Techniques
Chevy-GMC Pickup Repair
Chrysler Engine Swapping Tips & Techniques
Engine Swapping Tips & Techniques
Ford Pickup Repair
How to Build a Street Rod
Land Rover Restoration Tips & Techniques
MG 'T' Series Restoration Guide
MGA Restoration Guide
Mustang Restoration Tips & Techniques
Performance Tuning - Chevrolets of the '60s
Performance Tuning - Pontiacs of the '60s

MOTORCYCLING

BROOKLANDS ROAD TEST SERIES

AJS & Matchless Gold Portfolio 1945-1966
BSA Twins A7 & A10 Gold Portfolio 1946-1962
BSA Twins A50 & A65 Gold Portfolio 1962-1973
Ducati Gold Portfolio 1960-1974
Ducati Gold Portfolio 1974-1978
Laverda Gold Portfolio 1967-1977
Norton Commando Gold Portfolio 1968-1977
Triumph Bonneville Gold Portfolio 1959-1983

BROOKLANDS CYCLE WORLD SERIES

Cycle World on BMW 1974-1980
Cycle World on BMW 1981-1986
Cycle World on Ducati 1982-1991
Cycle World on Harley-Davidson 1962-1968
Cycle World on Harley-Davidson 1978-1983
Cycle World on Harley-Davidson 1983-1987
Cycle World on Harley-Davidson 1987-1990
Cycle World on Harley-Davidson 1990-1992
Cycle World on Honda 1962-1967
Cycle World on Honda 1968-1971
Cycle World on Honda 1971-1974
Cycle World on Husqvarna 1966-1976
Cycle World on Husqvarna 1977-1984
Cycle World on Kawasaki 1966-1971
Cycle World on Kawasaki Off-Road Bikes 1972-1979
Cycle World on Kawasaki Street Bikes 1972-1976
Cycle World on Norton 1962-1971
Cycle World on Suzuki 1962-1970
Cycle World on Suzuki Off-Road Bikes 1971-1976
Cycle World on Suzuki Street Bikes 1971-1976
Cycle World on Triumph 1967-1972
Cycle World on Yamaha 1962-1969
Cycle World on Yamaha Off-Road Bikes 1970-1974
Cycle World on Yamaha Street Bikes 1970-1974

MILITARY

BROOKLANDS MILITARY VEHICLES SERIES

Allied Military Vehicles No.2 1941-1946
Complete WW2 Military Jeep Manual
Dodge Military Vehicles No.1 1940-1945
Hail To The Jeep
Land Rovers in Military Service
Military & Civilian Amphibians 1940-1990
Off Road Jeeps: Civ. & Mil. 1944-1971
US Military Vehicles 1941-1945
US Army Military Vehicles WW2-TM9-2800
VW Kubelwagen Military Portfolio 1940-1990
WW
2 Jeep Military Portfolio 1941-1945

RACING

Le Mans -The Jaguar Years- 1949 -1957

2396

CONTENTS

Page	Title	Publication	Date	Year
5	Munich Masterpiece	Motor Sport	May	1976
6	BMW's New Coupé	Autocar	Mar 13	1976
8	A New Coupé from BMW	Road & Track	June	1976
12	New BMWs	Wheels	May	1976
15	A BMW to Boast About	Car	April	1976
19	BMW 630CSi	Road & Track Special		1977
20	BMW 633CSi Road Test	Autocar	Oct 16	1976
26	BMW 630CSi Road Test	Road & Track	June	1977
30	BMW 630CSi Road Test	Road Test	July	1977
34	Near Perfection: The BMW 633CSi Road Test	Motor Sport	Aug	1977
36	Luxotouring Comparison Test Jaguar XJ-S vs. Mercedes-Benz 450SLC vs. BMW 630CSi	Car and Driver	Dec	1977
42	Horses for Courses Mercedes-Benz 450SLC vs. BMW 635CSi	Motor	Nov 11	1978
44	BMW Alpina 630 Road Test	Motor Trend	Jan	1979
47	Almost a Batmobile 633CSi	Autosport	July 5	1979
48	BMW 635CSi Road Test	Autocar	Jan 6	1979
54	Understated Elegance 633CSi	Sports Car Graphic	Nov	1980
58	BMW 628CSi Road Test	Motor	Dec 27	1980
62	BMW 633CSi Road Test	Road & Track Special		1982
66	Any Other Business 628CSi	Car	Jan	1982
67	Better BMW 635CSi	Autocar	June	1982
68	The Sporting Life 635CSi Road Test	Autosport	Sept 9	1982
70	BMW 633CSi Turbo Road Test	Motor Trend	Sept	1982
71	BMW 635CSi Driving Impressions	Road & Track	Oct	1982
74	BMW 6 and 7 Series Buying Secondhand	Autocar	July 9	1983
78	BMW 635CSi Coupé Automatic Road Test	Car South Africa	Sept	1983
82	Graceful Racer 635CSi Track Test	Autosport	Nov 17	1983
84	No Ordinary Day 635CSi Race Test	Autosport	Nov 17	1983
86	M-azing M635CSi	Autosport	Nov 17	1983
87	Flying the Coupé 635CSi Road Test	Drive	May	1983
91	BMW M635CSi	Car and Driver	July	1984
92	Big Black Bimmer 628CSi	Sporting Cars	Dec	1983
96	BMW 635CSi Automatic Road Test	Autocar	April 28	1984
98	BMW M635CSi Road Test	Autocar	April 28	1984
104	BMW 628CSi Road Test	Motor	Sept 15	1984
106	BMW M635CSi Road Test	Motor Trend	Feb	1985
109	BMW 635CSi Road Test	Car and Driver	Feb	1985
113	BMW M635CSi	Performance Car	April	1985
114	Test Match BMW 630CSi vs. Jaguar XJ-S Comparison Test	Fast Lane	June	1985
118	Inside JPS-Team BMW	Modern Motor	Sept	1985
123	Power-Fix 6 M635CSi	Sports Car World	Aug	1985
124	Quandary at Furnace Creek Comparison Test BMW 635CSi vs. Jaguar XJ-S vs. Mercedes-Benz 560SEC	Road & Track	July	1986
132	Duel of the Titans Comparison Test BMW 635CSi vs. Mercedes-Benz 560SEC	Wheels	Sept	1986
138	The M6 Alternative M635CSi	Car	Oct	1986
141	Sisters Under the Skin? M635CSi and 507 Road Test	Motor Sport	Feb	1987
145	4 Seasons Test BMW 635CSi Road Test	Automobile Magazine	Aug	1987
148	BMW M6 M635CSi Road Test	Motor Trend	Sept	1987
152	BMW 635CSi Road Test	Autocar	April 20	1988
156	Power Trip! BMW M6 vs. 928 S4 Comparison Test	Motor Trend	April	1988
164	BMW M635CSi Road Test	Autocar	Jan 18	1989
168	Practical Supercar	Classic Cars	Aug	1989
170	BMW 635CSi	Popular Classics	July	1990
172	Waving the Wand Alpina 635CSi	Modern Motor	Jan	1989

ACKNOWLEDGEMENTS

BMW's reputation picked up enormously during the Sixties after the New Class 1500 saloons and their successors demonstrated the depth of the company's engineering expertise, and by the mid-Seventies there was no doubt that this once-ailing concern could attempt to challenge Mercedes-Benz for the domination of the Autobahns. One of the ways in which it did so was by building a top-of-the-range, high-performance luxury coupé - the 6-series - which is as widely respected today as it was when new. It was an obvious subject for us at Brooklands Books to cover in our Gold Portfolio series.

Regular readers of Brooklands books will know that books like this one are an attempt to create a living archive and to make available to enthusiasts material about their cars which has become hard to find. In putting such books together, we depend on the generosity and understanding of those who originally published the copyright material we reproduce. For material in the present volume, our thanks therefore go to the owners of *Autocar, Automobile Magazine, Autosport, Car, Car and Driver, Car South Africa, Classic Cars, Drive, Fast Lane, Modern Motor, Motor, Motor Sport, Motor Trend, Performance Car, Popular Classics, Road & Track, Road Test, Sporting Cars, Sports Car Graphic, Sports Car World* and *Wheels*.

R.M. Clarke

BMW's 6-series coupés (the E24 models) were announced at the Geneva Show in March 1976. They had a shortened 5-series saloon floorpan with a modified rear suspension, and stylist Paul Bracq's sure touch produced an elegant car with a broad and purposeful stance and a strong family resemblance to the E9 CS coupés it was to replace. With the exception of additional spoilers and a modified rear bumper wraparound (the latter in 1982), that styling remained unchanged during a production run of 16 years - and the 6-series looked just as much the wealthy man's grand tourer in 1989 as it had done on its introduction.

The first cars were carburetted 630CS and injected 633CSi models, which used engines seen in the older E9s, but from 1978 there was also the formidable 635CSi, a car designed specifically to out-perform the latest 5-litre Mercedes coupés. The carburetted 630CS was replaced in August 1979 by the similarly powerful injected 628 CSi. The 3.5 litre models meanwhile showed up well in European Touring Car championship events, and prompted BMW's Motorsport division to announce the special high-performance M635CSi derivative in 1983. Final changes then came in May 1987, when the 628CSi disappeared from the range and the remaining cars had a mild front-end facelift.

In the US, the dollar exchange rate pumped up the already high prices of the 6-series cars. The first cars on sale there were 630CSi models - essentially the European 630CS with extended "safety" bumpers and with fuel injection to give better control of exhaust emissions. These arrived in 1977 and lasted just one year, being replaced by the more torquey Federalised 633CSi. Sadly, American never did get the 635CSi, however.

The last 6-series coupés were built early in 1989, but BMW did not replace these big Autobahn-stormers directly. There was a decent interval before the E31 8-series coupés arrived in 1990 - and they were rather different cars in a number of ways. So different, in fact, that the 6-series were slow to lose their market values and very quickly became accepted as enthusiasts' cars. The articles in this book help to explain just what was - and still is - so special about these big coupés.

James Taylor

Munich Masterpiece

BMW introduce the 6-series coupé

FOR SUCH an expensive car, BMW's trend-setting, six-cylinder, two-door coupé series, from 2.5CS to 3.0CSi and CSL, enjoyed extraordinary success, selling 45,000 examples, helped by its image as the touring car racing King of Europe. Production ended last December, rather prematurely we thought, until we were introduced to its 6-series successor in Munich and Marbella. Aesthetically and mechanically the new model shows a worthwhile improvement and keeps BMW well in contention with the opposition, from which, like Jaguar, they single out the Mercedes 350/450SLC as arch-enemy.

Of the two versions of the 6-series coupé, the 3-litre, carburetter 630CS and the 3.2-litre, fuel injection 633CSi, only the latter will be available in Britain—and not until September. Pound and Deutschmark permitting it is likely to cost some £12,000, inclusive of air-conditioning and electric front windows, optional in other markets. In essence, the 6-series follows the same straight-six, independently-sprung path as before, but with many improvements. The body is entirely new, built by Karmann, who'll also be assembling the cars at the rate of 8,000 per annum, 500 of which will head for the UK. The 3-litre saloons continue unchanged.

Nose styling of the slightly wedge-shaped body comes from the mid-engined Turbo prototype. The vast window area is of Parsol bronze-tinted glass; the side areas are no longer pillar-less, the gap being split by a roll-over bar, which helps give better window sealing. The body is designed with safety crush zones and strengthening members and gains 69% in torsional rigidity over the old model. There is some improvement in interior space and the boot gains an extra 3 cu.ft. Although the frontal area change is fractional, aerodynamics are much improved.

The longer-stroke 3,210 c.c. straight-six of the 633CSi gives an identical 200 b.h.p. at 5,500 r.p.m. to the 3.0 CSi. Torque is improved from 200.4 lb. ft. at 4,300 r.p.m. to 209.7 lb. ft. at 4,250 r.p.m. The compression ratio is reduced from 9.5 to 9.0 to 1, but improved efficiency comes from combustion chamber modifications and a change from D-Jetronic to L-Jetronic Bosch fuel-injection. Transistorised ignition is fitted. An extra 40 kg. in weight will probably outweigh any performance gains.

On the chassis side, the track is increased front and rear. At the front the caster angle is reduced and kingpin inclination increased for better direction stability. Instead of separate coil springs and dampers at the rear, spring/damper struts are fitted, along with an anti-roll bar. The ventilated front discs are of 11.0 in., instead of 10.7 in., diameter.

By far the most important chassis improvement is the use of the new ZF ball-and-nut hydraulic steering system, with power assistance varying with engine speed—more power at low speed and vice versa. An appreciably higher ratio is used, too.

Inside, the now familiar BMW curved facia includes an all-systems-go check device : press a button and seven lights should light up if fluid levels and so on are all right.

All seats have headrests, the rear ones including lidded oddments trays. The driver's seat adjusts for height. An electrically controlled door mirror is a very useful gimmick.

On the German autobahn or in the hills of southern Spain there was no doubting the stability, excellent braking and the high performance of this smooth, 135 m.p.h. straight-six. The handling was excellent, but the suspension too soft in the mountains (there is a stiffer option pack) and we'd have appreciated a limited slip differential, which will be standard on UK cars. Most impressive of all was the precision and feel of the power steering, though a shade more caster return wouldn't go amiss. Minor modifications to the linkage have improved the gearbox, too.

We look forward to publishing a full road test when this impressive coupé appears on the British market.—C.R.

New at Geneva

BMW's new Coupé

No new technical barriers broken by new coupé to replace defunct 3.0 CS model. General all round improvement based on brand new body, designed by BMW to be built by Karmann, as was the predecessor. Steering features progressive assistance and small but telling differences to track and suspension improve roadholding and ride. Comprehensive check of major systems available through facia-mounted test panel. Certain to be more expensive than the earlier car; deliveries should start to the UK by the end of the summer

by Andrew Shanks

AFTER SPENDING the latter end of the last decade in developing their engines and running gear, BMW have been amongst the most prolific introducers of new cars in the first half of this decade. First the 500 Series reintroduced the company in the medium/large sector of the family car market and then the 300 Series gave a fresh new competitor in the stylish 2-door family car range in which they had been so strongly served by the 1600/2002 coupé for many years. Now we have a replacement for the least dated of the familiar range, the 3.0CS coupé.

The new 600 series coupé comes in two model versions, the 630 with the same engine as the superceded 3.0CS and the 633 which shares its 3.2-litre engine with the 3.3Li long wheelbase version of BMW's 4-door large car range. Unlike the previous big coupé, the new car is the work of BMW's own styling studios and the striking design is surprisingly different from its predecessor while offering very similar accommodation.

Clearly, the first question to answer is why the 3.0CS model required replacement at all, considering its continuing sales success. The main reason is one of passenger safety legislation for the previous pillarless design was unsuitable for the passing of the stringent US Federal rollover and side impact standards. These have been met in the new design by the adoption of study B-post and of horizontal strengthening beams in the doors. But while seeking to improve in this most important of areas, the BMW engineers have incorporated a number of improvements in the detail design too.

In place of Bosch D-Jetronic fuel injection, the new 633CSi has the later and more efficient L-Jetronic system which as well as possessing superior potential fuel economy is more efficient in emissions control as well. As a parallel aid for improved efficiency, the 633CSi has contactless ignition and a two-pipe exhaust system of greater volume than that of the old car. While no gain in peak power output has been achieved (or sought), the torque is improved by 10 lb. ft. to 210 lb. ft. though the engine speed at which it is developed is down by a marginal 50 rpm to 4,250 rpm.

At this point, it is as well to explain that although the new 600 series includes the 630 and 633 models, BMW say only the latter will be sold in the UK and the two cars differ only in their power units and associated services though the lighter weight of the 3-litre engine does allow the smaller-engined car a higher maximum payload.

In this description, therefore, all engine details can be taken as applying to the 3.2-litre engine whose extra cubic capacity results from a lengthening of the stroke from 80 to 86 mm; the bore of 89 mm is unchanged and thus the engine remains over-square. An engine oil cooler is fitted with a pressure-sensitive by-pass to protect the cooler and its pipes when the oil is cold and thick.

Above: Three-quarter rear treatment is very different from superceded 3.0CS models. The inclusion of a substantial 'B' post has enabled the quarter panels to be usefully smaller

New transmission

Though the 4-speed Getrag manual gearbox is as before, an innovation is the ZF-BMW developed 3-speed automatic gearbox that replaces the Borg-Warner unit. Previously, automatic had only been available on the 180 bhp carburettor engine but now the ZF gearbox can be had on the top-of-the-range engine too. Old and new gearboxes use a torque convertor and the most important difference is the lowering of the gearing of Low and Intermediate to cope with the slightly increased weight of the new cars. Top gear is still direct and the final drive is unchanged and thus overall gearing (and, therefore mph/1000 rpm figures) is the same as that of the 3.0CSi, namely 3.25-to-1 (top gear) and 22 mph/1000 rpm. Completing the transmission details is a limited slip differential that is available as an option (in mainland Europe at least). This may yet form part of the standard UK specification when the cars come here later in the year.

Revised suspension details

As in any semi-trailing arm rear suspension arrangement, there are bound to be considerable camber changes during the movement of the wheels and the familiar BMW system is no exception. In order to lessen any undesirable tendencies of the basic design, the new car has a substantially lower rear roll centre than before which in theory should lead to less roll resistance while helping to avoid rear wheel "tuck under". To combat the former natural result, the 633CSi has a rear anti-roll bar and an increase in rear track of no less than 3 in. while to leave total wheel travel unaffected, both bump and droop travel are increased, the former by nearly 1 in. Conversely at the front, the increased frontward weight distribution has been countered by a narrowing of the front track whose effect is not completely countered by a marginal increase in front anti-roll bar section. The net result is to tidy up the handling especially over indifferent road surfaces and under fully laden conditions.

Progressive power steering

The new car features a highly effective progressive ZF steering whose assistance diminishes as road speed increases. In addition, the steering is higher geared and, without loss of essential assistance at parking speeds, it gives much improved road feel at speed. A welcome bonus is a reduction in the diameter of the steering wheel without noticeable increase in steering effort.

Again with the increased weight in mind, the diameter of the ventilated front disc brakes has been increased though that of the ventilated rear discs is unchanged. The system is of the ubiquitous dual circuit type, but in the BMW system, one system operates on all four wheels and the second one on the front discs only. A 9 in. dia. Mastervac duplex servo is the same as that of the superceded 3.0CSi.

The handbrake if of duo-servo type (making it equally effective whether it is used going forward or back) and it operates on small drums which form the centre of the rear discs. For added effect, the ratio of the handbrake mechanism has been increased by 20 per cent.

Roomier body

Important major details of the new body have been mentioned and to these should be added the new car's much improved accident "ride down" characteristics. These are accomplished by the familar shark-nose so beloved of BMW designers which enables initial retardation to be fairly gentle and controlled in a head-on collision.

New at Geneva

In what is becoming the norm, the body design is a three-box configuration with collapsible sections fore and aft and an immensely strong central passenger cell. The strength of the latter is greatly assisted by the adoption of the strong B-posts whose tops are linked by a substantial double-skinned compression panel. A double-skinned panel links the tops of the rear quarter panels and an immensely sturdy triple-skinned box section surrounds the windscreen. The design of this new body and the materials used in its construction result in improved torsional rigidity when compared to the old body and such improvements promise longer life for the body as well as improved crash performance.

Though outwardly similar in dimensions to the old body, slight differences have permitted considerable gains in accommodation. Headroom front and rear has been increased by nearly 1 in. and kneeroom for rear passengers goes up by over 1 in.

In overall dimensions, the new car is only 1½ in. longer but 2 in. wider and the latter dimension is made more noticeable by the adoption of more curved glass for the four side windows. Thus as well as improved interior accommodation, there is a useful improvement in boot space of no less than 23 per cent, though the fuel tank capacity is slightly reduced.

As well as the collision considerations, the adoption of a B-post has enabled the rear quarter panel to be reduced in size.

Sedentary systems check

Though various design exercises have threatened it, the BMW 633CSi becomes the first production car to offer a really comprehensive integrity check of the car's important systems. Pressing a "Test" button on a facia panel checks that the warning lights for the following systems are properly operating: engine coolant level; engine oil level; brake fluid level; brake pad wear indication; side lights; brake lights; windscreen washer bottle contents. In addition to these warning lights, there are also the more familiar ones for oil pressure, generator output, handbrake and foglamps (when these are fitted).

A hooded binnacle contains the speedometer in its centre with flanking rev counter to the right and a combined water temperature and petrol tank contents gauge to the left. The speedometer contains total as well as trip mileage recorders, the latter with a single press reset. It may be a sign of BMW confidence in the improved reliability of their cars that the total mileage recorder now reads to six digits on the total recorder rather than five of the earlier car.

In the left-hand corner of the shaped binnacle is the Test panel previously referred to, while the centre section of the binnacle contains the controls for the comprehensive heater control system as well as a BMW 500 Series-type clock contained in a surround that also operates the three-speed heater fan.

On top of the transmission tunnel and sweeping down from the centre section of the instrument binnacle is a central console into which a radio and/or cassette player can be mounted. The console has a deep stowage well at its front and the controls on the electric window lifts flank the gearlever at its rear end.

Improved heating

Though of the water-valve type, BMW have sought with the heating of the 633CSi to remove many of the usual shortcomings of the system. Firstly, it is possible to blow unheated fresh air from triple multi-position central air vents and fresh air is also available from vents at the facia outside edges. Warmed air can come through three vents serving windscreen demisting as well as through vents at the extremities of the fascia to demist the side windows. By providing a big enough heater matrix with very careful and progressive control of the water valve, the new system is an improvement on most water-valve system, even if still no match for a good air-blending system though, perhaps, cheaper to build.

The seating layout is shaped for four passengers with separate seat backs in the rear. There is space for a third adult in the back, but the padding in the centre, over the transmission tunnel, is rather hard. The driver's seat is adjustable for height and the steering column can be adjusted for reach, bringing the fingertip stalks with it when it is. Naturally, the front seat backs can be raked and they have adjustable head restraints set in their tops. Head restraints again appear on the rear seats and are similarly adjustable — a neat styling feature is the continuation of the rear head restraint shape across the rear parcel shelf. This gives two lockers beneath hinged lids in which a first aid box or similar might be placed out of sight and unable to slide across the shelf.

Seat belts for all four seats are of the inertia reel type with locking by vehicle retardation as well as reel pull-out speed. All four seats are of the lap-and-diagonal type and those in the rear are easy to put on and comfortable to wear.

External trim

A substantial rubber-faced rubbing strake continues the line of the bumpers down the sides of the car. The bumpers themselves are designed to absorb serious parking bumps without coming into contact with the bodywork and at both front and rear, they continue down the sides of the car to give good corner protection.

A new feature for the big BMW Coupés is the adoption of Mercedes Benz-style deep rain guttering up the screen pillars to prevent rain water from getting onto the side windows.

A small lip in the body sheet panelling below the front bumper considerably reduces front end lift at speed, BMW claiming that at the car's maximum speed, its presence reduces lift from 150 to only 40 kg. Incidentally, all the body is in steel sheet and there are no immediate plans to build a lightweight version to replace the 3.0CSL.

Driving impressions

Excellent all-round vision and the visible corners of the car give a commanding driving position. Though never over-light, even at parking speeds, the new steering soon makes the big car feel smaller than it is. This steering must surely be amongst the best available, since it allows effortless low-speed manoeuvring without any loss of feel at speed — the ideal arrangement.

Though small in character, the changes in suspension design have greatly improved the roadholding and ride compared with the superceded models. Even such a severe test as mid-corner braking fails to reveal any adverse reactions and this big car can be thrown around with confidence after the minimum of familiarisation. The suspension's ability to iron out indifferent surfaces and to ride well enough to allow the full roadholding, whatever, is most reassuring.

Outright performance testing will have to await a full Road Test, but there seems little reason to doubt the claims of a top speed of 133 mph and acceleration from rest to 100 mph in around 20 sec. A rather optimistic claim of 26 mpg (DIN consumption) would be excellent for such a powerful car, but if the performance of the Porsche models is used as a guide, then such consumption may well be possible — we must wait and see.

The only criticism that must be levelled at the new design is that it fails by only a little to provide real 4-seat comfort. In this respect it does better than our own Jaguar XJ-S but it is no match for the Mercedes Benz SLC models. One must say in defence of the BMW, however, that its close-coupled looks are an improvement on those of the Mercedes Benz but nonetheless, the ship is spoilt for a ha'porth of tar.

Depending on specification, the 630 and 633 models are priced in Germany at between £8,000 and £9,000. It seems likely, therefore, that it will compete less with the Jaguar XJ-S than with the thirstier Mercedes Benz 450 SLC. Present BMW 3.0CS coupé owners will be disappointed at such an increase over the new price of their cars but they should be placated by the knowledge that the new are is an all-round improvement and a definable step up when compared with the previous car.

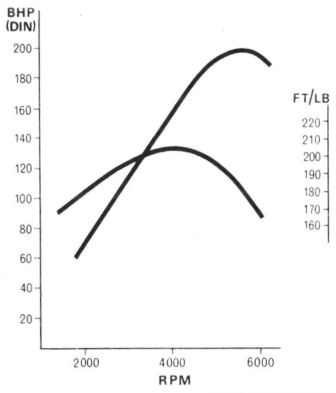

Below: Power and torque curves for the 3.2 litre fuel-injected engine show broad spans of useful power and torque. Incidentally, since BMW fitted rev limiters to their bigger engines, warranty and rebuilt costs for damaged engines have been cut by 70 per cent

Competitive comparisons

Model	Cylinders	Bore mm	Stroke mm	c.c.	Power	Torque rpm	C.R.	Induction
BMW 633CSi	6, in line	89.0	86.0	3210	200/5500	210/4250	9.0	Bosch EFI
Mercedes Benz 450SLC	8, vee	92.0	85.0	4520	225/5000	278/3000	8.8	Bosch EFI
Ferrari 365 GT4 2+2	12, vee	81.0	71.0	4390	320/6200	318/4000	8.8	Weber (6)
Jensen Interceptor	8, vee	109.7	95.3	7212	280/4800	380/3200	8.2	Carter
Aston Martin V8	8, vee	100.0	85.0	5340	Not known	Not known	8.3	Weber (4)
Jaguar XJ-S	12, vee	90.0	70.0	5343	285/5500	294/3500	9.0	Lucas EFI

Model	Wheelbase (in)	Length (in)	Width (in)	Height (in)	Weight (lb)	Tank (gal)	Tyres	Final Drive	mph/1000rpm
BMW 633CSi	103.4	187.2	68.0	53.7	3340	15.6	195/70 14	3.25	22.0
Mercedes Benz 450SLC	111.0	186.6	70.5	52.4	3635	19.8	205/70 14	3.07	24.0
Ferrari 365 GT4 2+2	106.0	189.0	70.5	52.0	3913	26.0	215/70 15	4.30(3.5)	22.8
Jensen Interceptor	105.0	188.0	70.0	53.0	3947	20.0	ER70 15	3.07	24.8
Aston Martin V8	102.8	191.8	72.0	52.5	3940	21.0	GR70 15	3.54(2.99)	26.9
Jaguar XJ-S	102.0	191.7	70.6	50.0	3852	20.0	205/70 15	3.07	24.7

A NEW COUPE FROM BMW

Technical analysis & driving impressions of the 630 CS and 633 CSi

BY RON WAKEFIELD
European Editor

THE OLD BMW coupe, known most recently as the 3.0 CS, had been with us in one form or another for 10 years, and though it was still a nice car its time had come. As retiring chief engineer Bernhard Osswald of BMW put it, "Let's face it, the old coupe was no longer at the forefront technically." It first appeared in 1965 as the 4-cylinder 2000 CS (for Coupe Sport), related to the 2000 sedan. A very glassy passenger cabin, much wood trim inside and clean body lines from the cowl rearward were its strong points; less impressive were its awkward front-end design and an engine that seemed too small for the car's weight.

When BMW's first postwar 6-cylinder came out in 1968 the coupe was adopted into the new series. It got a harmonious new front end, new front suspension and a 2.8-liter 6-cyl engine, although the old rear suspension and drum rear brakes remained, and thus became a much more attractive car. It has continued in essentially this form ever since. The engine grew to 3 liters, the rear wheels acquired disc brakes, and there were eventually variations ranging up to the wild, winged 206-bhp CSL and its racing counterpart—which, incidentally, is still winning races. But any good car is eventually overtaken by better technology, and the latest BMW sedans were better cars than the top-price coupe.

The Bavarian firm, which had a record year in 1975 and shows no signs of slowing down, has now corrected the situation with a handsome new coupe. Available presently in Europe in two versions, 630 CS and 633 CSi, it is not (as had been rumored) based literally on the 530i sedan, although it shares basic technical characteristics and the floor stampings immediately under the passenger compartment. It's more likely that the new coupe will have considerable component interchangeability with a new series of large BMW sedans, probably to come along sometime in 1977 to replace the current 3.0 Si and its siblings. Like its predecessor, the new CS will be assembled (and its body manufactured) in small quantity at the Karmann factory in Osnabrück rather than one of the BMW factories in the Munich area. As recently introduced BMW sedans like the 530i also outstripped the old coupe in assembly quality, it's to be hoped that updated production techniques used for the new one will bring it into line.

Built on the same 103.3-in. wheelbase as the old model, the new CS is predictably slightly larger: 4.5 in. longer, comparing European-specification cars new and old, and 3.0 in. wider. Thus in the normal sense the new car is a logical successor to the 3.0 CS, though in these days of ever-climbing fuel prices one might have hoped for a lighter, more compact car. In appearance it could be nothing but the new BMW coupe. The characteristically glassy look continues, even though pillarless coupes are now passé and the new one has a central roof pillar/rollbar. In fact, glass area is up 7 percent. Bronze-colored tinted glass is standard, and BMW people admit freely that this was chosen because it looks better: it is only 60 percent as effective as the usual green. It does look good; and the customer who must have green can still order it.

The shape is also related to the latest sedans, with its low and sloping hood, the virtually square BMW "kidney" grille center and raised hood section leading back from it, the wedge side profile and squared-off tail end. Sides are lean and clean and the windows drop below the main beltline. Only at the rear is the form somewhat disaccordant: the rear bumper looks unrelated to the front one. Here you see the chrome European bumpers; for the North American version there will be aluminum ones on hydraulic energy absorbers, protruding about 2 in. farther at each end and thus not nearly as prominent as those of the 530i or the 1974 3.0 CS USA. (No 1975 coupes were sold in North America.)

Inside, the concept of a luxurious 4-seater is continued but there is significantly more leg, head and shoulder room in the rear shell seats. As before there's a fold-down center rear armrest; new are adjustable head restraints, behind which molded shells blend into the package tray and lift up to reveal two small compartments for odds and ends. The front seats, with their widely adjustable head restraints, are as before generously dimensioned and comfortable; the driver's seat has inherited a manual height lever from the 530i. This 2-way lever allows not only the height but the angle of the whole seat to be set, and in combination with separately variable backrest angle and a telescopically adjustable steering column leaves little "room" for complaint about driving position.

The old coupe was rich in wood inside, with paneling across the dash and along the sides. In the new one there is no wood at all, but rather the latest in padded, ergonomic, plastic contours. The dash looks very much like that of the 3-series BMW (yet to appear in America as the 320i), its instrument-control board wrapped slightly so that the driver can reach everything easily, and the contours of this section also wrap around into the doors to give visual continuity. Above this are several air outlets for the heating-ventilation system, which is greatly improved. Directly in front of the driver are the three instrument dials, orange-illuminated at night, and the main warning lights. To the left is a new set of "monitor" warning lights, reminiscent of those in the Toyota Corona hardtop though not as complete. Here the driver can check the level of four important liquids—engine oil, coolant, brake fluid and windshield-washer fluid—plus the condition of the front brake pads and operation of brake- and taillights by pushing a single bar labeled "test." If anything is amiss (fluid low, pads worn too far, bulb burned out) its light does not come on. BMW's philosophy here is that if one of the warning lights itself is burned out the driver will be given the pessimistic message and therefore be warned of a possible problem. If the bulbs were to light only when there's trouble and one were burned out, he would be given a false sense of confidence.

BMW has given the optional air-conditioning system more attention than it got in the old coupe, not a bad idea as the old one couldn't keep front passengers cool at more than 90 degrees outside without sounding like a hurricane and never mind the folks in the rear. Though not a fully integrated system in the American sense, it is now controlled by the same temperature and blower controls as the heater and its cooling capacity is up from 4500 kilocalories per hour to 6200, a meaningful 37-percent increase. Air distribution is also much better: cooled air comes from outlets at both ends of the dash, the main center grille and a small center one directed at the driver instead of a single center one. BMW and Behr cooling engineers developed the new system partly in Texas and Arizona, so there is hope for it. Another item very popular in America these days, the sunroof, has been changed to the lift-or-slide type. And there is provision for four built-in stereo speakers.

The trunk, beautifully finished but not large in the old coupe, is just as well finished in the new and about 23 percent larger. On the underside of the trunklid one still finds the clever drop-down toolbox, an idea of former U.S. importer Max Hoffman, with its fascinating assortment of ordinary and special tools plus bulbs, fuses and sparkplugs. The fuel tank remains under the trunk floor, bucking Mercedes' trend to over-the-axle placement, and is actually slightly smaller than before.

Of course there are the usual safety improvements like longer "crush zones" front and rear and a stronger passenger cell. The rollbar, finished in matte black and thus de-emphasized in the car's styling, provides roof strength for U.S. safety rules as well as a better place to hang the front seatbelts, and as a side benefit stiffens the body. Although the pillarless CS was remarkably rigid for such a structure, according to BMW engineers the new one is 31 percent stiffer in bending and 69 percent better in twisting.

Engines & Transmissions

FOR THE European market there are two engine choices. The 630 CS has the familiar 2986-cc inline six with single overhead cam, now fed by a 4-barrel Solex carburetor like that of the Mercedes 280 instead of the previous Solex 2-barrels. This boosts the power from 180 to 185 bhp DIN, allows a lower fast-idle speed (and thus hopefully avoids the ominous noises carbureted BMW sixes make when started from cold) and reduces emissions. The 633 CSi engine is the 3210-cc injection six recently introduced in the 3.3 Li over here (why not, then 3.2 Li and 632 CSi?) and produces 200 bhp. When the coupe appears in America it will most likely have the same detuned 2986-cc injection unit now used in all North American 6-cyl models. The decision is not firm yet, but if so the name will be 630 CSi and the power 176 bhp SAE net.

The manual 4-speed gearbox is like that used previously, but the automatic is new and completes a transition from the Borg-Warner 65 to a new ZF type. Also a 3-speed unit with torque converter, the ZF is lighter, more compact and smoother-shifting thanks to its Simpson (instead of Ravigneaux) planetary gearset, an arrangement long since normal in American automatics. The

final drive ratios are 3.45:1 for the 3-liter, 3.25:1 for the 3.2 and probably 3.64:1 for the heavy-breathing American version.

Chassis

EVER SINCE BMW brought out the 4-cylinder 1500 model that effectively rescued the company from financial ruin in the early 1960s, BMWs have stuck with a basic combination of MacPherson-strut front and semi-trailing independent rear suspension. The new CS models hold to this formula with refinements. The front suspension geometry is altered for better straight-line stability and better behavior in the event of uneven braking, but the layout remains the same with simple lower lateral links, drag struts and the MacPherson coil-shock struts. The shocks are now easily exchangeable to simplify replacement; spring travel remains at 7.9 in. and the anti-roll bar is very slightly larger. As before it is mounted to the lateral links with rubber above and below it so that it does not come into play on small bumps or gentle road undulations. Only when the rubber is

The dash is similar to 3-series BMWs and uses orange light for illumination. At left is bank of monitor warning lights. The rear seats have more room than old CS and head restraints, which hide two small compartments. U.S. introduction will be this fall.

fully compressed does it begin to act as an anti-roll bar, thus avoiding some of the harshness a bar can introduce on one-wheel bumps.

At the rear, all the components are new and the coils and shocks are now concentric as in the 3- and 5-series. An anti-roll bar is standard here too, as it has always been on the U.S. version of the coupe. Springing has been softened slightly at the front and stiffened at the rear to get rid of the old BMW pitch, and there's now more spring travel at the rear: 8.5 vs 7.9 in. The rear track, formerly much narrower than the front, is now considerably wider.

Brakes have been improved in an evolutionary way too. The front (ventilated) discs are now 11.0 in. in diameter vs 10.7 before, and now all versions have the 10.7-in. vented discs at the rear. BMW has stuck with the same wheel and tire sizes: 195/70VR-14 on 6-in.-wide rims. The engineers say that at the moment they aren't charmed by wider tires, such as the 60- or 50-series, for this sort of luxury GT. But "at the moment" seems significant, perhaps hinting that a sports version like the 3.0 CSL is in the offing. Alpina alloy wheels are standard.

Power steering is now standard and has been changed to the recirculating-ball type used in the 530i. The ratio is thus reduced from 18.1:1 to 16.9:1 and the wheel turns lock-to-lock are cut from 3.7 to 3.5. More interesting is that the "assist curve" of the power-steering pump has been recalibrated so that power assist falls off with increasing engine speed. Up to about 2000 rpm it delivers 100-percent boost for easy parking and low-speed maneuvers. From there up to the engine's redline of 6500 rpm the assist is smoothly reduced to 70 percent of maximum, so that at high speeds the steering communicates a firmer feel to the driver. Up to now only Citroën had offered such a feature.

The new coupe is 110 lb heavier than the old, partly because there's more sound insulation and partly because it's bigger and stronger. It can also be expected to be more expensive and most likely will cost around $20,000 in the U.S. with full equipment (air conditioning, leather upholstery, stereo, electric front window lifts, etc).

Driving Impressions

ASIDE FROM audible air turbulence around the windshield and variable quality of sealing around the door windows, the old CS was an exceptionally quiet car, and with about 40 lb more sound-deadening one would expect the new one to be quieter. As far as wind noise goes it is, with the qualifier that if I lowered a front window after closing the door the rubber sealing no longer worked on the two prototypes I drove. Otherwise the new car seems to be, if anything, a bit noisier. The differential sang a hearty song in both cars, and in the fuel-injected model a lot of engine noise came through too, although it was the virile sort of noise an enthusiastic and fast driver can enjoy. Using the car speedometer, I timed the 633 CSi with manual gearbox as follows, shifting at 6500 rpm:

0-30 mph	2.5 sec
0-60 mph	7.2 sec
0-80 mph	13.5 sec
0-100 mph	17.8 sec

The shift linkage was a bit stiffer than in the older car, perhaps because of newness. In any case the BMW six and manual gearbox remain a sporty, satisfying way to get a lot of performance from a moderate-size engine in what is a relatively large car by European standards.

For contrast I tried the 630 CS with automatic, which probably approximates the performance Americans will get, and found it would attain 60 mph from rest in 9.4 sec with or without manual gear selection. This engine is quieter than the injection unit; it doesn't sound as "hard," but still a lot of exhaust noise came into the car just as in the 530i we tested last year. The new automatic is decidedly better than BMW and Borg-Warner's mediocre past efforts. On moderate acceleration it shifts very quickly out of 1st gear, like earlier BMW automatics, but it does stay in 2nd longer and thus gives a less strained effect. With the foot firmly planted on the floor it upshifts both times at 5700 rpm, making manual override unnecessary for maximum performance, and kickdown to 2nd for passing is available to 75 mph. With the 4-speed 633 CSi I reached an indicated 220 km/h (136 mph), with the automatic 630 CS 200 km/h (124 mph) on the long straights of BMW's bone-shaped high-speed track.

Going to the handling course, I wanted to see if the new power steering would make the car tricky to handle in the sort of hard driving on a curving road in which one changes gears often. I'd found the steering feel very good at high road speeds, though there was no crosswind to really test the car's stability, and by running along at steady speed in one gear and then shifting up or down I could feel the difference in power assist—but it was pretty subtle. Changing gears while turning the steering wheel failed to unearth anything that would come to consciousness if one weren't looking for it, so I'd say BMW has accomplished this improvement in power steering without introducing any dangerous or bothersome side effects.

BMW's characteristic final oversteer is there, but this is something one experiences only with energetic provocation. The handling course also includes a good sample of bumpy curves; their effect on the car is virtually nil even when cornering at the limit, the people platform remaining supremely stable. I found the brakes less impressive; they're powerful but too easy to lock up on a hard stop. Overall, the ride—which I checked over bumps, dips and cobblestones—is slightly better, mainly because the old tendency to fore-aft pitching is gone. Ride development seems to be on a plateau, although a very high one, with sophisticated independently sprung cars such as this one.

The old coupe's electric window lifts, particularly after 1973, were hopelessly slow. Unfortunately they now seem to be worse—I timed the meager 5-in. descent of the rear-quarter panes at what seemed an eternal 5.5 sec. No cars with front lifts were on hand. Otherwise, though, there's little to complain about in the CS's interior. It's roomy up front, usefully more spacious in the rear. Instruments and controls are laid out beautifully, the seat-height adjustment and telescopic steering column make it possible to find just the right combination, and you can impress everybody on the block with that panel of warning lights. Ventilation is superb with all those air vents and works very quietly, but I had no opportunity to try the air conditioning.

The new BMW coupe carries forward its predecessor's blend of performance and refinement with worthwhile improvements in comfort, styling and safety. If it seems large in the European context, in America it is a car of moderate size and weight considering how luxurious it is, and of course a chassis of this sophistication is simply not available in a domestic car.

BMW 630 CSi
SPECIFICATIONS
(Probable North American Version)

GENERAL

Curb weight, lb	3300	Gear ratios (automatic):	
Wheelbase, in.	103.3	3rd (1.00)	3.64:1
Track, front/rear	56.0/58.5	2nd (1.48)	5.39:1
Length	191.1	1st (2.48)	9.03:1
Width	67.9	Final drive ratio	3.64:1
Height	53.7		
Fuel capacity, gal.	18.4		

ENGINE

Type	sohc inline 6
Bore x stroke, mm	89.0 x 86.0
Displacement, cc/cu in.	2986/182
Compression ratio	8.1:1
Bhp @ rpm, net	176 @ 5500
Torque @ rpm, lb-ft	185 @ 4500
Fuel injection	Bosch L-Jetronic

DRIVETRAIN

Transmission	4-sp manual or 3-sp automatic
Gear ratios (manual):	
4th (1.00)	3.64:1
3rd (1.40)	5.10:1
2nd (2.20)	8.01:1
1st (3.86)	14.05:1

CHASSIS & BODY

Body/frame	unit steel
Brake system	vented discs, 11.0-in. front, 10.7-in. rear; vacuum assisted
Wheels	alloy, 14 x 6J
Tires	steel-belted radial, 195/70VR-14
Steering type	recirc ball
Overall ratio	16.9:1
Turns, lock-to-lock	3.5
Front suspension:	MacPherson struts, lower lateral links & drag struts, coil springs, tube shocks, anti-roll bar
Rear suspension:	semi-trailing arms, coil springs, tube shocks, anti-roll bar

NEW BMWs
Again, just that bit better...

Our European correspondent SLONIGER has an exclusive WHEELS pre-release drive of the stunning new BMW 630 CS and 633 CSi. Eat your hearts out, he says...

BMW FANS, eat your hearts out. Anything those wonderful 3.0 Coupes could do before is ancient history now as Munich releases the 630 CS and 633 CSi on a luxury market.

Oh, the new 2 plus 2 machines are based on their previous coupes and closely related to the 3 and 5-series sedans but they are also so thoroughly revamped and improved that you get more comfort, greater safety, lower exhaust emissions — and still have machines which in handling run rings around the old 3.0 CS. After all that, even the elevated prices seem reasonable.

Styling first: the two new coupes have a strong touch (in the grille) of the mid-engined Turbo design exercise (it too will appear later in limited series). They have more class and smoother lines but, in the process, BMW may have lost a hint of that solid Germanic identity. However, BMW has certainly tossed in all the trim, leather and gadgets an owner-dandy might want.

Seats not only recline, they move up and down, in conjunction with a steering column which goes in and out. Almost anybody can find comfort. In

back anybody with very short legs will be happy on individual seats with their own adjustable head rests.

Electric windows are standard in back and optional in front, and if you order the (optional) sliding roof (tilts up in back for use in rain) with electric action you find its motor has been moved to the boot to cut interior noise. In general, BMW can claim 20-30 percent less interior noise, thanks to new sound deadening material — and nearly twice as much of it.

All glass is tinted bronze unless you order green or the clear option. The bronze doesn't cut glare as much as green but it doesn't reduce vision as much either. Besides, they think it helps styling. For the same reason light metal wheels are standard on both cars, mounting 70-series steel radials. BMW doesn't think lower-profile tyres are suitable on the open road.

Body engineering apart, there is nothing entirely new about these exclusive (Carrera price class) coupes but little that has not been re-thought, strengthened, updated and generally projected into the eighties. We forget, after all, that the previous CS was seven years old and even based in large part on a decade-old 2000 Coupe.

Even a brief drive around the wintry Bavarian test track six weeks before the coupes were released made it clear that these are not only safer and friendlier to the ecology but — most important — aimed directly at the driver who wants the Grand in his Touring. Engine elasticity, spring rates, fittings and styling are all aimed at the sybarite. Yet the 633 CSi WHEELS drove (the photo car was a 630 CS) would out-run and out-corner any previous BMW coupe except the CSL homologation special.

It simply did this with less drama, less lean, less stiffness over cobbles and lower demands on driver skill. With a 6300 red line the car will turn 160 km/h at 4250 and see 6000 with a super-long straight. It will also pull from 1000 rpm in top gear.

Handling is near-neutral with a touch of understeer and total stability in high-speed bends where the 633 took on

Interior is combination of best BMW practice, with indirect orange lighting. There are ample air vents everywhere.

NEW BMWs

a mild lean and held that position all the way around a large skid-pan circle. Of course, the 633 is a bit stiffer-sprung than a 630. Gear speeds work out to roughly 60, 105 and 160 km/h using all the revs but that isn't the way to drive a 633. In hairpins you have ample torque to kick out the tail.

Steering is steady at all speeds and the shift pattern accurate if a little wide-spread. Their new gearbox (updated really) is, in fact, an improvement. Brakes are very easy to dose from high speeds and locked wheels totally absent. Despite the addition of a B-pillar the side windows are still frameless and produce some whistle at speed.

In short — a car that drives up to its price.

But the real changes are under the skin, in either of the new versions. Both naturally meet all foreseeable emissions standards with the carburetted 630 using improved combustion chambers from their 3-series and gaining slight extra power 200 rpm lower than the previous three litre coupe powerplant. The new carb is also more progressive.

The 633 model is not precisely the same as the 3.3 engine from the stretched sedan. In fact, it has less stroke and is really a 3.2 but does have Bosch L-Jetronic, an injection based on air-flow measurement and slightly more sophisticated than K-Jetronic (Porsche). BMW thus gets the power of an old 3.0 CSi with lower compression and more torque by boosting capacity for comfort.

Otherwise, both engines have larger generators, larger cooling fans and even an oil cooler for the 633 although this may not prove necessary. The 3.3 also has transistor ignition which BMW doesn't "consider necessary" below the 3000 cm³ limit.

While gear ratios are unchanged the box newer with more accurate syncro. The automatic option is all new — BMW had some delivery trouble with B-W and have switched the entire line to a German ZF with adjustments to fit each engine. Even within the coupe range the 630 and 633 have different torque converters. A separate oil cooler for gearbox fluid comes with the automatic. One nice touch is the indication of automatic gear position on the dash so you needn't study the tunnel for every shift.

The 6-series chassis is largely as before — plus improvements. There is slightly less track in front but this was done to suit the styling. Spring travel is the same 20 cm but no longer equal both ways: There is less bounce travel, more on rebound and the stabiliser is 1 mm thicker.

In back BMW has gone to spring legs, wider track, lower roll radius and more of both bounce and rebound, plus a new 16 mm stabiliser. The rear is harder, overall roll and dip reduced.

Steering has a new servo system with reduced input as speed increases to maintain feel. Brakes have more progressive boost, a balance valve and inner ventilation for the rear discs while the handbrake gives more leverage.

While some metal is carried over from the 5-series and larger sedans the chassis and body have been entirely re-computed to give 31 percent bending stiffness and 69 percent more torsional with a quarter more nose crush zone. Chief addition is the B-pillar, a profile unit which serves as integral roll-over bar. A spoiler integrated with the front bumper reduces lift at speed from 150 to 60 kg. Boot capacity is up by a quarter to near-sedan levels.

And the fuel tank rides under the boot, mounted from below and designed to drop out in a tail-end shunt. Engineering boss Oswald doesn't believe a tank above the axle is as safe and besides, since all tanks will leak at some time, he'd rather they did it entirely outside the body.

Dash and dials come from the 3-series with one new trick. Seven control lights for oil, water, brake and screen washer levels, brake and rear light function and brake pad wear are all activated by one test button before you drive off — a full safety check with belt fastened.

BMW's new Coupe(s) set the pace by doing all things just that little bit better; as only happens when you start with a good design and give it a thorough rethink. *

SPECIFICATIONS

ENGINE; Inline, single-OHC six with tri-hemisphere combustion chambers and double-downdraft carb (L-Jetronic injection). Bore/stroke 89/80 mm (89/86), capacity 2986 cc (3210). 136.1 kW/5500 rpm (147.1/5500), 260 Nm/3500 (290/4250), 9.0:1 compression.

DRIVE LINE: Membrane-spring clutch with hydraulic action. Fully-syncronised four-speed manual box, Getrag with Borg-Warner syncro. Ratios 3.855, 2.203, 1.402 and 1.0 with 3.45 final drive. Optional ZF 3-speed automatic with Fichtel & Sachs converter, ratios 2.478, 1.478 and 1.0 with 3.25 final drive. 25 percent limited slip diff optional.

BRAKES: Dual-circuit, all-disc with back-up circuit on front wheels, servo and balance valve plus pad wear warning. All four discs inner-vented.

WHEELS & TYRES: Alpina alloy rims 6 J 14 standard with 195/70 VR 14 radials.

DIMENSIONS: Wheelbase 262.6 cm, front/rear track 142.2/148.7, LxWxH 475.5 x 172.5 x 136.5, empty weight 1450 kg (1470), permitted load 380 kg (360). Tank 70 litre.

PERFORMANCE (works): Top speed "over 200 km/h". Acceleration 0-100 km = 8.9 s (7.9), 0-400 m = 16.3 s (15.8) — with manual box. Price: around $A12,000 in Germany.

A BMW TO BOAST ABOUT
.....so long as you can put upwards of £10,000 where your mouth is. By Mel Nichols

I AM NOT AN ARDENT BMW FAN. THE MILES HAVE TAUGHT ME to think of them as good cars, yes; but not exceptional cars. Refined cars, but not redoubtable cars. Efficient cars, but not outstandingly enticing. Adequate cars, but not overwhelmingly alluring. They have my respect, but not my reverence. It wasn't always that way though; there was a time, when the 2002 was new and I was younger, that I believed it to be the car for me. But then I came to know Alfa Romeos and the lesson was that one could have looks, performance and handling with character in equal proportion. Another time, long ago, I decided that a BMW 2.8 saloon was to be preferred to a Jaguar XJ6 4.2 because it had better rear accommodation, and because the Jaguar's power steering prevented it being driven as enterprisingly although in reality its abilities matched the BMW's and it felt rather more luxurious than the 2.8. But the 2.8/3.0 was much younger then, and the Jaguar has been exorcised; moreover, the S class Mercedes has arrived in the meantime and so has the Peugeot 604. Indeed, Ford's Granada is closer to the big BMW than most people would ever expect. And while it is easy enough to admire the chassis of the Five series cars — in extremis, their behaviour is magnificent — all the examples I have tried have suffered so badly from windnoise as to be rendered unacceptable at the price. Disappointment for me was spelled BMW 2.8CS, a car of considerably less ability and driver appeal than the saloon from whence its power plant came. The reason was that the coupe was based on the old 2000 saloon, not on the more modern 2.8 saloon. The 3.0CS was better but still a disappointment, although there was the time I drove a fully-winged homologation special CSL home to Munich and enjoyed its tautness and the way its appetite for miles was balanced against its thirst. I've yet to drive the 320i, and need more time with the 320. But while the sheer competence of the 316 is notable; especially its gentle, fail-safe handling, the advances made in these areas have come at the dilution of character, and I wonder whether I might find the car a little too bland were I to own it.

Now BMW have a new flagship, the 630CS and its more powerful sister the 633CSi, and as I walked through the snow and slush to drive them I couldn't help wondering if they were of similar nature to the Three and Five series cars; cars developed sensibly and meticulously. But inspiringly? Exceptionally? Enticingly?

An initial reaction to 630 series is that it is very handsome but conservative. It conforms to the styling trend begun with the Five series and continued with the Three series, and mates harmoniously with them. In the process it contrasts a little with the outgoing 3.0CS in style and in background; the new coupe was designed in-house, the old one by Karmann who built it as well.

Are you surprised that it is such a conservative-looking car? Well-balanced, tidy, attractive; but little more than a sleeker Opel Commodore coupe. Perhaps, if you were expecting something less closely related to the Five series and Three series, you were confusing the 630 with the mid-engined BMW coupe still to come. The one BMW have asked Lamborghini to build for them. The one powered by Munich's own and still secret V12; the one that will be a homologation car to enable BMW to go racing with a new vengeance. Yes, it will be possible to buy one. Five hundred will be built, but each will cost upwards of £17,000 on current values. Countach and Boxer competitors. The advent of this mid-engined V12 BMW, still some time away, has significant bearing on the 630 coupe in two ways. It means that BMW could design and engineer the 630 purely as a road car and forget the concessions necessary for a lengthy racing career. Anyhow, BMW believe a mid-engined car is the only way to be sufficiently competitive from now on. And, of course, the mid-engined car removes the need for them to launch a V12 onto the mass market, and especially to engineer the 630 to take it. However, it will be there, fully developed in a road-going vehicle, should they decide it is worth employing more generally in the future as the Jaguar/Mercedes putdown they originally planned. Why are they asking Lamborghini to build the V12 car in the face of a great deal of pressure to have such a prestigious machine kept all-German? They say they have looked around and they believe Lamborghini can do the job best. So the 630 can be a conservative, traditional three boxes notchback. Save for the nose, the ideas revealed in the BMW Turbo showcar several years ago are being kept for the wild one.

The Five series was initially to have more than just a styling relationship with the 630. At the planning stage, the idea was to build the coupe on the Five series' floorpan and to use its mechanicals. But there would have been too many compromises in what BMW wanted to be an especially singular coupe, the most expensive car they'd build until the mid-engined car came along. The 630, then, is a new car in its own right; a concept that employs various bits and pieces from the other BMWs but nevertheless has its own engineering identity. That identity, however, is almost as conservative — you could say well-judged — as the styling. And although the 630 is built by Karmann it is strictly an in-house BMW design, and more parts are being built by BMW and supplied to Karmann for assembly too. Torsional rigidity in the old CS was terrible, and so was passenger cell protection, BMW will now admit. Those factors alone, in the face of toughening legislation, were enough to prompt creation of the new car, and to explain the three-pillar aspect of its construction whereas its more striking predecessor was pillarless, and it's slightly greater (by 1.9 percent) 3190lb weight. It does not look a big car; but the length is 187in, the wheelbase 103in. And that makes it smaller than the 191.7in XJS but slightly bigger than the 186.6in 450SLC.

The 630's mechanical story in relation to the 3.0CS is one of considered improvement rather than radical change; a story of refinement. The base model 630CS has the familiar 2986cc inline six; the topline 633CSi has the 3210cc six and there are several changes in both power plants to distinguish them from their previous applications. The cylinder head on the 3.0litre engine benefits from the combustion chamber shape devised for the 316/320. It also has a new twin choke Solex carburettor and the results are an improved power figure of 185DINbhp at 5800rpm with 188lb/ft torque at 3500rpm. The 633's 3.2litre engine has the same combustion chambers as the 3.0CSi. But the fuel injection is now Bosch L-Jetronic which provides more accurate control of air intake and mixing. The 633's ignition is transistorised too. Peak power — 200bhp at 5500rpm — is unchanged but the torque is up 4.48percent to 209lb/ft and is now developed at 4250rpm instead of 4300rpm.

There is, as yet, no five-speed transmission. Getrag developed a new version of their 262 four-speeder for the 630 though, and the three-speed automatic optional on both models comes from ZF. Ratios in the manual transmission are the same for both coupes but the 633 comes with a 3.25 to one differential that is lower than the 630's 3.45 to one final drive. A limited slip differential with 25percent lockup is optional for both cars. So are firmer suspension settings.

The front suspension is essentially similar to the Three and Five series cars' — lower links, struts inclined inwards and to the rear at the top, and an anti-roll bar. The changes come at the rear: in place of the familiar coil springs and separate dampers, there are now struts there too. The benefits come not so much in greater roadholding and handling but in improved ride and noise suppression. Location of the independent axles is by the usual BMW trailing arms, and there's an

Without breaking any new ground, the 630's styling is both elegant and sporty (above and far right) with excellent proportions. The top-line 633CSi's engine (right) now has Bosch L-Jetronic injection and more torque. Finish throughout is impeccable, of course. Handling is magnificent, allowing long, glorious high-speed power slides like this (above right) whenever the driver wishes to indulge

anti-roll bar. It is, by Mercedes, Jaguar and Peugeot standards, a very simple suspension. But, as one soon discovers, an arrangement that is to be greatly respected. The specification sheet says only that the steering is, once again, by ZF recirculating ball. That's the dry side of the argument. BMW, taking a note from Mercedes' and Citroen's books, among others, have developed servo assistance for the steering that is related both to engine rpm and vehicle speed. The boost diminishes as these factors rise, so that when most steering feel is called for there is virtually no assistance. The front brakes are enlarged, and the rear discs are now ventilated too. Alpina alloy wheels are standard, and 6in wide; they carry 195/70VR14 high-speed radials, mostly Michelin XVRs (some aquaplaning problems in puddles say BMW, but the most consistent quality by a long way).

The construction of the body points up the main differences with the old CS. A strong crush-resistant passenger cell that's 60percent more rigid is the prime feature, and there are some very impressive examples of deformative engineering ahead of it and behind it. Worthy lessons learned with the Three series are applied to the full. Something else BMW have borrowed from Mercedes are the deep channels that run up the leading edges of the A pillars. They prevent wet weather much reaching the side windows. Munich did not, however, think Stuttgart's ribbed tail lights a good idea — too hard to clean after the car has stopped. The glass area is big; the glazing itself something called Parsol Bronze, which absorbs heat and cuts glare while offering greater transparency than the older green glass.

The interior is as efficient as the 320's, and it seems as likely to be imitated as some of BMW's earlier dashboards. There is the now typically BMW wrap around section in front of the driver that houses the instrument pod, its dials as big and clear and appealing as ever. Stalks are basic; efficient. BMW, considered at the forefront of ergonomics, do not however see the need to follow in the footsteps of Citroen, those fervent ergonomic pioneers. Perhaps, in the search for marque-to-marque uniformity they have a point. The steering wheel adjusts for reach, and the (new) seat is adjustable for height. There are other niceties. But are there overtones of Ford-style marketing slickness about a device called the 'interrogation unit?' This is a warning light system with a display panel on the outer edge of the fascia to cover engine oil level, coolant level, brake fluid level, washer reservoir level, brake pad wear, brake light and tail light fail-

Nigel Snowdon

Dashboard is new, a paragon of logic and efficiency although instrumentation is less attractive than in some of the older cars. Heating and ventilation is excellent, and there is a pleasant harmony about the way the dash meets with the door cappings. There is surprisingly good rear seat accommodation

ure. Fine, but rather than having the lights coming on when a system fails or a level falls too low, BMW couldn't resist involving the driver. He has to press a button on the panel to activate the check system; the light for the system in need of attention will not glow if something is wrong. There is a new heating and ventilation system — a cause of considerable complaint in all older BMWs — with outlets to keep the side windows clear (an Audi trick, most notably).

From the driver's seat — you can raise it in parallel, or at the front or at the rear — the 630 feels tidier and more solid than the old CS; smaller, and somehow more purposeful. Driving it leaves you with much the same impression. Although the 630's engine has no more torque than before and only another five horsepower, the curves are different and the result is that the 3.0litre car feels livelier. Again, more definite. More sporting. It has more identity. And yet, while immediately imparting this impression of greater sporting integrity than the old 3.0CS it is also more comfortable and rather more refined. Noise insulation is increased by 42percent, and that, in conjunction with the smoother and quieter rear suspension means that the 630 is rather more satisfying in this area alone than the 3.0CS was. Not that the ride is soft. Not at all. It is quite firm, but is especially nicely sorted, providing both excellent control as well as comfort. There is never trace of wallow, nor is there a suggestion of road surface irregularities overcoming the damping. There is an impressive and pleasing overall feeling of solidity about the car over bumps, none of which can move it off line or make it bump steer.

The new steering is extremely good. It is sharp and keen without being especially quickly geared (16.7 to one), smooth and progressive. The servo assistance is never particularly noticeable — even at parking speeds you note only that turning the wheel is no real effort. There is no super-light, watery feeling or the 'bounciness' that comes with some other systems. It is sufficiently good when the car is on the move for the driver never to think about it. The gearshift and pedals work with similar efficiency; a lack of flab that further increases the air of purposefulness about the car. The vision comes in here — it is excellent even though the A pillars are thick by the standards of, say, Opel — and the driving position is one of those that make you feel you're in an ideal environment in which to go about your work.

That work, in this new BMW, is, I am very pleased to report, quite delightful. The 630 gets up and goes hard without being outstandingly fast. There is a sense of eagerness, of willingness about it. The stability is notable, and there is not much noise. What there is comes from the trailing edge of the frameless door windows; there is none at the leading edges. BMW said the test cars had experimental sealing strips and that they were being changed for production, thus eliminating all windnoise. But having suffered in the Five series, and knowing how poor the outrageously expensive 3.3Li is in this respect, I'll reserve judgement until it is proven to me. However, the noise that did exist on these early cars was small; talk was easy at speeds well in excess of 120mph. There is practically nothing by way of road noise, and the suspension itself works extremely quietly. But the really outstanding thing about the 630 — beyond its eager performance, its clean and elegant looks, its comfort and the logic of its controls — is its handling. The roadholding is strong; I was expecting that. There is considerable precision to the handling; I was expecting that too. But I was impressed, more and more with every mile that I drove, by the 630's impeccable balance. Because it feels rather small and so taut around you, because it has a lively character, because it has that deft steering and because it responds to your inputs both at the wheel rim and on the throttle, it is very definitely a sporting car; a very sporting coupe with considerable comfort as a

side benefit. Now you might say that the old car was like that. But no, it was nothing like as good as this one. This one has real harmony, a matter-of-fact decisiveness. It feels closer to the Five than the 3.0CS.

Its abilities are exhibited best when the car is flicked, flat out in second or third, through ess bends so close and difficult that they demand the very best from car and driver. The 630 will be heeling hard onto one side (there is enough body roll to make you thankful for the big bolsters on the sides of the seats) when the time comes to change to the other lock. It snaps across to heel the other way in an instant, obedient to the extreme. There is no wallow, no hesitation. Just *balance*. There is considerable feel — or should I say conveyance of information? Enough so that, on a streaming wet road with icy patches, when the front does start to edge into understeer your own reactions seem all the sharper as you lift off, and the sliding of the front wheels stops. A prod on the power, as the nose comes tight again, scoots the tail out and the control is such then that it can be kept out for extraordinary long periods with lots of opposite lock cranked on, and it all held so precisely on the throttle. Behaviour into the bends under brakes and over crests, the aspect I have always thought to be the BMWs' greatest virtue, is even better than in the Three and Five series cars.

After an hour of sheer enjoyment in the 630 manual, and then a quick run in the 630 automatic to learn that the ZF automatic works flawlessly, I switched to the 633 with its firmer dampers and its limited slip differential. Impressed as I had been with the 630, and curious to see how the 633 could be better without being less comfortable or

A BMW TO BOAST ABOUT

relaxing, I soon found that there was, for me at least, no contest. The 633 is even tauter but just as comfortable; even more eager and notably more purposeful than the 630. This one screams out to get on with it. It bolts away where the little one merely gallops; savages the bends where the other one merely consumes them. It is, of course, far less forgiving to begin with. The combination of the slippery diff and more power means that the prod on the throttle that will merely bring the 630 to a neutral attitude sends the 633 CSi into lightning-fast oversteer, and calls for correction of the same order. The feel of the car is as close to that of a good racing saloon or coupe as I can recall experiencing, rather akin in sharpness to the Ford Escort RS1800. So you lift your driving to its standards and get cracking. Because it is tighter, sharper, more immediate — and faster, in every way — the actions of control are smaller, quicker and more precise by necessity. The balance is no better than that of the 630, but the time between extremes of body movement is less. You can really use this one to test yourself, and yet you need have no fear that it will betray you. It works with you all the way, not against you. The only flaw I could find in two hours of the hardest driving in my life — the others had gone home and I had BMW's test track to myself — was that there is just not quite enough feel in the steering to reveal front-wheel lock-up under full braking from high speed in the wet. Suddenly you have no steerage. It's a communications breakdown that comes as a great surprise after the data flow you've grown accustomed to. So you learn to allow for it, to brake earlier and adjust your line. To get back onto the throttle a little sooner and drive around with full traction, gradually increasing pressure on the throttle so that the car pivots around its nose tucked tightly in there at the apex. A touch of opposite lock. Ah yes, hold it at that. Sweep through. Wind it out, out, out to the edge of the track as the bend becomes the straight. Bring on full power as the offside rear wheel touches the side of the tarmac. Snap the next gear. Set up the next bend, and the next, with throttle alone. Get all that power down again. And sweep it through and out and on in yet another of those long power slides for no reason other than to indulge in the sheer pleasure of holding a car at 30 or 35 or even 40 degrees for as long as possible. Magnificent! *What control!* Even the section with the ripples mid-bend fails to wrest this car from you.

Have I made the 630 sound too good? It *is* good. A sporting coupe with handling second to none; better and more refined and more enjoyable than any other BMW yet. This one, for me, is the definitive BMW, the epitome of the character that the Bavarian company set out to develop and to market. They have done their job well this time. And yet, I find it extremely hard to accept that it is any more thorough than all cars should be. Nor, despite certain trappings, does the 630 have the air of luxury about it that one expects to find visible at this price, and indeed gets in the cold quality of a 450SLC or the quietness and smoothness of the XJS. Or did I simply enjoy the driving of the thing too much to notice anything else? Is this one really more of a Porsche competitor? An unbridled driver's car? I know only that for the first time among current BMWs my feeling goes beyond mere respect; this one beckons tantalisingly. But could I possibly pay something like £12,500 for the 633CSi; £3000 more than for an XJS, more than for either of the Ferrari Dinos, more than for a Urraco 3.0 or a Jarama or a Bora. More than for a Carrera, and almost as much as for an Espada. Yes, it is good, this new BMW. But I can't bring myself to believe it's *that* special. ●

BMW 630CSi

WE HAVEN'T had a chance to road test the 630CSi yet as it arrives in the U.S. just about the time this magazine is printed. We have, however, driven the European version in Germany and, frankly, we're impressed.

A few fortunate automobile manufacturers are able to build extraordinary cars, hang the expense, and a few fortunate people are able to buy them. The 630CSi is one of the finest European touring cars we've had the pleasure of meeting and, of course, its price means it's for the fortunate few.

The handsome styling is immediately impressive. It's refined and subtle while allowing good outward vision. But you have to know the 630CSi for a while before you realize it is also practical, efficient and sensible.

Being a BMW, it is great fun to drive. The handling and performance are impressive. It stops when it's told to–disc brakes are at all wheels. The same willing single-overhead-cam inline 6-cyl engine found in the 530i is in the 630CSi. You have your choice of transmissions: 4-speed manual by Getrag or 3-speed automatic by ZF. Fuel injection is by Bosch L-Jetronic. The cars we drove in Germany had excellent finish and high-quality interior materials. The seats are infinitely adjustable and the steering column moves in and out. The 49-state 4-speed gives an EPA rating of 23 mpg on the highway.

A specialty of the 630CSi is a test panel that monitors brake lining wear, brake fluid level, radiator water level, windshield washer water level, engine oil level and brake and taillight operation. Of course, the famous BMW tool kit is standard.

Welcome to the U.S., 630CSi.

SPECIFICATIONS

Basic price (est)	$24,000	Tires	195/70-14sbr
Country of origin	Germany	Curb weight, lb	3470
Body/seats	2D/4	Wheelbase, in.	103.4
Engine position/drive	F/R	Track, f/r	55.0/58.5
Engine	sohc inline 6	Length	192.7
Bore x stroke, mm	89.0 x 80.0	Width	67.9
Displacement, cc	2986	Height	53.7
Equivalent cu in.	182	Turning circle, ft	32.2
Compression ratio	8.1:1	Turns, lock-to-lock	3.9
Bhp @ rpm, net	176 @ 5500	Steering type	recirc ball
Torque @ rpm, lb-ft	185 @ 4500	Fuel capacity, U.S. gal.	16.5
Transmission	4M, 3A	Fuel economy (est), mpg:	
Final drive ratio	3.45:1	U.S.	21
Suspension, f/r	ind/ind	Calif.	21
Brakes, f/r	disc/disc		

AutoTEST
BMW 633 CSi

Efficient engine and reasonably high gearing give rare combination of performance with economy.
Ride rather firm and lively; handling and steering very good.
Excellent seating, comfort and finish.
A fine car, but a formidable price.

The CSi has electric window lifts and electric adjustment for the exterior mirror, which works very effectively. The front windows do not go completely down, and rear windows open only about half way

ACCORDING to the British BMW importers there is an annual market for over 3,000 top-price luxury cars in Britain, and it is expected that this country will be the second biggest export market for their new coupé, after the United States. Whether this prediction will hold good in the face of the current disastrous state of the economy and the exchange rate for sterling, remains to be seen. The Press release giving the price of the car as £13,599 had no sooner arrived than it was followed by another recording a substantial increase on account of the collapsed pound.

Be this as it may, one can only say that the BMW 633CSi deserves to do well; it is a magnificent car. It continues the traditions set by the former 3.0CSi, and offers the same standards of driving pleasure and enthusiast appeal in a roomier and more comfortable body, while the larger engine gives even more performance.

The power unit is the six-cylinder 3,210 c.c. ohc engine with Bosch L-Jetronic fuel injection, as fitted in the big 3.3Li saloon. Peak power is 200 bhp (DIN), developed at 5,500 rpm, exactly the same values as were quoted for the 3.0CSL lightweight coupé. The maximum torque shows a small increase, the value being 210 lb. ft. instead of 200. The new Coupé is quite a big car, so its weight of 29.3cwt at the kerb is heavy only in relation to the lightweight coupé, which was some 3½cwt lighter.

Comparisons with this former CSL model show the sort of moderate decline in all-out acceleration that is to be expected, the difference being only a second or two for all speeds up to 70 mph. At higher speeds, the appreciably larger body begins to tell, and the new 633CSi takes 42.4sec to reach 120 mph from rest, compared with 35.1sec for the CSL. Fairer comparison, though, is with the former CSi Coupé — since the csl lightweight Coupé was something of a special — and here it is seen that the 633 is quicker than its predecessor on all counts.

Our maximum speed runs were made in perfect conditions of still air, on a straight amply long enough for the ultimate speed to be stabilised at 131 mph. This speed corresponds to 5,800 rpm and so, in typical German fashion, is fairly well over the engine revs at which peak power is developed.

The CSi keeps a very level attitude when cornered hard. The front screen pillars are rather thick and tend to obstruct cornering vision, and the centre pillar also blocks the driver's over-the-shoulder view

AutoTEST
BMW 633 CSi

This 3.3-litre engine is a superb unit, and with Bosch L-Jetronic injection it seems to be even smoother and more responsive than before. The car trickles along smoothly with the traffic at about 30 mph in top gear, but then makes a wonderful change of character when gears and throttle are used to zoom the car forward to the front of the column. In full power getaways there is a quick screech of wheelspin and marked squat of the tail as the car surges forward. The changes up to second and third gears again bring a pronounced dip of the tail as the torque reaction compresses the springs, although the rear suspension has anti-squat geometry.

For once we have enjoyed a BMW without feeling that it is too low-geared. Top gear on the 633 is low enough to give lusty acceleration if one is in idle mood and not inclined to change down to third, yet it allows really relaxed cruising at three-figure speeds. It was a great delight to include a long Continental trip in the Road Test mileage, and the 633 settled down effortlessly at an indicated 110 mph for hour upon hour of motorway travel. Later speedometer correction showed this to be a true 105 mph. Particularly good is the ease with which the car resumes this sort of cruising speed without any need to change gear, after there has been need to shed 20 or 30 mph for slower traffic.

Transmission; economy

Automatic versions will become available later, but so good is the manual transmission, with its delightfully light yet positive change, that we certainly would not hurry anyone into choosing the automatic model. The gears are well-spaced, and thanks to the free-revving nature of the six-cylinder ohc engine, a really wide span of performance is available in each ratio. First gear takes the car to more than 30 mph, and yet is low enough for a very easy restart to be made on a 1-in-3 hill. There is no complaint if an attempt is made to start from rest in second gear, yet this ratio allows a maximum of 65 mph.

Equally impressive is third gear, which takes the car from as low as 10 mph (with only a trace of vibration) all the way to its 103 mph maximum.

In the indirect gears the 633 is extremely quiet, to the extent that it is easy, on occasion, to forget that third is still engaged and that a change up to top is due. A faint whine is audible from the back axle under light throttle loads in top gear.

A little care is needed to achieve quiet selection of reverse gear. Perhaps because of the fairly high (800 rpm) engine idling speed, it is necessary to keep the clutch depressed for some time before knocking the lever across to the left and into reverse; or the technique of baulking the layshaft by momentary selection of first gear before going into reverse can be used to avoid an undignified crunch of gears. The gear lever knob itself is neatly made in stitched leather, and the lever is encased in a little gaiter with a zip fastener — nice, if old-fashioned, touches of refinement.

Clutch take-up is smooth in normal conditions; a suggestion of judder came only when making the restart on the 1-in-3 test hill. In standing start take-offs it absorbs full power, causing controlled wheelspin. A limited slip differential is standard on manual transmission models. The clutch operating load is on the heavy side by today's standards, at 35lb, but this is not objectionable even in tedious traffic work.

Automatic mixture enrichment for cold starting is provided by the Bosch injection, and it compensates without any of the problems of hunting, stalling, or causing the engine to race unnecessarily when starts are made after a short spell of idleness, or before full warm-up is reached. In the first minute or two of running with a cold engine there is some combustion noise from the exhaust hot-spot, which diminishes quickly and disappears altogether as soon as the engine has warmed enough for the gases to be sent straight down the pipe instead of having to heat up the manifold.

To judge from the fuel consumption recorded, with an overall figure of 20.6 mpg for the complete test, it is clear that BMW have managed to improve the efficiency of this car — certainly when compared with the CSL, which returned only 16.7 mpg. The ability to better 20 mpg while regularly putting more than 90 miles into every hour in motorway travel is

Specification

ENGINE	
	Front engine, rear-wheel drive
Cylinders	6 in line
Main bearings	7
Cooling	Water
Fan	Viscous
Bore, mm (in.)	89 (3.5)
Stroke, mm (in.)	86 (3.39)
Capacity, cc (in³)	3,210
Valve gear	ohc
Camshaft drive	Chain
Compression ratio	9-to-1
Octane rating	97.5 RM
Carburation	Bosch L Jetronic fuel injection
Max power	200 bhp (DIN) at 5,500 rpm
Max torque	210 lb. ft. at 4,250 rpm

TRANSMISSION	
Type	Manual
Clutch	Diaphragm spring MF 240

Gear	Ratio	mph/1000rpm
Top	1.0	22.5
3rd	1.402	16.05
2nd	2.203	10.21
1st	3.855	5.84

Final drive gear	Hypoid, with limited slip differential
Ratio	3.25 to 1

SUSPENSION	
Front—location	Independent, oblique MacPherson struts
springs	Coil, eccentric
dampers	Telescopic
anti-roll bar	Yes
Rear—location	Independent, semi-trailing links
springs	Coil
dampers	Telescopic
anti-roll bar	Yes

STEERING	
Type	ZF worm and roller
Power assistance	Standard
Wheel diameter	14.7 in.

BRAKES	
Front	ATE 11 in. dia. ventilated disc
Rear	ATE 10.9 in. dia. ventilated disc with inner drum for hand brake
Servo	Vacuum

WHEELS	
Type	Alloy
Rim width	6J
Tyres—make	Michelin
—type	XWX radial
—size	195/70 VR 14

EQUIPMENT	
Battery	12 volt 66 Ah
Alternator	75 amp
Headlamps	4 x 55-watt Halogen
Reversing lamps	Standard
Hazard warning	Standard
Electric fuses	17
Screen wipers	Two-speed plus intermittent
Screen washer	Electric
Interior heater	Water valve
Interior trim	Leather and cloth seats, fabric head-lining
Floor covering	Carpet
Jack	Screw pillar, winding handle
Jacking points	Two each side under sills
Windscreen	Laminated
Underbody protection	Valvoline Tectyle ML

MAINTENANCE	
Fuel tank	15.4 Imp. galls (70 litres)
Cooling system	21 pints (inc. heater)
Engine sump	8.8 pints SAE 20W/50
Gearbox	1.9 pints SAE 80
Final drive	2.6 pints SAE 90 Hypoid
Grease	No points
Valve clearance	Inlet 0.013 in. (hot) Exhaust 0.013 in. (hot)
Contact breaker	Contactless transistorised ignition
Ignition timing	0 deg. BTDC (static) 22 deg. BTDC (stroboscopic at 1,800)
Spark plug—type	Bosch W175T30
—gap	0.028 in.
Tyre pressures	F33, R30 psi (normal driving)
Max payload	740 lb (336 kg)

A large and clearly-marked speedometer is flanked by fuel gauge and thermometer in a round dial to the left, a rev counter to the right. There are individual warning tell-tales for low fuel level or high coolant temperature. An illuminated button switch for the heated rear window is to the left of the central facia vent. Although the console is angled towards the driver, it is still very difficult to read the clock. Four thumb pads in the steering wheel spokes sound the deep, windtone horns. The lights switch is beside the steering column, and there are lever switches for indicators and lamp flashing or dipping (left) and wipers (right) including an intermittent wipe position. Electric window switches are beside the gear lever

Maximum Speeds

Gear	mph	kph	rpm
Top (mean)	131	211	5,830
(best)	132	212	5,850
3rd	102	164	6,400
2nd	65	105	6,400
1st	37	60	6,400

Acceleration

True mph	Time secs	Speedo mph
30	2.9	33
40	4.4	43
50	6.1	53
60	8.1	63
70	10.7	74
80	13.7	84
90	17.4	94
100	21.9	104
110	27.6	115
120	39.4	125

Standing ¼-mile: **14.9 sec**, 83 mph
Standing kilometre: **30.3 sec**, 112 mph

mph	Top	3rd	2nd
10-30	—	6.1	3.7
20-40	8.9	5.6	3.4
30-50	8.7	5.4	3.6
40-60	8.5	5.4	3.6
50-70	9.4	5.4	—
60-80	9.5	6.6	—
70-90	9.8	7.3	—
80-100	10.1	9.4	—
90-110	13.2	—	—
100-120	18.6	—	—

Consumption

Fuel
Overall mpg: **20.6**
(13.7 litres/100km)
Calculated (DIN) mpg: **25.5**
(11.0 litres/100km)
Constant speed (manufacturer's figures):

mph	mpg
30	31.0
40	32.8
50	32.1
60	30.3
70	28.0
80	25.5
90	21.9
100	19.3

Brakes

Fade (from 70 mph in neutral)
Pedal load for 0.5g stops in lb

	start/end		start/end
1	25-20	6	30-30
2	30-20	7	30-30
3	30-20	8	30-30
4	30-30	9	30-30
5	30-35	10	30-30

Response (from 30 mph in neutral)

Load	g	Distance
20lb	0.30g	100ft
35lb	0.95g	32ft
Handbrake	0.30	100ft

Max. gradient 1 in 3

Clutch Pedal 35lb and 6in.

Autocar formula
Hard driving, difficult conditions
18.5 mpg
Average driving, average conditions
22.6 mpg
Gentle driving, easy conditions
26.8 mpg
Grade of fuel: Premium, 4-star
(98 RM)
Mileage recorder: 1 per cent over reading

Oil
Consumption (SAE20W/50)
1,000 miles/pint

Test Conditions

Wind: 5-10 mph
Temperature: 18 deg C (65 deg F)
Barometer: 29.5 in. Hg
Humidity: 63 per cent
Surface: dry asphalt and concrete
Test distance: 1,307 miles

Figures taken at 3,500 miles by our own staff at the Motor Industry Research Association proving ground at Nuneaton, and on the Continent

All Autocar test results are subject to world copyright and may not be reproduced in whole or part without the Editor's written permission

Regular Service

Interval

Change	5,000	10,000
Engine oil	Yes	Yes
Oil filter	Yes	Yes
Gearbox oil	No	Yes
Spark plugs	No	Yes
Air cleaner	No	Yes
C/breaker	No	Yes
Total cost	**£6.45**	**£24.73**

(Assuming labour at £4.30/hour)
Note: Initial service and first 5,000-mile service free

Parts Cost
(including VAT)

Brake pads (2 wheels) — front	£8.01
Brake pads (2 wheels) — rear	£6.27
Silencers	£43.02 (front)
	£45.36 (rear)
Tyre — each (typical advertised)	£44.00
Windscreen	£105.86
Headlamp unit	£11.15
Front wing	£95.47
Rear bumper	£76.19

Warranty Period
12 months/unlimited mileage

Weight

Kerb, 29.3cwt/3,280lb/1,490kg
(Distribution F/R, 56.6/43.4)
As tested, 31.4cwt/3,520lb/1,560kg

Boot capacity: 14.6 cu. ft.

Turning circles:
Between kerbs
 L, 34ft 1in; R, 34ft 4in.
Between walls
 L, 36ft. 7in. R, 36ft. 10in.
Turns, lock to lock 3.6

Test Scorecard

(Average of scoring by *Autocar* Road Test team)

Ratings: 6 Excellent
 5 Good
 4 Better than average
 3 Worse than average
 2 Poor
 1 Bad

PERFORMANCE	5.33
STEERING AND HANDLING	4.66
BRAKES	5.20
COMFORT IN FRONT	4.42
COMFORT IN BACK	3.14
DRIVERS AIDS	4.38
(instruments, lights, wipers, visibility etc.)	
CONTROLS	4.75
NOISE	4.67
STOWAGE	4.67
ROUTINE SERVICE	4.60
(under-bonnet access: dipstick etc.)	
EASE OF DRIVING	4.91
OVERALL RATING	**4.55**

Comparisons

Car	Price £	max mph	0-60 sec	overall mpg	capacity c.c.	power bhp	wheelbase in.	length in.	width in.	kerb weight cwt.	fuel gall	tyre size
BMW 633CSi	13,980	131	8.1	20.6	3,210	200	103	187	69	29.3	15.4	195/70 VR 14
Aston Martin V8	14,836	145	7.5	11.7	5,340	—	103	184	72	35.6	21.0	GR 70/V15
Ferrari 365 GT4 2+2	15,719	150	7.0	11.2	4,390	340	106	189	71	29.6	26.0	215 VR 15
Jaguar XJS	10,507	153	6.9	15.4	5,343	285	102	192	71	34.8	20.0	205/70 VR 15
Mercedes-Benz 450SLC	12,874	136	9.0	14.1	4,520	225	111	187	71	32.9	19.8	205/70 VR 14

BMW 633 CSi

much appreciated, as is the resultant ability to cover the better part of 300 miles between fuel stops. Tank capacity, at 15.4 gallons, is the same as for the 525/528. With moderate use of the performance, owners should quite readily stretch the economy to 25 mpg.

Steering, ride and handling

One is never inclined to enthusiasm on seeing the words "worm and roller steering" in the specification of a car, but as applied to the 633 the system gives little cause for complaint. It is not as precise as a good rack and pinion system, and leaves one sometimes a little doubtful of the ability to shoot a narrow gap or to counter the wind-spill from a heavy lorry. But there is little lost movement or wander at speed, and we felt very confident of the car's directional stability when making our maximum speed tests. The ZF power assistance is standard, and makes light work of manoeuvring or tugging the car round tight corners, while remaining unobtrusive when accurate response is needed. The assistance provided diminishes as speed increases. The 34ft turning circle measured is quite fair for a car of this wheelbase.

MacPherson struts are used for the front suspension, with eccentric arrangement of the coil springs to prevent "stiction"; and at the rear the 633 has the familiar BMW layout of independent semi-trailing arms with coil springs. A very substantial-looking anti-roll bar is fitted at each end, and it takes really hard cornering to make the car lean over noticeably on the outside front wheel. The degree of understeer is perhaps less than a 57/43 weight distribution would suggest and, with plenty of power applied, the car takes fast bends in a very reassuring manner. Tyre squeal gives plenty of warning before the adhesion limit is reached on a dry road, and grip is also good in the wet, though the power then has to be used with moderation to avoid provoking a tail slide. Like

Bumpers are rubber-faced and have energy-absorbing mountings. A gutter on the screen pillar helps to keep the side windows clear

A full-length rubber-faced guard rail protects the sides from car park knocks. Alpina alloy wheels are standard

most of the BMW range, the 633 lets the driver provoke a slide fairly readily but, when he does, it is easy to maintain control because of the good steering response.

On the firm side, as to be expected, the suspension gives a comfortable ride and has a good range of vertical travel to cope with uneven sections of motorway at three-figure speeds. It is surprisingly upset by transverse ridges such as are encountered on concrete surfaces, provoking a sharp thump and slight lurch. The ride becomes impressively better the higher the speed, and the standard of comfort is generally very good indeed. There is just an occasional tendency to wallow and rear-end floating, which makes one feel that slightly harder damping would be an improvement.

Noise; brakes

Insulation of road noise is excellent, whatever the surface; but we were a little disappointed in the sealing of the side windows. When we first tried the car, there was excessive wind whistle around the triangular post which serves as a steady for the leading edge of the frameless door window. After drawing attention to this it was adjusted on the driver's side, but we then became aware of excessive wind noise at the trailing edge on the passenger's side. Owners would be wise to be fussy on this point, and insist on the dealer getting it right, because it is clear that the car *can* be really quiet if wind noise is not allowed to spoil the general air of mechanical refinement. However, BMW should perhaps examine an Opel Commodore Coupé to

see how big, frameless windows can be made to be completely silent and trouble-free.

Ventilated disc brakes are fitted both front and rear; the hydraulic circuit to the front brakes is duplicated, and there is a pressure-limiting valve in the line to the rear brakes. In general use, the impression given is of excellent brake response, due to the vigorous servo assistance. On test, however, it proved rather too easy to lock the front wheels, with only 40lb pedal load, leaving heavy black lines on the road and reducing efficiency from its peak of 95 per cent to 85. Slightly less pedal load gives quicker stopping, without marking the road, but is rather difficult to achieve accurately in an emergency.

Fade tests produced no noticeable increase in pedal load, but dampened the slight tendency for the brakes to "wrap on" once they had heated up after the third or fourth application. The handbrake operates separate drums within the rear discs, and is capable of 0.3g efficiency. The lever does not need to be pulled up particularly hard to hold the car securely on 1 in 3, facing up or down.

Comfort, fittings and equipment

Although it is the superb engine and responsiveness of the 633 which will delight the owner once he has bought the car, it is probably the impressive interior finish and layout that will sell it to him in the first place. The seats are upholstered in cloth on the wearing surfaces, with the edges, sides, and backs trimmed in leather with piped joints.

As often the case with German cars, the first impression is that the seats are rather hard for a person of light build but, in the test of long hours at the wheel, their correct shaping, with adequate lateral and squab support, ensures that they remain comfortable and that occupants step out without any aches and pains after a long journey. The rear seats are similarly opulent and, providing the front seat occupants

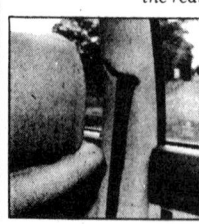

Inertia reel seat belts are neatly built into the side pillar. A release lever on the squab of each front seat allows it to tip forward for access to the rear

Really comfortable seating is provided in the rear, though front seats have to be moved forward a little for adequate legroom. Rear seat belts are included in the standard BMW delivery charge

co-operate by moving their seats forward an inch or two to provide adequate rear legroom, one can travel in the back very comfortably indeed. Forward view from the rear seats is spoilt by the large front headrests, which cannot, it seems, be removed.

The rear seats also extend up into large headrest structures, the backs of which can be tilted upwards to reveal useful out-of-sight stowage for cameras or other possessions. One of these contains a Hartmann first aid kit. There is also a spacious, lockable, drop-down compartment beneath the facia on the passenger side, and an open cubby in the centre of the console, ahead of the gear lever and ashtray. There is also a recess for oddments on the top of the facia panel. A rechargeable (12-volt) torch is fitted beside the glove box.

The facia itself is a neat, one-piece moulding across the full width of the car, neatly marrying up with matching black cappings to the doors. Instruments are well sited

Although the engine is rather buried beneath its fuel injection equipment and air trunking, access to all service points is good, and the layout is orderly. There is a light on the inside of the boot lid, wired through the side lamps

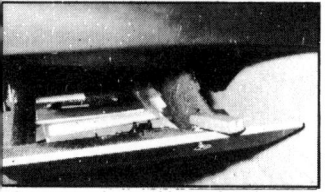

Above: A rechargeable torch is fitted at the side of the under-facia glove box
Below: Lift up panels on the rear shelf behind each back seat headrest reveal useful stowage compartments, one of them being occupied by the standard first aid kit

for the driver to see through the upper half of the four-spoke steering wheel, and comprise a large central speedometer with at-a-touch trip milometer reset, and a slightly smaller circular dial on each side. That on the right is the clearly marked rev counter, and on the left, the dial is divided between fuel gauge and coolant thermometer, each with its own relevant warning tell-tale in the base of the instrument. Above these dials are warning tell-tales in two groups of three, with space for two spares.

Ventilation is all too often rather a weak point on BMWs and, on the coupé, even with the provision of five outlets for cool air, the flow of air tends to be inadequate unless the rear side windows are opened. This immediately reduces the pressure and allows air to flow, but results in excessive wind noise and buffeting at speed. The heater has vigorous output, and is a little more responsive to its temperature control than is usually the case when a water valve system is used. The booster speed is infinitely variable.

A comfortable driving position is assured by the ability to raise or tilt the cushion of the driving seat, using the lever to the left of the driving seat. The steering column is also telescopically adjustable. A lever to the right of the seat allows backrest adjustment through fine notches, down to full recline, and there is a release near the front outside edge for to-and-fro adjustment of the whole seat. An additional release lever on the side of the squab allows either backrest to tip forward to give access to the rear seats.

The doors require a hefty slam to close them fully — otherwise they stop on the safety catch position. A master key works all the locks on the car, and there is another key for attendant's use, which does not open the glove box or boot. The passenger door can be locked from outside by pushing down the button before closing the door.

Test at a touch

Typical of the attention to detail and the emphasis on safety is the systems check panel to the right of the facia. A small touch-bar marked TEST is pressed, and five labelled green tell-tales light up to confirm that all is well with the engine oil level, brake fluid, coolant, screen-washer contents, and disc brake pads. Failure of any light to come on reveals a fault — except in the case of the engine oil light, which comes on only if the test is made

Above: Systems check on the right of the facia allows all important services to be confirmed at a glance before moving off

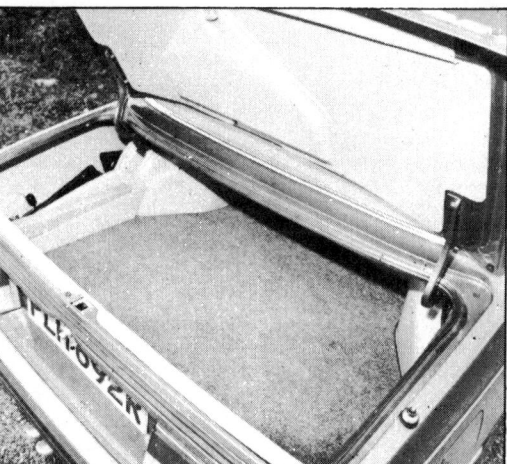

Left: Compartments at the side of the boot give stowage for the toolkit and a hazard triangle, and a magnificent toolkit is built into a drop-down tray in the under-surface of the boot lid

when the engine is not running. Two more tell-tales at the same time check the rear lamps and stop lamps, but these are energized only when these lamps are on.

It is less complicated than it sounds. Before moving off, switch on side lamps, put a foot on the brake pedal and touch the button; if seven tell-tales come on, you know all is well.

Further touches of refinement are the standard fitting of electrically-operated windows with switches on the console around the gear lever; an electrically-adjustable, door-mounted outside mirror; and the provision of interior lights under the bonnet and in the boot. The engine room light is wired through the ignition for security reasons, and have an isolator switch. The driver's window on the test car tended to open jerkily, especially if the glass was wet.

Where it fits in

In Germany there is also a carburettor version of this car, the 630CS, but the importers have wisely decided to concentrate on the 633 with injection and the slightly larger engine capacity. Standard equipment for the UK includes tinted glass, four automatic seat belts, and a delay switch for the interior lights, which keeps them on for some five seconds after closing a door. The price puts the 633 in that realm of the fantastic, where the odd thousand pounds here or there is of little consequence, and its chief competition is from the Jaguar XJS, Mercedes-Benz 450SLC, Aston Martin V8, and the Ferrari 365 GT4. Even amongst such formidable rivals, the 633 has a great deal to offer, particularly on the score of appearance both inside and out, and economy.

Conclusion

BMW are, with this new model, moving into the sort of market where buyers will justifiably expect great things. We can only say on conclusion of a test we have thoroughly enjoyed, that owners are not likely to be disappointed. It is a very impressive car indeed.

MANUFACTURER:
Bayerische Motoren Werke AG; München, West Germany

UK CONCESSIONAIRES:
BMW Concessionaires GB Ltd., 991 Great West Road, Brentford, Middlesex.

PRICES
Basic	£11,948.70
Special Car Tax	£995.73
VAT	£1,035.56
Total (in GB)	**£13,980.00**
Seat belts (front and rear) Licence, number plates, Delivery charge, full tank of fuel, and anti-freeze	£183.00
Total on the Road (exc. insurance)	**£14,163**
Insurance	On application

EXTRAS
None listed.

TOTAL AS TESTED ON THE ROAD £14,163

BMW 630CSi

Will the Bavarians' new Valkyrie carry off the XJS and 450SLC to Valhalla?

MOST CAR ENTHUSIASTS know the initials BMW stand for Bayerische Motoren Werke (or Bavarian Motor Works). But few are aware that BMW traces its roots back to a company called the Bayerische Flugzeug Werke (or Bavarian Aircraft Works), founded in 1916 by Gustav Otto, son of August Otto, inventor of the 4-stroke-cycle engine. The aircraft heritage explains where the circular blue and white quartered emblem comes from. Blue and white are the Bavarian state colors and the badge symbolizes propeller motion. The company's first product, a 6-cylinder single-overhead-camshaft engine, immediately set a world record for high-altitude flight, and another early aircraft engine powered Germany's famous Red Baron.

When peace came to Europe in 1918, BMW began to diversify, producing its first motorcycle in 1923. Five years later, BMW began building cars, an additional project to their aircraft work which went on until the end of World War II. Success quickly came on the road, just as in the air, and a factory team of three BMW 3/15s (a derivative of an English Austin Seven originally called the Dixi) took the team prize in the 1929 5-day Alpine Trial. This first taste of victory helped to orient BMW to sporting competition, to the development of numerous racing machines and production cars designed for the person who loves to drive, a tradition carried on by the latest in a long line of sporting coupes, the 630CSi. (For additional historical perspective on this latest coupe plus styling analyses and driving impressions of some of its important predecessors, please refer to Jon Thompson's styling analysis in the June 1977 issue of R&T.)

Ron Wakefield gave a technical rundown on the new coupe in June 1976 and Dottie Clendenin drove the European 633CSi in Germany two months later. Now the coupe has come to America and we've had a good opportunity to extensively drive and test it on familiar roads in its U.S. form.

In the upper stratosphere of car prices in which the 630CSi competes, how much a car costs is not simply a question of design sophistication and standard equipment but of what the market will bear. And when you start talking about spending around $20,000, a car becomes a relatively price-insensitive commodity. In other words, someone who's rich enough to afford a $20,000 car can in most instances as easily spend another $5000 or even another $10,000 if he or she wants a particular car badly enough. With that as a background, it's not surprising that BMW should ask close to $24,000 for the 630CSi. After all, the Jaguar XJS lists for $20,000 and a Mercedes 450SLC for $27,000. It isn't surprising, but it is dismaying as it puts this car way out of the reach of even a semi-affluent buyer. But it does promise the person who can afford it a certain air of exclusivity. And to some buyers this is all important.

Adding insult to injury, BMW has the audacity to charge extra for the radio. BMW believes a person who spends $24,000 would want to personalize the car with his own sound system. This might be true, but the least BMW could do would be to tell a 630CSi owner to install the system of his choice (up to a maximum of $500 seems fair) and send the factory the bill. Admittedly, the base price includes several items—air conditioning, electric window lifts, fuel injection—not standard on the previous coupe, the 3.0CS, but still, charging extra for a radio on this type car seems a bit much.

The trend in interior car design these days is toward safety and function. The old coupe's interior was rich in wood, with paneling across the dash and along the sides. In the new one there's no wood at all, but rather the latest in padded, ergonomic, plastic contours with a black finish that is functional and glarefree but lacking in warmth. One exception is the top of the dash which reflects into the windshield if the sun strikes the car at a certain angle.

Although a few drivers mentioned lack of lateral support during spirited cornering, overall the front seats score high marks. The driver's seat has the usual rake and fore/aft adjustments plus a 2-way lever that allows not only the height but the angle of the whole seat to be varied. Combine this with a telescopic steering wheel and you've got a choice of driving positions to fit virtually any driver.

Directly in front of the driver are the eminently readable instrument dials, bathed in easy-on-the-eyes orange light at night, and the main warning lights. To the left is a test panel that allows the driver to monitor engine oil, brake fluid, coolant and windshield-washer fluid levels, plus the condition of the front brake pads and operation of brake- and taillights by pushing a single bar labeled "test." Above the curving dash that wraps into the doors are the outlets for the heating-ventilation-air conditioning system. These four vents in conjunction with one additional outlet directly above the logically arranged and easy-to-understand temperature, fan speed and air distribution dials provide 630CSi occupants with appreciably better control of their environment than was possible with the 3.0CS. For around-town driving you need to have the fan going to boost ventilation, but the outlets are more directable than on any previous BMW and fan noise is reasonably subdued. If you set the blower motor to the high position in the old coupe, it sounded like a hurricane blowing at full gale.

Backseat accommodations are just about what you'd expect from a coupe. There's more room than in the 3.0CS but we still wouldn't recommend them for long journeys. Between the seats is a fold-down center armrest and above each is an adjustable head restraint. Molded into the package tray behind the restraints are small compartments. Lift the lid and you've got space to store odds and ends. For larger items there's the expected voluminous glove box up front containing a cordless plug-in trouble light, plus the expected snap-shut pockets in each of the doors.

Powering the 630CSi is the 3-liter single-overhead-camshaft inline six. We've said it before but it's worth repeating: This engine is without a doubt the most sophisticated production inline six in the world. Although the effects of retarded ignition timing and exhaust-gas recirculation rob it of some of its around-town response, when driven hard it frees up, smooths out and goes nearly as well as the 3.0CS we tested in 1973. The 630CSi is quicker from 0–60 mph, only a tick slower in the quarter mile and only begins to lose out to the less tightly controlled 1973 version in the illegal speed ranges. That's quite an accomplishment when you consider the much stricter emission regulations the current

We tested the 4-speed 630CSi but used an automatic for photography.

engine (a 49-state version) has to meet. Give credit to the Bosch L-Jetronic for the engine's exceptional driveability. The engine lights instantly when cold, can be driven away immediately without stumbling or stalling and warms up quickly, albeit with a bit of roughness when the enrichment first shuts off. And the engine returned 18.0 mpg in our fuel economy test, quite a respectable figure for a car with a curb weight of 3510 lb and enough performance to propel it to a speed of 60 mph in less than 10 sec.

The 4-speed gearbox is a real delight. The light clutch and crisp, precise gear changes are perfectly matched to the sporty character of the car and really let the driver get a lot out of what is a relatively small-displacement engine.

The new coupe is quieter at all speeds than its predecessor, but that old BMW bugaboo—wind noise—is still present. If a front side window is lowered and then raised after closing the doors, the electric motors can't crank out enough power to properly seat the glass into the rubber seals. The noisy differential Ron Wakefield noted in the two cars he drove in Europe wasn't present in the one we tested. In its place was a low rumbling vibration most evident around 40–50 mph that seemed to be emanating from the center driveshaft bearing. It's nothing we'd expect to be characteristic of the car, however.

To appreciate how BMW can even begin to justify charging $24,000 for this car, one must drive it . . . and drive it hard. Out on a twisty road you discover the combination of ride, handling, braking and steering that make the coupe one of the world's best road-going GTs. On a skidpad the 630CSi generates 0.754g, higher than all but a handful of all-out sports cars, and the way it

hangs on during fast transitions as in our slalom test is confidence inspiring. BMW's characteristic final oversteer is there, but there's no abrupt transition from understeer to oversteer if you back off the throttle or stab the brakes when cornering hard as is sometimes the case with semi-trailing-arm rear suspension. You have to work really hard to provoke the 630CSi into anything unexpected. The ride is comfortable and soft, so soft that some drivers would equate it with some American luxury cars. Don't be deceived. The ride is not only soft but well controlled and is accompanied by superb handling that is achieved by all-around independent suspension, anti-roll bars, exceptionally responsive yet compliant Michelin radial tires and plenty of suspension travel.

Power steering similar to the recirculating-ball type used in the 530i is standard. More interesting is the variable nature of the assist. Up to about 2000 rpm the pump delivers full boost for easy parking and low-speed maneuvers. From there up to redline the assist is reduced to 70 percent of maximum, so that at high speeds the steering communicates a firmer feel to the driver. The coupe's steering is wonderfully exact and accurate, giving the driver a computer-like ability to instinctively crank in precisely the right amount of lock regardless of the speed or the tightness of a turn. Even the impressive Mercedes steering, which we have considered the world's finest, can't top the BMW's in overall road feel, effort and precision response. Marvelous.

Vented discs are fitted front and rear. During panic stops there's a tendency for the fronts to lock but the ease of pedal modulation makes for straight and undramatic stops. In normal driving they're even better; pedal effort is just about ideal and the linear relationship between effort and deceleration rate makes for smooth comfortable stops, something your passengers will appreciate.

As the ultimate BMW currently built for the U.S., the 630CSi answers the question: "What do I buy if I want to outdo my friends who drive Jaguar XJSs and Mercedes 450SLCs?" The price virtually assures you won't be seeing another one in the same neighborhood, much less on the same block, and its driving characteristics place it among the world's best GTs. But we're left with one nagging thought. A 530i doesn't give up much to a 630CSi, has four doors, comfortable seating for four (or a friendly five) and costs about half as much. Whether the coupe's distinctive styling and slightly better overall performance are worth an extra $11,000 is something only your ego and your pocketbook can decide.

ROAD TEST
BMW 630 CSi

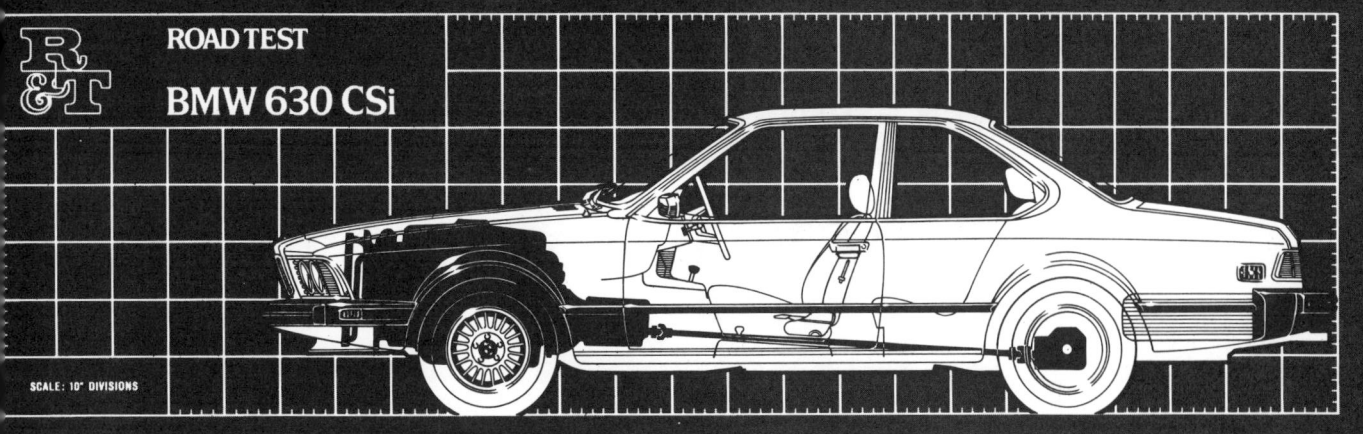

SCALE: 10" DIVISIONS

PRICE
List price, east coast......... $23,600
List price, west coast......... $23,700
Price as tested, east coast.. $24,175
Price as tested includes metallic paint ($325) dealer prep ($150)

IMPORTER
BMW of North America, Inc
BMW Plaza
Montvale, N.J. 07645

GENERAL
Curb weight, lb 3510
Test weight 3630
Weight distribution (with driver),
   front/rear, % 56/44
Wheelbase, in. 103.4
Track, front/rear 56.0/58.8
Length 192.7
Width 67.9
Height 53.7
Ground clearance 3.7
Overhang, front/rear 42.3/47.0
Usable trunk space, cu ft 15.7
Fuel capacity, U.S. gal. 16.5

ENGINE
Type .... sohc inline 6
Bore x stroke, mm .... 89.0 x 80.0
  Equivalent in. .... 3.50 x 3.15
Displacement, cc/cu in. ...2985/182
Compression ratio 8.1:1
Bhp @ rpm, net .... 176 @ 5500
  Equivalent mph 118
Torque @ rpm, lb-ft ... 185 @ 4500
  Equivalent mph 96
Fuel injection .... Bosch L-Jetronic
Fuel requirement ...... regular, 87-oct
Exhaust-emission control equipment:
  air injection, thermal reactor, exhaust-gas recirculation

DRIVETRAIN
Transmission .... 4-sp manual
Gear ratios: 4th (1.00) 3.45:1
            3rd (1.40) 4.83:1
            2nd (2.20) 7.59:1
            1st (3.86) 13.32:1
Final drive ratio 3.45:1

ACCOMMODATION
Seating capacity, persons 4
Seat width, f/r, in. .. 2 x 21.5/2 x 20.5
Head room, f/r 37.0/35.0
Seat back adjustment, deg 70

CHASSIS & BODY
Layout .... front engine/rear drive
Body/frame .... unit steel
Brake system . 11.0-in. vented discs front, 10.7-in. vented discs rear, vacuum assisted
Swept area, sq in. 547
Wheels .... cast alloy, 14 x 6J
Tires .... Michelin XVS, 195/70HR-14
Steering type .... recirc ball, power assisted
Overall ratio 16.9:1
Turns, lock-to-lock 3.6
Turning circle, ft 37.4
Front suspension: MacPherson struts, lower lateral links, compliance struts, coil springs, tube shocks, anti-roll bar
Rear suspension: semi-trailing arms, coil springs, tube shocks, anti-roll bar

MAINTENANCE
Service intervals, mi:
  Oil change 6000
  Filter change 6000
  Chassis lube .... none
Tuneup 12,500
Warranty, mo/mi 12/unlimited

INSTRUMENTATION
Instruments: 140-mph speedo, 7000-rpm tach, 999,999 odo, 999.9 trip odo, coolant temp, fuel level, clock
Warning lights: oil press., brake system, alternator, coolant temp, low fuel, rear-window heat, thermal reactor, exhaust-gas recirc, test panel (see text), seatbelts, hazard, high beam, directionals

CALCULATED DATA
Lb/bhp (test weight) 20.6
Mph/1000 rpm (4th gear) 21.4
Engine revs/mi (60 mph) 2800
Piston travel, ft/mi 1470
R&T steering index 1.35
Brake swept area, sq in./ton .. 301

RELIABILITY
From R&T Owner Surveys the average number of problem areas for all models surveyed is 12. An average of 7 of these problem areas is considered serious enough to constitute reliability areas that could keep the car off the road. As owners of earlier-model BMWs reported 9 problem areas and 3 reliability areas we expect the overall reliability of the BMW 630CSi to be better than average.

ROAD TEST RESULTS

ACCELERATION
Time to distance, sec:
  0-100 ft 3.6
  0-500 ft 9.5
  0-1320 ft (¼ mi) 17.5
Speed at end of ¼ mi, mph 81.0
Time to speed, sec:
  0-30 mph 3.2
  0-40 mph 5.2
  0-50 mph 7.2
  0-60 mph 9.7
  0-70 mph 13.2
  0-80 mph 17.0
  0-100 mph 31.0

SPEEDS IN GEARS
4th gear (5700 rpm) 122
3rd (6400) 94
2nd (6400) 61
1st (6400) 35

FUEL ECONOMY
Normal driving, mpg 18.0
Cruising range, mi (1-gal. res) .. 279

HANDLING
Speed on 100-ft radius, mph .. 33.6
Lateral acceleration, g 0.754
Speed thru 700-ft slalom, mph..55.5

BRAKES
Minimum stopping distances, ft:
  From 60 mph 160
  From 80 mph 275
Control in panic stop .... very good
Pedal effort for 0.5g stop, lb 25
Fade: percent increase in pedal effort to maintain 0.5g deceleration in 6 stops from 60 mph..32
Parking: hold 30% grade? .... yes
Overall brake rating .... very good

INTERIOR NOISE
All noise readings in dBA:
Idle in neutral 57
Maximum, 1st gear 80
Constant 30 mph 63
         50 mph 68
         70 mph 72
         90 mph 77

SPEEDOMETER ERROR
30 mph indicated is actually .. 30.0
50 mph 49.5
60 mph 59.5
70 mph 69.5
80 mph 78.5
Odometer, 10.0 mi 10.0

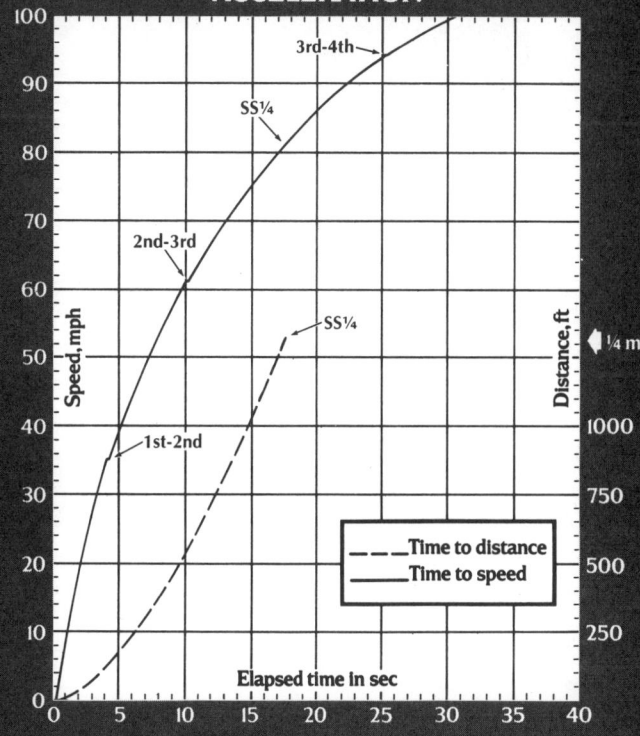

ACCELERATION

BMW 630CSi

This car's worth depends upon whether you're looking for a car, or for a status symbol.

ROAD TEST

There it stands, not quite hat-in-hand and coaxed to its spot by ambitious parentage, at the portals of a very exclusive club: Cars Costing Over $20,000. It's a snobby joint and you don't get invited; you have to qualify all by yourself and if your folks were just a little too optimistic when they pushed you there you will have a tough time of it. This club is only for the real cars, the ones actually in some sort of real production, and so, excluding the non-qualifiers like Rolls-Royce, Ferrari and the like, the current members are Mercedes-Benz, Porsche, and Jaguar, seasoned veterans of the Beverly Hills marketing wars. And standing at the portal, not *quite* hat-in-hand, is the new member, the BMW 630CSi, wondering about being accepted.

That this car even exists should tell you a lot about the hopes and aspirations of the parent corporation, and that it exists in the U.S. should tell you a lot about the marketing

PHOTOGRAPHY: LARRY GRIFFIN

hopes of BMW of North America, the importing arm of the company. Because it isn't *just* twenty thousand bucks; by the time our proud owner has ordered a normal complement of options it will be closer to *twenty-five*. Think about that.

Now there are, obviously, several ways to look at this. For most of us—those to whom this kind of car is only a photograph on a page or a blur on a highway—it would seem obvious that for twenty-five grand it ought to be a hell of an *automobile*. But that type of thinking is found only among those who must buy—at least to an extent—for value simply because their bank books won't allow frivolity or displays of excess social status. On the other hand, folks who can actually afford that kind of money for a car are more concerned that it be a *hell* of an automobile, and it's important to not lose sight of that difference. You see, probably the most important point of a twenty-five grand car is that it actually cost twenty-five grand, and until you understand that you will never be on the same wave length with the potential buyers of the 630.

And that is the only way the 630 holds up, because judged on its merits as a car and balanced against that price tag, it just doesn't make it. If you're after twenty-five grand worth of automotive pleasure there are better ways to spend your money.

One of those better ways would be to go down to the BMW showroom, walk past the 630 and buy the 530i for about half the money, leaving some twelve thousand to blow at Malibu Grand Prix for a dollar a lap or maybe to spend on a winning Super Modified to race on half-mile dirt ovals on Saturday nights. Either way, you'd still have the 530i.

And that's the rub. Because the 630CSi is really a two-door version of the four-door 530i. Both cars use the same engine and transmission. Except for minor details the chassis is shared, including the MacPherson strut front and semi-trailing arm rear suspension. Underneath the skin the only worthwhile change is that the four disc brakes on the 630 are vented and those on the 530i are not. The real difference is in the sheet metal; off came the 530i's smart, attractive four-door body and on went the 630 CSi's smart, attractive two-door body. In the bargain the 630 gained about 200 pounds and lost a lot of interior room, mainly in the head and leg areas and especially in the back seat (which is now +2 at best).

Add it up. For twice as much money you get a car with practically identical running gear, less interior room and more weight. And it goes slower.

So, obviously, it will have to sell to those who don't buy cars based upon value alone.

Now, deciphering that kind of buyer is kind of tough, especially if you've never been one. But, in the end of town where Mercedes 450s are referred to as Beverly Hills Volkswagens the 630 will have some attributes. First, the price. Like we said, just costing that much is a positive point in that market, because you can't have your neighbors thinking you're on the down and out when you roll a BMW into your driveway; they might think it's like the 320i they gave their kid for his sixteenth birthday. But that should only be a concern for the early buyers, because once word gets around, the town will know it's o.k.

And being fair, you also get a very nice car. In fact, judging it as Very Nice is perhaps the most accurate evaluation. It is, first and foremost, extremely handsome in a very refined, civilized way. It has a balance of form, shape and line rarely achieved. You look at it and little things keep catching your eye, like the subtle and deft creases that highlight the shape perfectly, and the louvers on the B-pillar, gently tapered together at the top, the line of each one of them extending into space and converging at a common point with the lines of the front and rear roof pillars, things you don't notice at first, all combining to make one very handsome car. It settles in that there is nothing silly or

childish here, and so another word for the car comes to mind: Adult.

It is also very well executed. Panels all fit properly. There is integrity in the feel of the mechanism. The interior is full of the smell of real leather. The finish was the best we have seen in a long, long time. But then, for twenty-five thousand bucks it ought to be pretty good.

And so those things, the things in the 630 that the owner and his neighbors can see and touch and feel and smell, will perhaps combine to help give the buyer what he really needs in a car: status in his peer group. And you better believe it, in this market that's what counts.

And our buyer better stay with those parameters, because trying to justify it for its value as a car will leave him stranded. Performance? Middle-of-the-pack, with 0–60 in 10.6, (accompanied by enormous amounts of rear wheel hop under hard acceleration, courtesy of the semi-trailing arms), emergency stopping distances from 60 mph averaging 160 feet (with immediate and uncontrollable front-wheel lock up), interior noise level at 70 mph of 73 dBA (a partial result of an improperly sealing driver's side window, which was probably unique to our test car but there just the same), and lateral acceleration a respectable 0.75g, a respectability unfortunately not matched by the subjective feel and open-road behavior which was characterized by a little tail happiness at the limit and a good dose of trailing throttle oversteer, typical of all those cars similarly suspended. The engine, like all BMWs, performed flawlessly, with immediate starts in all weather and smooth, quiet running all the way to redline. Just like it does in the 530i.

The interior too is good—respectable—but not particularly well-done or luxurious. Modern and German-looking, it has definite family resemblance to the 530i and 320i. But the family resemblance means that, except for a monitoring system of lights on the left hand side of the dash the controls and instruments are almost all straight from the lower-priced little brothers, complete with little lapses, like the clock-with-fan-speed-switch that seems so neat in the other cars but which is angled down in the 630 so you can't see the top of the clock face. Seating comfort is only mediocre, and for a BMW—where it is usually excellent—mediocre isn't very good. True, the seat will adjust up and down, back and forth, will tilt the cushion and adjust the backrest and the steering wheel moves in and out. But if you're very tall—and you don't have to be *very* tall to count—your crown will be friendly with the headliner. Rear seat comfort, for adults at least, is minimal. Looking at the 630CSi another way, what you get for $25,000 is a relatively cramped sedan with six-cylinder engine and Bosch injection, four-speed transmission, MacPherson struts and semi-trailing arms and a host of warning lights, a description that also fits the Datsun 810.

And so the BMW 630CSi is an enigma at best. Exactly what was in mind when the car was conceived is unclear, and that only adds to the puzzle. Originally, it was to have cost around twenty, not twenty-five, and while easier to take that still seems like *way* too much; on a value basis the 630 seems to us to be a real nice $15,000 car, no more. Is it an example of over-reaching, with insufficient hardware to really pull it off? Maybe the car's tie to the 530i should be compared to that between the Nova and Camaro in an absured automotive version of the you-can't-take-the-country-out-of-the-boy syndrome. Maybe somebody at BMW, flushed with recent marketing success, decided to charge hell-for-leather after the big boys, a charge unfortunately inspired by inflated opinions of relative worth and value.

Because when taken as a $15,000 car it could be judged on its merits, and in that arena it would have worth and value; it would appear as a reasonable alternative to, for instance, an XJ6 Jaguar or the upcoming 280 Mercedes coupe. But there is no forgetting that extra ten thousand bucks before you can drive away with one, and that changes things. Judged on value a twenty-five thousand dollar car ought to do a lot and the 630CSi doesn't, not with performance comparable to a properly-optioned Buick Le Sabre Sport Coupe, for instance. So it has to be judged on status and status alone. That leaves only one answer: That a search for status is the only plausible reason BMW decided to charge twenty-five thousand dollars for the 630CSi. ∎

road test data

BMW 630CSi

SPECIFICATIONS

ENGINE
Type	SOHC L6
Displacement, cu in	182
Displacement, cc	2986
Bore x stroke, in	3.50 x 3.15
Bore x stroke, mm	89.0 x 80.0
Compression ratio	8.1:1
Hp at rpm, net	176@5500
Torque at rpm, lb/ft, net	185@4500
Carburetion	fuel inj.

DRIVELINE
Transmission	4-spd manual
Gear ratios:	
1st	3.86:1
2nd	2.20:1
3rd	1.40:1
4th	1.00:1
Final drive ratio	3.45:1
Driving wheels	rear

GENERAL
Wheelbase, ins	103.4
Overall length, ins	192.7
Width, ins	67.9
Height, ins	53.7
Front track, ins	56.0
Rear track, ins	58.5
Trunk capacity, cu ft	NA
Curb weight, lbs	3470
Distribution, % front/rear	56/44
Power-to-weight ratio, lbs/hp	19.7

BODY AND CHASSIS
Body/frame construction	unit
Brakes, front/rear	vented disc/vented disc
Swept area, sq in	NA
Swept area, sq in/1000 lb	NA
Steering	recirc. ball
Ratio	NA
Turns, lock-to-lock	3.6
Turning circle, ft	37.7

Front suspension: Independent, MacPherson struts, coil springs, tubular shocks, anti-roll bar
Rear suspension: Independent, semi-trailing arms, coil springs, tubular shocks, anti-roll bar

WHEELS AND TIRES
Wheels	14 x 6.0
Tires	195/70 HR 14 Michelin XVS

INSTRUMENTATION
Instruments: 0-150 mph speedo, trip odo, 0-7000 rpm tach, fuel, temp, clock
Warning lights: directionals, high beam, amps, oil press, brake, fuel, temp, engine oil, brake fluid, brake lights, coolant, washer fluid, rear lights, brake lining, reactor, EGR, hazard, rear defog, seat belts

PRICE
Factory list, as tested: $24,325
Options included in price: Metallic paint—$325; limited slip differential—$250; dealer prep—$150

TEST RESULTS

ACCELERATION, SEC.
0-30 mph	3.9
0-40 mph	5.3
0-50 mph	8.5
0-60 mph	10.6
0-70 mph	13.7
0-80 mph	16.6
Standing start, ¼ mile	17.7
Speed at end ¼ mile, mph	79.6
Avg accel over ¼ mile, g	0.20

SPEEDS IN GEARS, MPH
1st (6400 rpm)	34
2nd (6400 rpm)	60
3rd (6400 rpm)	98
4th (5600 rpm) (observed)	119
Engine revs at 70 mph	3300

SPEEDOMETER ERROR
Indicated speed	True speed
40 mph	40 mph
50 mph	49 mph
60 mph	59 mph
70 mph	69 mph
80 mph	79 mph

INTERIOR NOISE, dBA
Idle	55
Max 1st gear	84
Steady 40 mph	65
50 mph	68
60 mph	70
70 mph	73

BRAKES
Average stopping distance from 60 mph, ft	160
Avg deceleration rate, g	0.75

FUEL ECONOMY
Overall avg range	15-20 mpg
Range on 16.5 gal tank	330 miles
Fuel required	regular

HANDLING
Avg speed on 100-ft rad, mph	33.5
Lateral acceleration, g	0.75
Transient response, avg speed, mph	24.1

RATING

PERFORMANCE/ECONOMY
*Acceleration	4
*Fuel Economy	3

RIDE/HANDLING
*Lateral Acceleration	4
Subjective handling	4
Predictability	3
Ride	3
Steering	4

ENGINE/DRIVETRAIN
Starting	5
Throttle Response	4
Noise/Vibration	3
Shifting Action	3

BRAKES
*Stopping Distance	4
Fade Resistance	4
Subjective Feel	3

COMFORT/ERGONOMICS
*Interior Noise	3
Controls/Instruments	3
Visibility	4
Entry/Exit	3
Front Seat Comfort	3
Rear Seat Comfort	2
Space Utilization	4
Interior Environment	3

QUALITY
Assembly	4
Finish	5
Hardware/Trim	4

TOTAL	**89**
Percentile rating	**71**

*Denotes recorded data

5=Excellent, 4=Above Average, 3=Average, 2=Below Average, 1=Poor, 0=Unacceptable.

Test Equipment Used: Testron Fifth Wheel and Pulse Totalizer, Lamar Data Recording System, Esterline-Angus Recorder, Sun Tachometer, EDL Pocket-Probe Pyrometer, General Radio Sound Level Meter.

ROAD TEST
Near Perfection: The BMW 633 CSi

The best of sporting BMWs from two eras. The 633 CSi paused with a 328 at Oulton Park. Handsome and compact lines are obvious in this side view.

THE BMW COMPANY of Munich may bring out an almost bewildering number of new models – you will soon be hearing about the great top-range Series-7 cars – but it retains the basic formula which has elevated BMW to a supreme rank among the World's best. BMW continue to place their faith in an in-line six-cylinder single-overhead-camshaft engine, in varying capacity sizes, with carburetters of fuel-injection to provide the power they require. They use suspension which permits of very "forgiving" fast cornering, allied to supremely accurate steering. The outward appearance of every model reflects good taste without ostentation, and all the models in the wide range are unmistakably modern BMWs.

There can be few less-ostentatious fast motorcars than the BMW 633 CSi coupe. It tends to attract appreciative looks only from the knowledgeable, the ordinary observer hardly associating this compact, handsome, 8 ft. 7 in.-wheelbase car with a top speed of over 130 m.p.h. in manual-transmission guise, and acceleration in the order of 0-60 m.p.h. in some eight seconds and a standing-start ¼-mile devoured in less than 15 seconds. Yet such are the capabilities of this top model BMW (until the 7-series), which uses an 89 x 86 mm. (3,210 c.c.) version of the o.h.c. 6-cylinder power unit, with Bosch L Jetronic petrol-injection, to produce no less than 200 (DIN) b.h.p. at a modest 5,500 r.p.m. Suspension is by coil springs, and is all-independent, using MacPherson struts at the front, semi-trailing links at the back. To arrest this very quick but unobtrusive car BMW have gone to ventilated disc brakes front and back, of 11 in. and 10.9 in. diameter respectively, vacuum servo-assisted, the system being ATE.

The body is a two-door, full four-seater coupe, with a luggage-boot which is truly capacious. As expected of a BMW, the interior is quietly appointed, again without a trace of ostentation. Comfort, rather, is the key-note. The present price here of this 633 CSi, which can be regarded as a more civilised edition of the famous lightweight 3.0 CSL coupe, is £14,799 – I may as well get this over with, before describing the car's near-perfection! This price applies to both the 4-speed manual and the Automatic-transmission version of the 633. For test I had the former, another example of which we have "on the strength", sucessor to a 3.0 CSi.

I was frustrated at the last moment of testing this fast BMW 633 abroad. But it gave me some very enjoyable motoring in restricted, speed-conscious Britain. I was immediately enamoured by the cloth-cum-leather-upholstered seats, hard in the comfortable BMW fashion, and adjustable for the driver for height and tilt-angle as well as reach and back-rest angle. The windows are electrically controlled, the action somewhat sluggish. If it is necessary to move a window-glass when the ignition-key is not available, this is done by opening a door – an excellent arrangement. The three main instruments live in a binnacle before the driver – big central speedometer, smaller tachometer, and heat/fuel meter. The clock on the left, angled towards the driver's sight-line – I thought the instruments slightly less easy to read than those on the Editorial 520i BMW. Neat warning-light windows are set above the main instruments and to the right of the facia you have an ingenious warning panel, on which lights come on if the services they cover are in order, at the pressing of a single button. Thus this panel tells of correct oil-level, brake-fluid level, coolant content, screen-washer content, and the brake pads being in order, if each green tell-tale comes on, and if the engine is started and the lights and brakes put on, it also tells if oil-pressure and the stop-lamp and rear-lamp bulbs are functioning properly.

The steering wheel has four horn pushes, there are the usual BMW heater and ventilation controls, in the form of rotatable selectors, illuminated at night, and the electric window switches are down on the central console. All four seats have head-restraints and luxury touches include the provision of a re-chargeable torch in the cubby-hole, a first-aid kit under a lift-up

The ergonomics of the 633 CSi are excellent. The ingenious tell-tale warning panel is hidden on the right of the facia.

panel on the rear-window shelf, and the expected BMW tool-kit in its compartment on the underside of the boot-lid. The o/s driving mirror adjusts electrically. But the real reason why I describe this 633 CSi BMW as coming near to perfection is the combination it gives of very great performance, comfort, docility, and fine engineering. I liked its modest outward demeanour – apart from the styled "multi-spoke" road wheels; they were shod with Michelin XWX tyres, great 195/70 VR14 radials. I liked its logically arranged, straightforward controls. I especially commend the excellence of BMW power-steering, which imperceptibly but most usefully raises its ratio as less lock is used. It is geared 3½ turns, lock-to-lock, and is among the finest of its kind, by ZF, with a 14.7 in.-diameter steering wheel.

In spite of the very impressive speed and acceleration, this BMW is the personification of docility, the engine pulling smoothly from very low r.p.m. Yet it can be safely extended beyond the peak rev. limit when maximum performance is called for, when it is as smooth as a turbine and about as quiet. It starts and idles impeccably, hot or cold. And it gave me an overall fuel consumption of 21.3 m.p.g. (The tank holds just less than 15½ gallons.) And BMWs never seem to use any oil.

Here I want to digress to remark that BMW steering not only functions so well but is as long-wearing. The manual steering on the aforesaid 520i shows no free-play at a mileage of over 54,000, in spite of the heavy loads imposed on it when parking the car – so heavy that, after driving other cars, I have started to move off in the BMW and have been convinced, erroneously, that a front tyre has gone flat. The accuracy, too, is just as good as when this L-registered vehicle was new. It is this combination of steering accuracy and forgiving cornering that makes any BMW so pleasant to drive quickly, whether a 633 or a 320. These qualities are pronounced in the 633, although the suspension, attempting to give good road-clinging without discomfort, has a curious floating feel over some surfaces, while being quite harsh over unmade roads. Maybe this is why Tom Walkinshaw, 530i driver in the Saloon Car Championship races, has specified modified suspension on the recently-announced "limited-edition" 633 CSi (price £21,300), as well as engine and other modifications. I think I might prefer this.

The brakes, of course, function very effectively but there is an unusual item to the hand brake, which works on inner drums on the rear discs – the car rolls a few inches after applying the brakes; unimportant, but disconcerting when first experienced. Another small idiosyncrasy was that so good is the door-sealing of the Karmann body that you either have to slam the doors heavily to get them to close, or have to first open a window – as on my 1955 VW Beetle. Cool-air ventilation wasn't 100%. The car I drove had the Getrag four-speed manual gearbox. If anything, it is even better than those in other BMW's, the gears seeming to feed themselves into mesh at the bidding of the stubby central lever.

The clutch is rather decisive in action. Revised rear suspension and a limited-slip differential aid adhesion when using the 633 hard in the lower gears. In fact, the only real irritation about this excellent near-perfect BMW that I can recall was the graunching of the wiper blades – I have suffered this on the old 520i, inspite of BMW servicing, so I assume it is attributable to the kind of wiper-blades the Germans fit. It would take more than that to mar the pleasure to be derived from using this splendid BMW 633, in which I enjoyed nearly 700 very restful, mostly rapid, miles. — W.B.

The 3,210 c.c. straight-six o.h.c. engine produces 200 b.h.p. DIN with the aid of Bosch L-Jetronic fuel-injection.

The rear seats offer just adequate space for two adults, although head room is modest.

Luxotouring

Starring the Jaguar XJ-S, Mercedes-Benz 450SLC and BMW 630CSi.

As the punishment should fit the crime, so the test should fit the tested.

BY STEVE THOMPSON

• We'd been calling our projected comparison test of three highline GTs "the luxotour" for so long that when it came time to sit down and plan the story, there suddenly seemed no reason not to do just that. Luxotour, that is. Grab a handful of staffers, plot an enjoyable route through some pretty countryside, stuff CB radios in each car and go. Manufacturers use the same technique when previewing new models—they call it "ride and drive"—so we were all amazed that we didn't think of it sooner. Especially since the very best way to compare what a selection of cars does to you is to sling each down the road in back-to-back testing. No memory-fade, no puzzling over notes. If one seat fits and the other doesn't, you know it pronto.

Obviously with a one-day jaunt (a necessity considering the huge variety of ingenious excuses *C/D* staff members manage to concoct for flying away every week to race tracks, new-car introductions and other suspect activities), the route is critical. I finally settled on a 300-mile combination of roads that would give us the best possible approximation of all the kinds of driving anyone buying a new Jaguar XJ-S, Mercedes 450SLC or BMW 630CSi could find himself doing. From the Red Ball Garage, the plan was to head west through the Lincoln Tunnel, across the awful congestion of industrial New Jersey on NJ 3 and US 46 (both known hereabouts as "franchise freeways"), and then pick up northbound NJ 23, a four-lane highway that gradually transforms itself from an urban artery to a pleasantly rustic road that dips gently through a number of equally pleasant little New Jersey towns. (And yes, Virginia, there really *are* pleasant little New Jersey towns.) That would take us to Port Jervis, a fascinating village strategically placed on the Delaware River at the tri-corn junction of Pennsylvania, New York and New Jersey. And from there, we'd shoot northwest on the Delaware River Road (otherwise known as NY 97), a two-laner that follows the snaky undulations of the Delaware on an alluvial plain with absolute fidelity; when the river twists, the road twists. Its attraction is that the road seems to be one of the best bits of paving this side of the Alps; gorgeous country looming on either side, it wriggles delightfully, with turns just slow enough to keep you busy but just fast enough to make you sweat.

At a minuscule hamlet called Hankins, we planned a turn to the northeast on a gravel-and-asphalt one-and-a-half laner that demands so much concentration to drive you might as well be herding a Ferrari P3 down the tortuous lanes and cart-tracks of the Targa Florio. If the cars were to bottom and do other awful things, this is where they'd do it.

Assuming we'd survived chasing each other for fifteen miles of that, we'd pick up NY 17 at Roscoe. That would put us on a major expressway, known to many as the way to Watkins Glen from the New England states, and to others as a freeway with endless numbers of frost-heaved concrete slabs. Even they had a useful job, since they would tell us how each car's suspension would work on a cross-country trip. (In case you haven't noticed, some cars develop sickening ride motions when subjected to rhythmic bumps.) We'd be southbound on 17 for quite awhile, picking up US 6—a two- and three-laner—for the southeast connection to the Palisades Interstate Parkway, which would then run us

LUXOTOURING

Mercedes: high steering effort, solid feel.

Jaguar: low effort, little direct feedback.

BMW: good steering, lots of body lean.

onto the crowded confusion of the George Washington Bridge and finally back down the Harlem River and FDR Drives to roost, tired but happy, at the Red Ball once again.

That was the plan, anyway. What actually happened was that once we'd painfully arranged to have David E. Davis, Jr., Terry Cook, Pat Bedard, Don Sherman and Zora Arkus-Duntov (we figured we'd better have an industry heavy along, both as an advisor and to lend the affair some semblance of dignity) at the Red Ball one fine August morning, only two of the cars were present. The 450SLC was still somewhere in New Jersey, having a floppy throttle linkage tightened.

Since we'd arranged five stops to compare notes, play musical drivers and cast critical eyes over the cars, and since each of these stops was mathematically calculated to insure fair distribution of driving time, we were already messed up. So after scurrying over to Montvale to pick up the Mercedes, we had to press on with an already confused program. The convoy reassembled itself at Point Charlie One (the Tip Top Diner on Route 23), and we set off again in pursuit of truth, beauty and the right line.

You're probably wondering, at this point, about the cars themselves. And well you might, since each is worthy of a good deal of wonder. We'd put them head-to-head on this exercise because by various means they'd done just that on the market; they were all expensive, luxurious Grand Touring cars which always seem to be compared in buying discussions. The new Jaguar XJ-S combines the poke of its magnificent 5.3-liter V-12 this year with a GM Turbo Hydra-matic 400 three-speed automatic transmission, a change demanded by the sloppy, lengthy changes of gear in the older Borg-Warner unit. Our blinding-red test car was the first one in the test fleet to arrive with this gear carton, so we were extremely anxious to see if the hands-across-the-sea mixture would eradicate a well-known flaw in the XJ-S character.

The Mercedes 450SLC, on the other hand, has had no such discernible flaws since its inception in 1972, so the car we had was simply the 1978 version of a proven winner. It finally arrived under the porte-cochere of the palatial M-B headquarters in Montvale in a gleaming coat of German racing silver, looking confident and aloof even next to two other potent luxocars. The only change even contemplated by the Mercedes planners for this car in the next handful of years will be the replacement of its current 4.5-liter V-8 with a lighter 5-liter unit. But that will happen only in Europe at first, and there is no word on when—or even if—that motor will appear in American cars. The 450SLC as it is today seems to fit its clientele so well that M-B clearly isn't about to change just for change's sake.

The BMW 630CSi came to us painted a marvelous metallic blue, and with considerably less of a reputation than either the Jaguar or Mercedes. Ever since we'd discovered its peculiar combination of beautiful styling, comfortable interior, willing-but-strained engine and enormous price tag in our May '77 test, the car had been the object of a great deal of controversy. Our inclusion of it in this comparison would, we hoped, define its niche for us once and for all.

Each car cost over $20,000 and less than

Each checkpoint was chosen for its celebrity.

thirty. Each car had independent suspension. Each was at least a nominal two-plus-two. The Jaguar and Mercedes had automatic transmissions, but the 630CSi had a four-speed manual, which we thought might do a bit to offset the power difference between the XJ-S's V-12, the 450SLC's V-8 and the little three-liter straight six in the BMW. As far as we were concerned, the cars were fairly matched in terms of market. It would be up to our actual back-to-back experiences to determine their relative stature.

Which brings us back to meeting the Michael Jordan-piloted Porsche 924 photo/chase car at Point Charlie One. Having used the CBs to good effect in finding Jordan and photographer Sutton, we set off down Route 23 a good two hours late, which meant that we were now firmly embedded in the rush-hour traffic that seems to orbit endlessly around New York City. The crush made it hard to maintain the tidy flight orders (the Jag was supposed to lead, followed by the Mercedes and BMW so that the driver changes could be made without confusion), but we finally carved our way into the countryside and settled down to some serious motoring.

The next five hours and 275 miles were the combination of high comedy and blistering driving you'd expect if you knew anything about the *C/D* style. First there was the goat that suddenly appeared in front of the XJ-S as it was leading the convoy through the picturesque series of esses you see in the lead photograph to this story. Being a pseudo-mountain goat, it simply stepped off the rocky wall adjacent to the road and landed square in the center stripe, seemingly munching away on some straggly grass growing there without even noticing the high-velocity stream of steel bearing down on it. I was driving the Jag, and while I wrenched the big coupe onto the shoulder, Group Captain Davis got on the R/T and ordered the flight to evasive action: "Red Leader to flight: Break right, break!" We must have all seen the same movies, because one after another, the Porsche, Mercedes and BMW howled around the bend and neatly dove to the side of the casually grazing goat.

And then there was the Amazing Vomiting Gas Tank. When we pulled in to Barryville for a driver change and fuel stop, Bedard dutifully rattled the fuel filler into the Jag's thirsty neck and began pumping expensive unleaded into it. Soon it had filled and automatically shut the fuel filler off. But before he could remove it, a whooshing jet of gasoline came pouring back up out of the pipe, and continued to flow for a healthy period until at last it stopped with maybe three gallons spread over the car and pavement. "Hmmm," said Bedard, as we frantically pushed the XJ-S and its super-heated catalytic converters away from the puddle of fuel.

But most of the trip was simply fun, if "simply" describes having the time of your life with high-buck GT cars on virtually deserted roads. Over freeways and rutted dirt tracks we hustled, mile after mile until at last we rolled, hot and tired, to a stop at the Alpine Lookout over the Hudson River, where ace-lensman Sutton wanted to do some final shots before the sun went down. It was an opportune time to reflect on the day's events, but something seemed missing. At last Sherman put his finger on it: The enforced discipline of the convoy had made it impossible to have The Race. Surely The Race was not essential to our understanding of the cars, but each of them would certainly get thrashed hard in its owner's hands at least once, right? As Humphrey finished his shots, we settled on the final act: Last one back to the Red Ball buys the beer.

Suddenly, nerves slackened from a day's effort were taunt again as drivers waited for their cars, and from the unofficial starting gate of the toll plaza, *C/D* Red Flight broke up into individual teams struggling to reach home base first. The traffic was heavy, so no outright speeding was needed; only nerves of steel and split-second timing, along with a willing engine and good brakes. The Mercedes led most of the way,

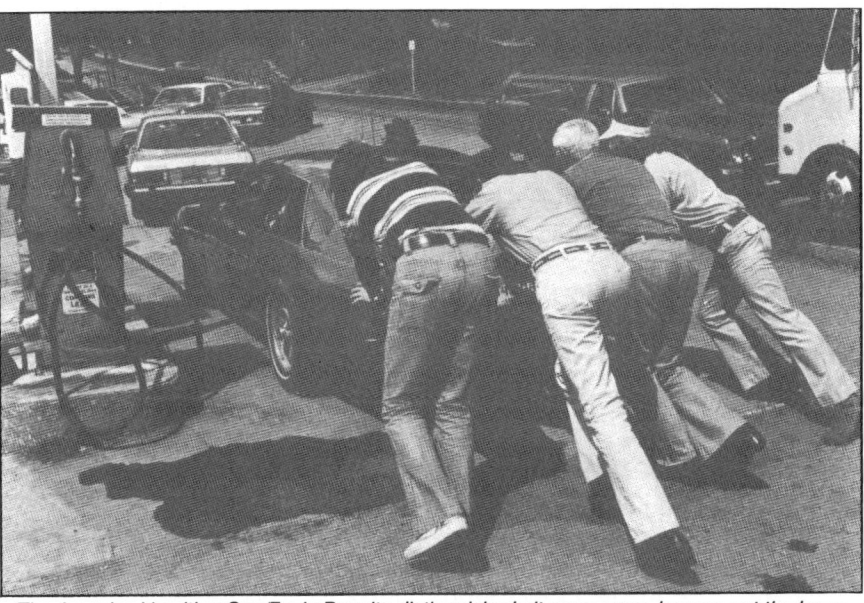

The Amazing Vomiting Gas Tank: Despite distinguished pit crew, even Jaguars get the burps.

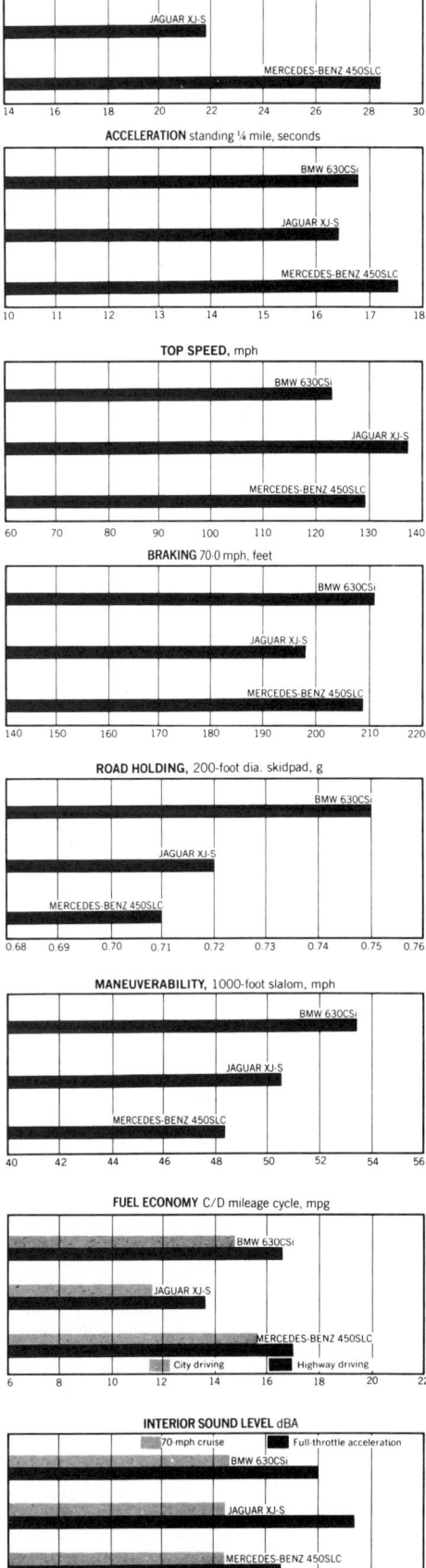

Mercedes: efficiency, excellent ergonomics.

Jaguar: comfy seats but low-ball upholstery.

BMW: tomorrowland dash, slippery seats.

but the BMW, under the guidance of Quarter-Mile Cook, screeched to a stop in the Red Ball well before either the 450SLC or the Jaguar. I bought the beer.

Sitting comfortably around the tables in the Good Times café, the cars' flaws and strengths began to emerge in congenial debriefing. We'd decided to look for two kinds of winner in this comparison: an overall best car that would be judged on a numerical rating system designed by Don Sherman, and the emotional winner, the car we'd most like to have regardless of faults or expense—in other words, the one with the highest lust factor. Each car could score a maximum of 55 points per driver, based on Sherman's system of eleven areas rated one through five points per area.

The Mercedes scored the highest on everyone's sheets, by a healthy margin. It rang up 207.5 points out of a possible 275, proving that even this disparate staff agrees that the word "engineering" could have been coined to describe the impressive abilities of the 450SLC. Second was the XJ-S, which gathered 192 points, losing mostly in the convenience areas, but doing well in performance figures. And third came the BMW 630CSi with 179.5, far behind because of its lack of power and odd ride sensations.

The lust-factor winner was another story. Only one staffer opted for anything else but the XJ-S here, and nobody thought that the car just voted most capable—the Mercedes—was emotionally as satisfying as the XJ-S. It seems that for all its ergonomic faults and unhelpful shifting tendencies, the feel of sheer svelte power the XJ-S imparted was a powerful enough stimulant to win it friends and lovers. If the Mercedes won the minds of the staff, the Jaguar could fairly be said to have won their hearts.

You can find out in more detail how the cars fared under "Scorecard," where each participant has his say. Naturally, there wasn't much complete agreement on any aspect of the luxotour results. Except, perhaps, that we need to do it more often.

But next time, having seen how hard it is to come to grips with evaluating cars as sophisticated as these on a single day's driving, we're going to schedule something a little more appropriate than New Jersey. Something, perhaps, like a month among the castles and crags of the Alps. No real luxotour is complete, after all, without mountain goats. ●

The BMW 2002
cabriolet saloon ti, tii touring turbo
a comprehensive guide to the classic sporting saloon

THE BMW 2002 - A COMPREHENSIVE GUIDE to the classic sporting saloons. This title written for Brooklands Books by James Taylor, motoring journalist, and Mike Macartney, a well-known BMW specialist, is a must for all 2002 enthusiasts. Chapters on: BMW before the 2002, History of the 2002, Selling the 2002, Press reports, The tuned 2002, A buyers guide, Running a 2002, Beneficial modifications, Restoring a 2002, The 2002 in competition, Technical specifications and more. Models: Cabriolet, Saloon, ti, tii, Touring, Turbo. 176 pages, 20 page colour supplement, 120 black and white illus.

Ref. A-BMW02   (ISBN 185520 3340)

From specialist booksellers or, in case of difficulty, direct from the distributors:
Brooklands Books Ltd., PO Box 146, Cobham, Surrey KT11 1LG, England Phone: 01932 865051
Brooklands Books Ltd, 1/81 Darley St., PO Box 199, Mona Vale, NSW 2103, Australia Phone: 2 9997 8428
CarTech, 11481 Kost Dam Road, North Branch, MN 55056, USA Phone 800 551 4754 & 612 583 3471
Motorbooks International, Osceola, Wisconsin 54020, USA Phone 715 294 3345 & 800 826 6600

MERCEDES-BENZ 450SLC

Importer: Mercedes-Benz of North America
One Mercedes Drive
Montvale, New Jersey 07645

Vehicle type: front-engine, rear-wheel-drive, 4-passenger coupe

Price as tested: $28,394
(Manufacturer's suggested retail price, including all options listed below, dealer preparation and delivery charges, does not include state and local taxes, license or freight charges)

Options on test car: base Mercedes-Benz 450SLC, $27,090; electric sunroof, $604; alloy wheels, $700.

ENGINE
Type: V-8, water-cooled, cast-iron block and aluminum heads, 5 main bearings
Bore x stroke ... 3.62 x 3.35 in, 92.0 x 85.0mm
Displacement ... 276 cu in, 4520cc
Compression ratio ... 8.0 to one
Carburetion ... Bosch K-Jetronic fuel injection
Valve gear ... chain-driven single overhead cam, hydraulic lifters
Power (SAE net) ... 180 bhp @ 4750 rpm
Torque (SAE net) ... 220 lbs-ft @ 3000 rpm
Specific power output ... 0.65 bhp/cu in, 39.8 bhp/liter
Max. recommended engine speed ... 5800 rpm

DRIVETRAIN
Transmission ... 3-speed, automatic
Max. torque converter ... 1.96 to one
Final drive ratio ... 3.07 to one

Gear	Ratio	Mph/1000 rpm	Max. test speed
I	2.31	10.3	60 mph (5800 rpm)
II	1.46	16.3	95 mph (5800 rpm)
III	1.00	23.9	129 mph (5400 rpm)

DIMENSIONS AND CAPACITIES
Wheelbase ... 111.0 in
Track, F/R ... 57.2/56.7 in
Length ... 196.4 in
Width ... 70.5 in
Height ... 52.4 in
Curb weight ... 3860 lbs
Weight distribution, F/R ... 53.5/46.5%
Fuel capacity ... 27.2 gal

SUSPENSION
F: ... ind, unequal-length control arms, coil springs, anti-sway bar
R: ... ind, semi-trailing arm, coil springs, anti-sway bar

STEERING
Type ... recirculating ball, power-assisted, centerlink damper
Turns lock-to-lock ... 3.0
Turning circle curb-to-curb ... 38.3 ft

BRAKES
F: ... 10.9-in dia vented disc, power-assisted
R: ... 11.0-in dia solid disc, power-assisted

WHEELS AND TIRES
Wheel size ... 6.5 x 14-in
Wheel type ... aluminum alloy, 5-bolt
Tire make and size ... Michelin XVS, 205/70HR-14
Tire type ... steel-belted, radial ply, tubeless
Test inflation pressures, F/R ... 32/36 psi
Tire load rating ... 1490 lbs per tire @ 36 psi

PERFORMANCE

Zero to	Seconds
30 mph	3.6
40 mph	5.5
50 mph	7.3
60 mph	9.6
70 mph	12.0
80 mph	15.5
90 mph	19.7
100 mph	26.5

Standing ¼-mile ... 17.5 sec @ 85.9 mph
Top speed (observed) ... 129 mph
70-0 mph ... 209 ft (0.78 g)
Fuel economy, *C/D* mileage cycle ... 15.5 mpg, urban driving
17.0 mpg, highway driving

JAGUAR XJ-S

Importer: British Leyland Motors Inc.
600 Willow Tree Road
Leonia, New Jersey 07605

Vehicle type: front-engine, rear-wheel-drive, 4-passenger coupe

Price as tested: $21,900
(Manufacturer's suggested retail price, including all options listed below, dealer preparation and delivery charges, does not include state and local taxes, license or freight charges)

Options on test car: none

ENGINE
Type: V-12, water-cooled, aluminum block and aluminum heads, 7 main bearings.
Bore x stroke ... 3.54 x 2.76 in, 90.0 x 70.0mm
Displacement ... 326 cu in, 5340cc
Compression ratio ... 8.0 to one
Carburetion ... Lucas electronic fuel injection
Valve gear ... chain-driven single overhead cam
Power (SAE net) ... 244 bhp @ 5250 rpm
Torque (SAE net) ... 269 lbs-ft @ 4500 rpm
Specific power output ... 0.75 bhp/cu in, 45.7 bhp/liter
Max. recommended engine speed ... 6500 rpm

DRIVETRAIN
Transmission ... 3-speed, automatic
Max. torque converter ... 2.50 to one
Final drive ratio ... 3.31 to one

Gear	Ratio	Mph/1000 rpm	Max. test speed
I	2.49	9.2	60 mph (6500 rpm)
II	1.49	15.4	100 mph (6500 rpm)
III	1.00	22.9	138 mph (6050 rpm)

DIMENSIONS AND CAPACITIES
Wheelbase ... 102.0 in
Track, F/R ... 58.0/58.6 in
Length ... 192.2 in
Width ... 70.6 in
Height ... 49.6 in
Curb weight ... 3930 lbs
Weight distribution, F/R ... 56.2/43.8%
Fuel capacity ... 23.0 gal

SUSPENSION
F: ... ind, unequal-length control arms, coil springs, anti-sway bar
R: ... ind, 1-transverse link, 1-halfshaft and 1-trailing link per side, 2-coil springs per side, anti-sway bar

STEERING
Type ... rack-and-pinion, power-assisted
Turns lock-to-lock ... 3.2
Turning circle curb-to-curb ... 39.0 ft

BRAKES
F: ... 11.2-in dia vented disc, power-assisted
R: ... 10.4-in solid disc (inboard), power-assisted

WHEELS AND TIRES
Wheel size ... 6.0 x 15-in
Wheel type ... aluminum alloy, 5-bolt
Tire make and size ... Dunlop SP Sport Super, 205/70VR-15
Tire type ... steel-belted, radial ply, tubeless
Test inflation pressures, F/R ... 24/26 psi

PERFORMANCE

Zero to	Seconds
30 mph	3.3
40 mph	4.8
50 mph	6.3
60 mph	8.3
70 mph	10.5
80 mph	13.2
90 mph	16.2
100 mph	21.8

Standing ¼-mile ... 16.4 sec @ 90.6 mph
Top speed (observed) ... 137 mph
70-0 mph ... 198 ft (0.83 g)
Fuel economy, *C/D* mileage cycle ... 11.5 mpg, urban driving
13.5 mpg, highway driving

BMW 630CSi

Importer: BMW of North America, Inc.
Montvale, New Jersey 07645

Vehicle type: front-engine, rear-wheel-drive, 4-passenger 2-door coupe

Price as tested: $24,000
(Manufacturer's suggested retail price, including all options listed below, dealer preparation and delivery charges, does not include state and local taxes, license or freight charges)

Options on test car: base BMW 630CSi, $23,600; limited slip differential, $250; dealer prep, $150

ENGINE
Type: 6-in-line, water-cooled, cast-iron block and aluminum head, 7 main bearings
Bore x stroke ... 3.50 x 3.15 in, 89.0 x 80.0mm
Displacement ... 182 cu in, 2986cc
Compression ratio ... 8.1 to one
Carburetion ... Bosch L-Jetronic fuel injection
Valve gear ... chain-driven single overhead cam
Power (SAE net) ... 176 bhp @ 5500 rpm
Torque (SAE net) ... 185 lbs-ft @ 4500 rpm
Specific power output ... 0.97 bhp/cu in, 58.9 bhp/liter
Max. recommended engine speed ... 6400 rpm

DRIVETRAIN
Transmission ... 4-speed, all-synchro
Final drive ratio ... 3.45 to one

Gear	Ratio	Mph/1000 rpm	Max. test speed
I	3.86	5.4	35 mph (6400 rpm)
II	2.20	9.5	61 mph (6400 rpm)
III	1.40	14.9	95 mph (6400 rpm)
IV	1.00	20.9	123 mph (5900 rpm)

DIMENSIONS AND CAPACITIES
Wheelbase ... 103.4 in
Track, F/R ... 56.0/58.5 in
Length ... 192.7 in
Width ... 67.9 in
Height ... 53.7 in
Curb weight ... 3420 lbs
Weight distribution, F/R ... 56.4/43.6%
Fuel capacity ... 16.5 gal

SUSPENSION
F: ... ind, MacPherson strut, coil springs, anti-sway bar
R: ... ind, semi-trailing arm, coil springs, anti-sway bar

STEERING
Type ... recirculating ball, power-assisted
Turns lock-to-lock ... 3.6
Turning circle curb-to-curb ... 37.7 ft

BRAKES
F: ... 11.0-in dia vented disc power-assisted
R: ... 10.7-in dia vented disc, power-assisted

WHEELS AND TIRES
Wheel size ... 6.0 x 14-in
Wheel type ... aluminum alloy, 5-bolt
Tire make and size ... Michelin XVS, 195/70HR-14
Tire type ... steel-belted, radial ply, tubeless
Test inflation pressures, F/R ... 33/30 psi
Tire load rating ... 1340 lbs per tire @ 36 psi

PERFORMANCE

Zero to	Seconds
30 mph	2.8
40 mph	4.4
50 mph	6.2
60 mph	8.4
70 mph	11.6
80 mph	15.1
90 mph	20.3
100 mph	28.7

Standing ¼-mile ... 16.8 sec @ 84.1 mph
Top speed (observed) ... 123 mph
70-0 mph ... 211 ft (0.78 g)
Fuel economy, *C/D* mileage cycle ... 14.5 mpg, urban driving
16.5 mpg, highway driving

CONTINENTAL DIARY

by Paul Frère

HORSES FOR COURSES
Paul Frère tries two of probably the finest four-seater coupés in current production

NEITHER the Porsche Carrera, nor the Turbo were ever serious competitors for the Mercedes 450 SLC or BMW's big Coupés. They were out and out "driving machines" with scant regard for interior space. But the Porsche 928, though not a full four-seater, and getting nowhere near the Turbo in terms of performance, is much more of a brick in the Mercedes and BMW gardens.

Though slightly less roomy, the 928 is faster, more modern and just as refined as anything either of its German competitors could offer at the time of its announcement. They had to do something about it and their answers are the "lightweight" 5-litre version of the Mercedes-Benz 450 SLC, and the BMW 635 CSi. I have tested all three cars in the last nine months, covering a considerable mileage across France, Germany and Italy, and was lucky to be able to swop the Mercedes for the BMW, with no other car in between, thus allowing an almost perfectly direct comparison.

An important difference between the Mercedes and the BMW is their age. The 450 SLC was developed from the original 350 SL, first announced in April 1971, and itself based on the now obsolete W 114 series (200 to 250 models) launched nearly ten years ago, but fitted with the V-8 engine. The BMW is a much newer design, announced early in 1976 as the 633 CSi, in which little else than the straight six engine is retained, in developed form, from earlier models, though the general design philosophy has remained the same as before.

The Mercedes SLC series was developed from the SL which is basically a 2-seater convertible. For the four-seater Coupé, the wheelbase was lengthened, but the very stiff platform structure required for the convertible was retained. Up to last year, the most powerful version was the 450 SLC using the 4.5-litre sohc V8 engine developing 225 bhp in the European K-Jetronic injected version. This is the car which won the recent marathon road race around South America. It comes with the Daimler-Benz three-speed automatic only, as the manufacturer has no manual gearbox able to stand the torque of the 4.5-litre engine and considers the demand to be too small to justify the cost of producing one.

Meanwhile the range of V8 engines (except for the 6.9) has been developed to use an aluminium cylinder block (the heads were already aluminium) in view of future models to come, which will be lighter than the existing ones in order to reduce the fuel consumption. In these engines, which will come in 3.8 and 5-litre sizes, the iron-coated alloy pistons run directly on the bores, the block being made of an aluminium alloy with a high content of silicium (Reynolds Aluminium patents, also used in the Chevrolet Vega engine and in the Porsche 928). As the 5-litre is 90 lb lighter than the 4.5-litre, and is also more powerful (240 bhp DIN), it was just what was required to make the 450 SLC a real challenger to the Porsche 928. But to make it noticeably faster than an ordinary 450 SLC, more work was required and the Daimler-Benz designers decided to redesign the entire floor pan and dispense with the stiffening that was not required for the fixed head Coupé. They also decided to make the luggage compartment lid and the bonnet in aluminium sheet and to fit light alloy wheels as standard equipment. Altogether this has taken some 265 lb off the car's weight, equivalent to about an 8 per cent reduction.

In this performance race, BMW were handicapped from the start because they don't have an engine as large as either the Porsche or the Mercedes V8s. But they had to produce something faster than the 633 CSi which is a larger and heavier car than the late, lamented 3.0 CSi, while its detoxed 3210cc engine isn't any more powerful than the old D-Jetronic 3-litre. The answer came from the racing department where long ago they had increased the engine capacity to over 3.5-litres, while maintaining reliability, even with much higher power output than could be contemplated for normal road use. To make the 93.4 mm bores acceptable for production, however, the cylinder centres had to be rearranged, while the stroke was actually reduced from 86 to 84 mm to improve smoothness and quietness at high engine speeds, though this still gave a capacity increase of nearly a quarter litre to 3453 cc, which raised the power from 200 bhp at 5500 rpm to 218 bhp at 5200 rpm and the torque from 210 lb/ft at 4250 rpm to 229 lb/ft at 4000 rpm.

The rear axle ratio was reduced from 3.25 to 3.07 to 1 and a Getrag five-speed gearbox was added for good measure, fifth being direct (no overdrive), according to normal BMW practice, with the lower gears closer together than in the four-speed box. To match the increased performance, the car was given stiffer springs and dampers and stiffer anti-roll bars to which wider wheels (6½ instead of 6 in) were also added, while a deeper front air dam and a rear spoiler reduce the high speed lift. All these modifications add up to making the 635 CSi a worthy substitute for the old 3.0 CSi, which it will see off while providing roomier and more comfortable accommodation and, at any rate, much better power steering.

No attempt has been made to lighten the car (it weighs about the same as the Mercedes) which comes as standard with the same full de Luxe equipment, including electric window lifters. Our Mercedes, however, was heavier than the "basic" lightweight "5.0", as it had the optional air conditioning equipment and an electrically-operated sliding roof. It also had electrically-heated front seats. Both cars had a central door locking system.

I suppose most people would guess that the BMW is the model which would best fill the needs of the enthusiast, but this is not necessarily so. If the BMW is just what I expected it to be, I really got the surprise of my life when I first drove the "5.0" at the Hockenheim track at the end of 1977, and ever since I have wanted to use one for some time on the road, to see if it is really as good as it felt on that occasion — and it is. Neither does it look or feel in any way dated, though it has been with us — basically — for more than seven years. Even the facia and the interior fittings are just what you expect in a modern car, though there is one — really stupid — exception: that huge steering wheel. There is absolutely no justification for it with the power steering and the too high leverage necessary spoils the feel. As it is, the power steering does not feel progressive enough; the weight does not increase enough as the cornering force increases. A smaller wheel would probably do the trick.

In this sphere, the BMW is very much better: you never notice there is a power steering until you need it, when parking or making a sharp turn, for instance. Recent BMW power steerings are — with the Lancia Gamma's — the best I know.

The handling of the Mercedes was quite a surprise. There is nothing soft about it: it is beautifully comfortable for high speed driving, though the low speed ride is harder than I expected, and in the same league as the BMW's with its uprated suspension. But the real surpise comes from the exceptionally quick response to steering impulses. That front end is really good, even though Daimler-Benz consider it an obsolete design, inherited from the W 114 series. Early 350 SLs used to understeer too much for my liking, but with the latest 450 SLC, this is certainly not so: the Michelin XDX-shod test car felt beautifully neutral up to the very high limit where oversteer could be provoked by using full throttle. With the anti-squat reaction arms used on European 450 rear suspensions, abruptly lifting off in a fast corner and even braking produces no drama, just some slight and desirable tuck-in. The brakes, too, were very efficient and fade free.

Fast driving and cornering with the BMW is just as enjoyable — even more so, thanks to the remarkably good steering — though I don't think it handles better than the Mercedes, which would indeed be difficult, at least for a car of that size and weight. At the limit, there is slightly more oversteer, but no more than necessary to make the car real fun to drive where conditions permit. Lifting off creates a slightly sharper reaction when cornering hard, but it is easily controlled. Here it should probably be mentioned that both cars were on Michelin VH-class radials, but the Mercedes was on the slightly more comfort-oriented XDX than the XWX fitted to the BMW.

Though the BMW is not by any means noisy for a high performance car, its engine is heard more, specially when it is pushed, and its exhaust note is of a higher tune than the Mercedes. But purists will say that the BMW "six" makes music, not noise, and however you rate it, it is a long way from being obtrusive. At high speed, say above 125 mph, it is difficult to say in either car whether it is the engine or the wind level which is worse, but it is never so high that you cannot listen to the radio in comfort. As far as wind noise is concerned, the Porsche 928 is better than either of the two cars under review. It is even quieter at speed, though I would say that its V8 engine is not quite as smooth as either the Mercedes or the BMW six.

The big difference between the two cars is in their respective transmissions. The Mercedes is automatic (with the best selector ever devised) and cannot be had with a manual, and the BMW is a 5-speed manual and cannot be had as an automatic. This is how the 3.5-litre from Munich can more or less keep up with the 5-litre from Stuttgart which it easily outdrags, except in maximum speed and at the top of the acceleration curve. Both cars accelerate from 60 to 120 mph in almost exactly the same time.

Our standing start figures on the 450 were taken by starting in position 2 and holding intermediate up to the 6200 rpm rev limit, but the gain is very small, amounting to only 1/10 of a second over the standing start kilometre. There is a higher (5800 rpm) change-up speed from 1st to 2nd in position "2" and this is very useful in mountain driving, as the kick-down speed is also raised, enabling the car to rocket out of hairpins without having to touch the lever.

With the BMW, rapid progress requires more work and you can keep up with the Mercedes only by making good use of the (many) intermediates of which 3rd gives you 80 mph and 4th 113 mph.

Both cars will cruise with no apparent effort at anything between 125 and 135 mph indefinitely, and driving the Mercedes from Nice to Paris on the "Autoroute" and the BMW back from Paris to Nice with complete disregard for the market speed limit of 130 km/h (81 mph) took virtually the same time, a drive slightly marred by the peak hours at which Paris was reached and left, corresponding to an average of just about 100 mph. These drives made an interesting comparison in fuel consumptions: 14.5 mpg for the BMW and 12.9 mpg for the Mercedes. Bearing in mind the greater losses in the automatic transmission, this is a good reminder that fuel consumption is much more closely related to the performance achieved than to the size of the engine. Though the Porsche 928 happily digests "regular" grade fuel requiring a comparatively low compression ratio, the 12.55 mpg I obtained under very similar conditions with the manual 928 seems rather high in comparison. But in spite of the the efforts of BMW and Daimler-Benz to match the Porsche's performance, the 928 is still, by a very small margin, the fastest of the three, both in acceleration and maximum speed, its biggest advantage being top speed (145.5 mph) when compared to the BMW's 139.2 mph. The Mercedes, however, can still claim to be the fastest automatic coupé, though the 6.9-litre four-door saloon from the same stable still has a marginal edge on it. In any case, the difference in performance between the three coupés is unlikely to carry more weight in the decision of prospective owners than inside room, preference for manual or automatic transmission and personal taste.

And please, don't ask me which I would choose! I have enough problems to solve for myself without adding yours!

PERFORMANCE DATA

	Mercedes 450 SLC — 5.0	BMW 635 CSi
Maximum speed	143.5 mph	139.2 mph
0-50 mph	7.0 s	5.6 s
0-60 mph	9.0 s	7.3 s
0-70 mph	11.4 s	9.1 s
0-80 mph	14.0 s	11.3 s
0-90 mph	16.8 s	14.5 s
0-100 mph	20.4 s	18.3 s
0-110 mph	24.3 s	22.5 s
0-120 mph	31.0 s	29.0 s
Standing ¼ mile	15.9 s	15.3 s
Standing kilometre	28.6 s	27.95 s
Fuel consumption, overall	12.9 mpg	14.5 mpg

Above: BMW's be spoilered 635

BMW Alpina

A quantum leap in the machineries of joy

by Peter Frey

We walked out the back door of the Petersen Building into the early morning sun. We stood a moment, blinking in the sudden brightness as our eyes adjusted. Sitting in a corner of the concrete parking pad, bathed in its own special shaft of sunlight, was a flowing exercise in functional aesthetics.

It was a cresting wave, thunder and fury still an instant away, frozen in time and turned to steel. It was a feline predator, crouched to spring, caught in the instant before motion, turned to pewter, and offered up to us. A sensual bond was formed in that instant, a bond that grew stronger with each step we took closer. We were enchanted, and the courier who delivered the spell to us spoke, telling us the story of how it came to be.

In 1962, a gentleman named Burkard Bovensiepen formed the tuning and high-performance firm of Alpina. In the years since, the firm has grown to be the largest high-performance and research center in Europe. They enjoy an extremely close working relationship with the BMW factory, to the point of exerting considerable influence in the design of the 3.0 CSL.

All the parts they manufacture, and conversions they do on complete cars, are engineered and built to the rather exacting standards of Germany's answer to the EPA, called T.U.V. The quality of their work was rewarded with considerable success in both the race tracks and marketplaces of the automotive world. As successful businesses are wont to do, they decided to expand, and Alpina/West came into existence in California. They funneled Alpina parts from Germany into the U.S. performance market.

Just recently, the company went through another evolutionary step. The people who were running the West Coast operation dropped the Alpina name and are now calling it Hardy and Beck Performance, Inc. (H&B), 1799 Fourth St., Berkeley, CA 94710, (415) 526-5489. They still import and use some Alpina parts but have also started designing some of their own pieces and are having them manufactured locally. The reason given for this is that some of the European parts don't conform to the U.S. version of the BMW. They flatly state that they will make no concessions in the quality of the parts, wherever they may come from.

In this case, the object of their attentions—and our affection—was one of the most capable, elegant touring cars to grace the automotive landscape in recent years, the BMW 630 CSi. In its stock form, the car is lean and handsome and a pleasure to drive. Its road manners are impeccable, but to the seat of the pants of an enthusiastic, capable driver, it is apparent that there is a level of performance, just out of reach, that was sacrificed in the name of comfort.

What H&B has done is to replace some of the stock parts with high-performance pieces specifically designed to make that other level of performance attainable. Having catered to the needs of velocity, they then tickle the visual, tactile and auditory senses with other equipment designed to make the car a joy to look at and comfortable to drive, and the cockpit a chamber of musical delights.

And this is how they do it.

The suspension, which is MacPherson strut in the front and Chapman strut in the rear, retains its basic configuration, but some of the parts are replaced. H&B high-performance coil springs, which are both stiffer and shorter than the stock springs, are fitted to lower the car an inch and a half. The shock absorber cores of the front struts are replaced with Bilstein units, and in the rear a complete Bilstein strut, with a provision for adjusting the ride height, replaces the stock one.

The anti-sway bars are replaced with larger-diameter versions. The front bar has nylon bushings, and the rear bar is completely adjustable by means of Heim rod ends. The sophistication of the rear suspension, with both adjustable ride height and roll stiffness, allows considerable fine tuning of the handling characteristics of the car.

The standard wheels and tires are exchanged for Pirelli P7s on one-piece, light-alloy rims. What were originally Michelin XVS 195/70 HR 14s on 6x14-inch rims become 205/55 VR 16s in the front and 225/50 VR 16s in the rear, all mounted on 8x16-inch rims.

The engine, an in-line, overhead-cam 6-cylinder, displacing 2986 cc and producing 176 horsepower at 5500 rpm, remains stock. So does the 4-speed manual transmission, except that the shift

BMW Alpina

knob is replaced with one made of Brazilian rosewood. The car is presently in the process of a turbocharging program, but customer availibility of that particular option remains in the future.

Apart from the wheels and tires, the only changes that are visible externally are the front air dam and the stripes. The air dam is a new design made of marine-quality fiberglass. It serves to smooth out the airflow in the front of the car and produces both improved fuel economy and high-speed handling. The stripes are of an Art-Deco design in the Alpina colors, an electric blue and green. They are by no means subtle, but they suit the car perfectly and set off the superb silver paint.

The remaining changes take place in the compartment that houses the ultimate judge of all these attentions, the driver. The factory seats are replaced with orthopedically designed Scheel driving seats, upholstered in a cloth that bears the Alpina colors. They are nothing short of perfect, offering both long-term comfort and lateral support. They also, praise be, slide back far enough to let even the long of leg get comfortable. The rear seats also are upholstered in the striking Alpina fabric.

H&B chose a sound system capable of delivering performance on a par with the car itself. What they picked was a Blau-

punkt Berlin, the famous one with the War-of-the-Worlds control head growing out of a flexible stalk. This is plugged into a Fosgate amplifier, and the sound flows out of KLH speakers. The system is strictly state-of-the-art and produces a sound of crystal clarity. Music of that quality goes a long way toward making driving a pleasure instead of a chore.

The final interior touch is a leather-rimmed Alpina sport steering wheel.

The courier finished his tale, handed us the keys and told us to have a good time, knowing there was not the slightest

doubt that we would. He departed, and we were left standing, just looking, afraid to break the spell the car was casting. The keys to the sterling fantasy grew warm in our hands.

We had the car for about 10 days and enjoyed every minute we spent in it. A long-traveled route from home to office that involved a lengthy, bumper-to-bumper freeway battle was exchanged for a twisty length of 2-lane blacktop that winds through the mountains. It was twice as far and took twice as long, but every second of that time was spent in a state of high-velocity stimulation. We would arrive at the office with a grin that didn't begin to fade until noon.

The effects of the suspension changes make themselves apparent in the supple ride and a rock-stable feeling at any speed, whether in a straight line or floating through a corner. The Pirelli P7s provide the ultimate in adhesion for a street legal tire and make the already excellent, power-assisted recirculating ball steering feel that much better by transmitting just a bit more road feel. They also raise the steering effort slightly, making it feel

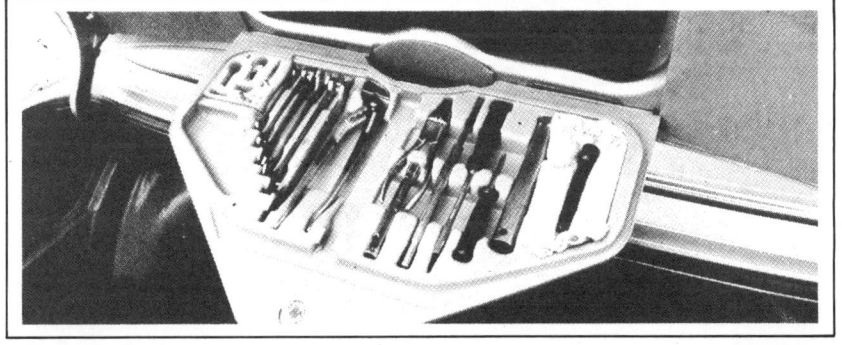

Alpina Components

Suspension	$696 plus 10 hours of labor
Wheels	$960 plus mounting & balancing (set of 4)
Tires	$225 each front, $265 each rear
Exterior group	
air dam	$150 plus 0.5 hours of labor
Deco stripes	$146 plus 6.0 hours of labor
Interior group	
steering wheel	$150 plus 0.5 hours of labor
shift knob	$9
sound system	$1580 installed
Scheel seats	$416 each plus 1.0 hour of labor

BMW Alpina

even more precise.

The seats, quite simply, hold you comfortably in place, even as the blood in your body rushes from side to side under the effects of the lateral g forces.

We put the car through every conceivable type of turn, several times, at speeds that left no room for error. It was under these extreme conditions that we discovered the car's only flaw. When cornering over uneven or undulating surfaces, there is a mild kickback through the steering. We attribute this to a slight incompatibility between the suspension geometry and the extreme sidewall stiffness of the P7 tires.

Under any other conditions, the manners of the BMW are absolutely impeccable—even when brought down to the less glorious task of slugging it out with merely mortal machines in the arena of "The Streets."

In short, the car is all we might wish for, and that's saying a lot. Our tastes have grown jaded and cynical over the years, and it takes something spectacular to excite us. We've also grown jaded to the prices of such machines. By the time you get your name on the registration of a car like this, you will have handed almost 30 thousand dollars to various people. When dealing from such rarified heights, you must keep in mind that if you want a car that does certain things and that makes you feel a certain way when you drive it, you must pay out a certain amount of money to own it. "Value" becomes an inoperative word.

In terms of physical destinations, any place you could go in the Alpina, you could go in any car—and, if you obey the law, at exactly the same speed. What makes the difference is that the BMW opens up a whole new category of destinations, located in that curious geography that lies between the ears of anyone who feels kinship with a machine. **MT**

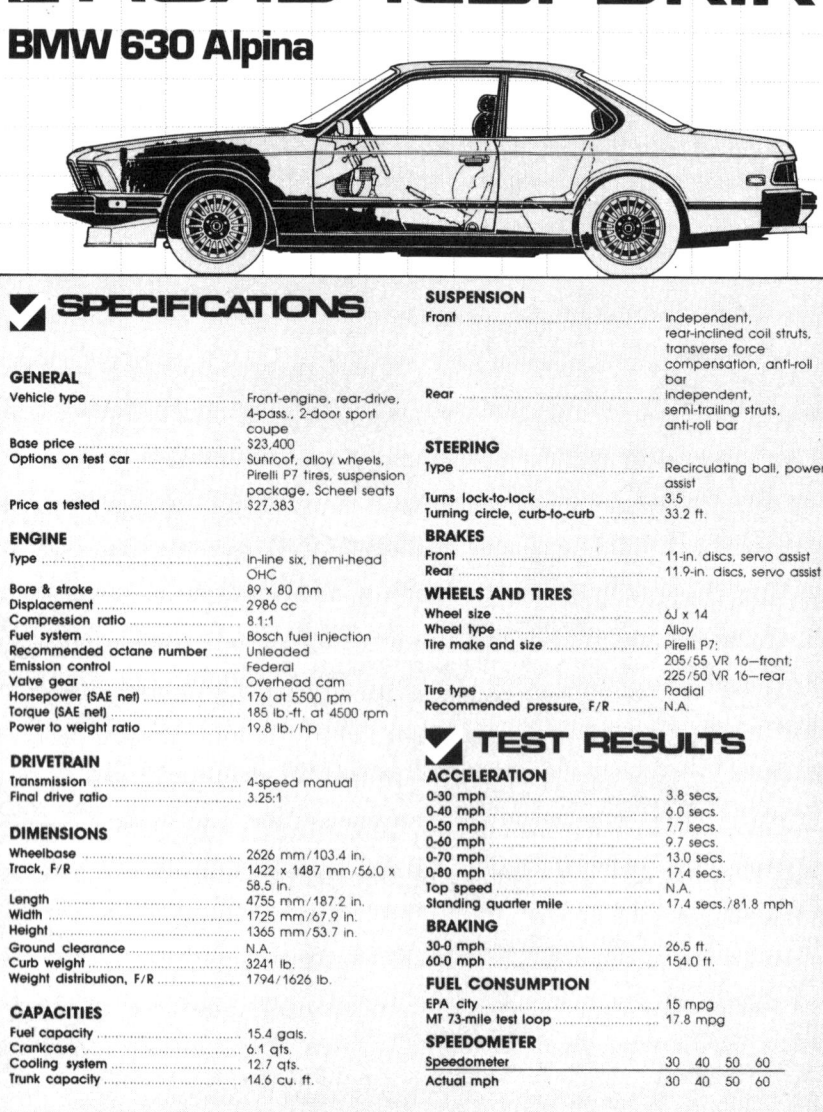

MT Staff Rating
(10=excellent, 1=poor)

Performance	8
Ride	9
Handling	8
Steering	9
Braking	8
Noise	9
Vibration	9
Instrumentation	8
Accessibility of controls	8
Transmission	9
Seating comfort, front	10
rear	6
Luggage capacity	7
Heating and ventilation	8
Vision, front	9
rear	9
sides	9
Exterior fit and finish	9
Interior fit and finish	9
Fuel economy	7
Ease of entry and exit, front	8
rear	5
Average overall rating	**8.2**

ROAD TEST DATA
BMW 630 Alpina

SPECIFICATIONS

GENERAL
Vehicle type	Front-engine, rear-drive, 4-pass., 2-door sport coupe
Base price	$23,400
Options on test car	Sunroof, alloy wheels, Pirelli P7 tires, suspension package, Scheel seats
Price as tested	$27,383

ENGINE
Type	In-line six, hemi-head OHC
Bore & stroke	89 x 80 mm
Displacement	2986 cc
Compression ratio	8.1:1
Fuel system	Bosch fuel injection
Recommended octane number	Unleaded
Emission control	Federal
Valve gear	Overhead cam
Horsepower (SAE net)	176 at 5500 rpm
Torque (SAE net)	185 lb.-ft. at 4500 rpm
Power to weight ratio	19.8 lb./hp

DRIVETRAIN
Transmission	4-speed manual
Final drive ratio	3.25:1

DIMENSIONS
Wheelbase	2626 mm/103.4 in.
Track, F/R	1422 x 1487 mm/56.0 x 58.5 in.
Length	4755 mm/187.2 in.
Width	1725 mm/67.9 in.
Height	1365 mm/53.7 in.
Ground clearance	N.A.
Curb weight	3241 lb.
Weight distribution, F/R	1794/1626 lb.

CAPACITIES
Fuel capacity	15.4 gals.
Crankcase	6.1 qts.
Cooling system	12.7 qts.
Trunk capacity	14.6 cu. ft.

SUSPENSION
Front	Independent, rear-inclined coil struts, transverse force compensation, anti-roll bar
Rear	Independent, semi-trailing struts, anti-roll bar

STEERING
Type	Recirculating ball, power assist
Turns lock-to-lock	3.5
Turning circle, curb-to-curb	33.2 ft.

BRAKES
Front	11-in. discs, servo assist
Rear	11.9-in. discs, servo assist

WHEELS AND TIRES
Wheel size	6J x 14
Wheel type	Alloy
Tire make and size	Pirelli P7; 205/55 VR 16—front; 225/50 VR 16—rear
Tire type	Radial
Recommended pressure, F/R	N.A.

TEST RESULTS

ACCELERATION
0-30 mph	3.8 secs.
0-40 mph	6.0 secs.
0-50 mph	7.7 secs.
0-60 mph	9.7 secs.
0-70 mph	13.0 secs.
0-80 mph	17.4 secs.
Top speed	N.A.
Standing quarter mile	17.4 secs./81.8 mph

BRAKING
30-0 mph	26.5 ft.
60-0 mph	154.0 ft.

FUEL CONSUMPTION
EPA city	15 mpg
MT 73-mile test loop	17.8 mpg

SPEEDOMETER
Speedometer	30	40	50	60
Actual mph	30	40	50	60

Depicted in front of the magnificent Goodwood House, a proud Derek Bell and his very special BMW 633CSi.

Almost a Batmobile

Continuing our occasional series about the cars racing drivers use off the track, we talk to Derek Bell about his BMW 633CSi.

It must happen to everyone at some stage in their motoring life — hurriedly selling a car, the loss of which you regret almost immediately. With some people, it is a vintage rarity, with others, a classic of the day. For Derek Bell, it was one of the latter — a left-hand-drive BMW 3.0CSL 'Batmobile'.

The white, 1974 two-door coupé — his second BMW, replacing a 3.0Si saloon — was sold when he joined the Broadspeed-run Leyland Cars European Touring Car Championship team to drive the Jaguar V12 coupé. Along with his Leyland contract came a Leyland car — an automatic Jaguar XJS. When the project was folded, Derek was offered the XJS, but he was keen to have a car with a manual gearbox. While remaining faithful to Leyland in one respect — he runs a Range Rover as a workhorse at his home in Sussex — he looked for a car with similar characteristics to his former Batmobile... and now he thinks he has found it.

Five months ago, Derek took delivery of a BMW 633CSi, the only manual version prepared by Tom Walkinshaw Racing to their 'Hallmark' specification, a conversion approved by BMW in England and Germany, but only available in this country. The engine of Derek's car has been blueprinted and the handling much improved with the fitting of high performance dampers, progressive rate springing and very special wheels and tyres: 7ins on the front and 8ins on the rear, with low-profile Pirelli P7s all round. A Hallmark front spoiler and twin hand-painted coach-lines make the car easily identifiable from the exterior while, inside, a BMW Motorsport steering wheel, Hallmark emblems on the headrests and a solid silver hallmarked (hence the name) ashtray lid complete the 'up-market' image. All that bumps the price of a 633CSi up to about £20,000.

Derek, who races very much for fun, said: "It is not that critical that I win everything, but I do like to have some fun. It is very much the same on the road. I like driving, and prefer not to use automatics. This 633 is verging on what my Batmobile used to give me. It's very positive in everything it does. It is very much a luxury car, but really responds when you give it some stick."

When he was racing for Ferrari, Derek used to drive himself all over Europe, his road car in those days being a Ferrari 275GT4. He said: "I used to love Ferraris, but they are not touring cars these days. I can travel hundreds of miles in the 633 and still feel relaxed. After finishing the Nürburgring 6 Hours a few weeks back, I left the circuit straight after the race to drive 490 miles to Le Mans. I averaged 84mph, including fuel and customs stops, but felt quite relaxed at the other end. I went straight to bed, and didn't even have to sit down and unwind — it was great."

Derek rarely takes his family with him to foreign race meetings, but reckons the 633 is ideal when he does want to. "The boot is so large, and the interior is quite big enough for Pamela and the two children — it is perfect. It is so comfy at 90mph and there's little draught noise."

Surprisingly, Derek has found the fuel consumption to be very reasonable: "I get about 20mpg if I'm cruising at speed on a motorway, and as much as 23mpg if I'm just pottering about in Sussex, or running backwards and forwards to London.

But, if any of you are trying to find Derek's 'phone number to buy this seemingly ideal car from him, don't bother. He says: "I've always wanted a 633. I thought they were a very attractive shape when they first came out. And I don't want to sell it." So there'!

One of the finest moments in Derek's long career was his win at Le Mans in 1975, driving a Gulf Mirage alongside Jacky Ickx. He was back in a Mirage this year with Vern Schuppan, but retired just before the finish.

BMW 635 CSi
Bavarian elegance

BMW 635
Latest version of 630 coupé range, announced with 3½-litre version of six-cylinder engine July 1978, with manual only gearbox. Original 630 and 633 replaced 3.0 CSi coupé at Geneva 1976; 633 CSi only available in Britain with automatic transmission since 635 launch.

MANUFACTURER:
Bayerische Motoren Werke AG, Munich, West Germany

U.K. CONCESSIONAIRES:
BMW Concessionaires GB Ltd, 991 Great West Road, Brentford, Middlesex TW8 9ED

SINCE, in spite of unnecessary overall speed limits in all European countries but enlightened West Germany, building fast cars is not yet illegal, it is encouraging to drive machines like BMW's 635 CSi. It is sad that, to some extent, legislation has driven manufacturers towards heavier, fatter cars — which need bigger engines to provide the same performance — a retrogression which began in BMW's case with the 633 CSi. But the 635 is nevertheless a highly satisfying return towards the standards set by the much loved 3.0 CSi, even if its greater frontal area and weight mean that it has lost an appreciable amount of the usual BMW economy.

The 633 CSi, which continues alongside the 635 CSi, is 6 per cent heavier and has roughly 10 per cent more frontal area than the previous 3.0 CSi. Compared with the 633, the 635 weighs 1cwt (5.1 per cent) more but has the apparent advantage of a fairly marked air dam front and a lipped tail, said to cut lift by 50 and 15 per cent front-rear. The engine is more different from the smaller-engined car than is immediately obvious. Where the '33 has a capacity of 3,210 c.c. produced with 89x86mm bore and stroke, the '35 has 3,453 c.c. and 93.4x84mm — it uses the big bore siamesed block design developed from the Group 5 racing 24-valve. In spite of the bigger bore and shorter stroke, the small increase in capacity might be expected to have slightly the opposite effect, the engine's performance peaks are both lower, thanks to different valve timing. Maximum power is 218 bhp at 5,200 rpm (633 has 200 at 5,500) and maximum torque is 224 lb. ft. at 4,000 rpm (210 at 4,250).

The four-speed gearbox of the '33 is replaced by a five-speed Getrag one, geared a little low as on most BMWs, at 23.55 mph per 1,000 rpm — a curious state of affairs here, when with such a wide spread of power it would have been so easy to have made fourth the true maximum speed gear, and fifth an economy overdrive, with small adjustments to the ratios. There is a 15 per cent bigger bore exhaust system, suitably altered Bosch L-Jetronic electronic fuel injection and a slightly higher 9.3-to-1 compression ratio, the car running on 98 octane as before.

Tyre sizes are unaltered (Michelin 195/70VR-14in. XDX on the test car) but the rim width on the standard BBS aluminium-alloy wheels is increased by ½in.

Above: Side view is one of the better aspects of the 635, recalling much of the grace of the old 3.0 CSi and diminishing the thickness of the new car, evident from three-quarter front. Chief identifying features are the plastic stick-on stripes, and front and rear aerodynamic aids
Below: Rear lip is soft-moulded for pedestrian safety; exhaust pipes are bigger for the larger engine compared with the 633 — up from 1.65in. to 2in. dia.

Performance
Smooth and zestful

Starting is typical of Bosch fuel injection cars. You are told not to apply any throttle from cold; turning the key for only a short period leaves the engine *just* turning over on its own, as if about to die — then, after a few seconds it picks up gradually to the right warm-up tickover speed, emitting that delightful and characteristic BMW high-pitched twitter from the exhaust. You get used to this behaviour after a while, since it is a reliable starter, but at first it is disquieting. Driveaway is excellent, and you have to restrain yourself from maltreating a cold engine which is so willing to go.

Not so once it's warm. This is the epitome of BMW zestfulness — that glorious smooth eagerness of a BMW six which is such a hallmark of the breed. There is one BMW that, up to 100 mph is quicker — the 323i — but what the 635 CSi loses to its big-engined little brother at the lower end, it more than makes up for in refinement and in the way it keeps going where the 2.3-litre falls off. The manufacturers claim the car to be fastest four-seat coupé made in Germany, with a top speed of 140 mph. The test car did exactly that in near still air, with the engine making a shade under 6,000 — 750 rpm (14 per cent) over its peak power. It is obvious that, if properly geared for maximum speed, the car would achieve perhaps 145 mph — of academic interest certainly, but if geared thus it would gain usefully in economy and unfussiness. For it is still fussier than some of its competitors especially the particularly quiet Jaguar XJ-S.

The advantage of undergearing which the German motor industry — and presumably the German driver — likes is of course easy top gear performance. The 635 has this in full measure, thanks to its generous spread of power. It will pull in top without protest from 19 mph — 800 rpm — and as the 20 mph increments show, it accelerates more or less evenly, give or take a half-second to 90 mph only gently falling off above that. This makes it a very relaxing car to drive generally.

Standing starts are best done using only around 3,300 rpm to produce mild wheelspin accompanied by some wheel-hop, any more wheelspin proved as usual to be slower. The Porsche-style gate pattern for the five-speed gearchange, with first on the dog-leg rather than fifth, makes the vital first-to-second change slower than usual. This is really a racing style arrangement, since first gear is only usually wanted once, on the start-line. For road use, the Alfa pattern with fifth on the dog-leg is better, since one rarely needs to change into top quickly whereas there are times when first is wanted fast.

Acceleration in any BMW is always exhilarating, and this one is no exception. There is a noticeable amount of attitude change — the car squats perceptibly on take off and after each gearchange. In spite of the awkwardness of first to second, there is a yelp of tyre squeal as the clutch bites again, and this is repeated on the faster change from second to third. Testing in December isn't always blessed with the right weather, and the inordinately high cross wind in which we were forced to figure the 635 has had its effect on the results. Nevertheless, as the times suggest, the car is a delight to drive fast as well as slowly. The gearchange takes some learning, requiring deliberate but light movements; heavy-handedness is encouraged by the notchiness of the change, but is punished by a tendency to baulk which disappears if you are gentler.

There's a rev-limiter at 6,350 rpm. In spite of the low peak power revs, we found it best to change up at 6,000 rpm. Response is superb, as usual on BMWs with the exception that on this particular car we encountered momentary bouts of total misfiring several times during hard acceleration — not we believe a typical fault.

Economy
What you'd expect

When comparing BMWs with competitors, economy has always been a strong point. But if engine size and weight and frontal area all go up, more energy is used to do the same job, and so it is not surprising to find the 635 CSi's consumption getting closer to its faster rivals, and well below (at 17.5 mpg overall) its 3.3-litre brother's 20.6 mpg. Nevertheless, there is still a worthwhile difference; among the comparison cars only the Lotus Elite, a smaller, lighter car, is better. To be fair, it should also be said that the 635's overall test mileage is less than the 633's so that the very thirsty MIRA test session distance occupies a larger proportion of the overall. The 15½ gallon tank fills easily except for the last half gallon which requires slow delivery because of blow-back from the breather system. The fuel warning lamp starts flashing at odd moments with at least two gallons left. BMW are, like Mercedes, delightful in detail; a pleasing one is the cup-like bracket in the fuel filler flap in which you can put the filler cap whilst re-fuelling.

Road behaviour
Total obedience

The car has that other hallmark of all BMWs — steering which at ordinary fast touring speeds is as near perfect a combination of total obedience and weight as can be expected in a road car. It is not matched by the car's ultimate cornering, which is slightly let down by the usual BMW trailing arm readiness to break away quite abruptly if the driver decelerates in the bend — but there is less of this than before which is a pleasant change. Naturally, it is easy to provoke a tail slide with the power available, but the fairly high-geared steering — 3½ turns for a handily tight 34ft lock — makes correction relatively easy. The steering's response is good, there is no serious amount of slop in the ball and nut steering gear, and the power assistance is strong enough to make parking easy but not so dominant that all feel is missing. Roll is limited to a comfortable and confidence-inspiring degree.

For all normal road use — and this includes fast, safe driving as opposed to track-style stuff — it is worth repeating that the response of the steering which is combined with the customary BMW readiness to turn in (to begin to turn) is something which makes the car particularly delightful to drive. On the other hand it is reassuringly stable, even in a cross wind at speed. We noticed, however, that at maximum speed the steering did feel a little lighter than before, suggesting that there is still some lift in spite of that modified front. Weight distribution is the same as that for the 633 CSi at with 56.6 per cent on the front with the car unladen.

The ride is a little abrupt at low speeds, rather more so than you might expect, but it improves to the usual BMW standard at speed. Road noise is there, slightly less than average, mostly made up of bump-thump, though there are enough vestiges of road roar to be noticeable. At 80 mph the volume of the excellent Philips stereo FM/AM radio must be turned up, but not to an uncomfortable level. Engine noise is there — a pleasing sound, typical of all six-cylinder BMWs, a sort of deep moan when you accelerate, which is subtly exciting — making the car only just quiet enough if judged absolutely. Wind noise on the test car came from around the very prominent door mirrors and, appreciably more so, from the optional electric sliding roof.

Brakes are just the right blend of power and low pedal effort. The ventilated discs resist fade convincingly and proved well up to the performance. The handbrake copes comfortably with the 1-in-3 test slope.

Behind the wheel
Clarity and simplicity

The driver's seat cushion can be adjusted fore and aft, up and down, and for tilt, so that together with the seat back rake adjustment there is little reason for discomfort. Cloth or leather seating is optional without extra cost; we found the leather lacked some sideways location. The steering wheel has fore and aft, but not angular adjustment, which is a small pity because some of our drivers felt that the wheel itself was a little high for them. Pedals are nicely arranged, and heel and toe changes are not at all difficult. The instrument panel remains an example to other manufacturers in its carefully placed reflection-free glass, the tastefully simple yet explicit markings and the neatness and position of dials with respect to the steering wheel, which does not obstruct one's view. We would continue to ask for an oil pressure gauge and an ammeter, in addition to the usual warning lamps in which the 6-series coupés

Above: Reflections seen here are caused bounce-flash lighting and are not normal seen by driver. Instruments, from left, are fuel and temperature gauges, speedometer with six-figure mileometer and press-zero trip, and revcounter. Warning lamps in centre are for (from left low fuel and high temperature, battery, indicators, oil pressure, rear fog lamps and, on right, high beam, fog lamp, handbrake and a spare. The six blank spaces across the top of the display are used on automatic 633 CSi as transmission indicators. Lamps are switched, pull-style, on left of column, fog lamps on right. Lamp check panel is on right. Heater controls comprise ambient air vents each side, temperature/air conditioning and distribution on left and right of quartz clock whose bezel is heater fan rheostat control.

Below: Lamp check panel enables a quic look at all major services before driving away

Impressive engine compart- ment. Note tidy, well identified fuse box with its transparent lid

abound. In addition to the main warning lamps in the instrument pod, there are as well a comprehensive panel of seven check lamps which on pressing a button tell you whether or not oil, coolant, washer and brake fluid levels, and brake lamps and linings are right.

The horn is controlled from thumb switches in the steering wheel spokes — rather too easily touched inadvertently and still not as ideal in an emergency as using the centre of the wheel. Electric

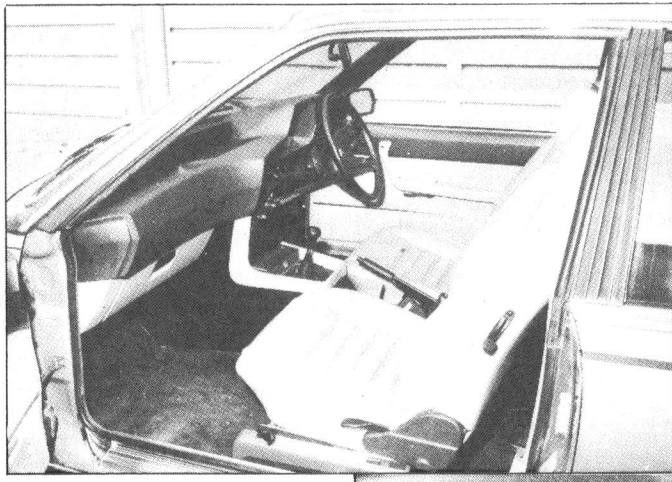
Below: Big wide doors make entry and exit easy. Cloth or leather (as here) are optional at no extra charge
Right: Rear seating is minimal, but adequate for short journeys if front seats are set fully back

windows are worked, rather slowly, with juddering, from switches on each side of the gearlever. The usual neat BMW heater fan rheostat is provided in the form of the bezel of the clock.

The heater itself continues to amaze us. How so many manufacturers, particularly makers of otherwise very good and expensive cars, can fail to take note of the simplicity, effectiveness and above all ideal temperature control of the air blending heaters used now for many years by Ford and Austin-Morris is beyond comprehension. That on the 635 is the depressingly usual water-valve controlled device one has learnt to expect from most Continental makers — it provides very poor temperature control, with little variation between full heat and none at all, plus slow response. Very tiresome on nearly £19,000's worth of motor car, especially when this price includes the boast of air conditioning (an extra) neatly switched from the same heater temperature control. One good mark is for the side window demisting

Drop-down bucket-type glove compartment includes a re-chargeable torch plugged in when not in use, on right

which is welcome in poor weather, and another goes to the ventilation which is adequate, at any rate in winter.

Visibility was a wonderfully strong feature of the elegant old 3.0 CSi body which had most pleasingly thin pillars on all corners and large amounts of glass. Presumably it was roll-over crush resistance and barrier impact test requirements which dictated the thicker pillaring of the replacement, but, although we would have thought that by using much thicker wall sections the windscreen pillars could have been kept to the same size as before, they

Extensions behind back seats on shelf cover useful concealed trays

are not too bad and the view is still good. You have got to remember that the size of those pillars is enough to hide a pedestrian momentarily however, and look more carefully. In the wet the proper British-biased screen wipers are appreciated by the driver, though there is a lot of area unswept on the passenger's side. Also in the wet, incidentally, after standing overnight in rain, there was a small leak from the front left hand corner of the sunshine roof on the test car.

Living with the 635 CSi

Overall one can sum up by saying that each of the Road Test team would very much like to. In addition to the overall pleasure of driving and running the 635, its detail design gives much satisfaction. True central locking is provided; turn the key or push down the sill pip (virtually thiefproof as it has no shoulder under which a wire could engage) in the driver's door, and both doors and the boot locked. Switches are all robust in feel and in the way they work. The bonnet is released by pulling an easily found lever under the dash on the passenger side — it is on the driver's side on left-hand-drive models and it is a pity that it isn't changed over for this country; but it works well, unlike those on some other cars.

Room in the back depends on how much the front occupants sacrifice their legroom, as you would expect in a coupé. The backs of the rear seats are continued neatly over the shelf behind in the form of lidded shallow compartments for

oddments. Other oddment space is fairly good; the lockable swinging-bucket type of glove compartment is big enough to carry a useful amount of guide books, maps and so on. The disadvantage of this type is that it becomes heavy to shut when full. At its side there is a handy rechargeable torch, permanently plugged into the car's electrical system but fitted with an automatic cut-off arrangement to avoid over-charging. There are small spring-loaded pockets in the doors. Door mirrors on the test car are the remote electric sort, controllable fortunately from the driver's side, since their ingenuity fascinates anyone unfamiliar with them. The driver's one is standard as far as British market cars are concerned, but not the passenger-side one.

The electrically motored sunshine roof moves slowly, and makes a tolerable increase in wind noise at speed when open. It can also be moved to hinge up at the back, for extra ventilation without being wide open. Boot space is generous, although the single skin sides must be respected with any large sharp-cornered load that can roll about if not restrained. The usual superb BMW toolkit is provided, neatly stowed out of the way under the bootlid. The aerodynamic rear lip is made of soft rubbery material to prevent damage.

If not quite in the Jaguar class, the underbonnet layout is good to look at. All major reservoirs are easily found; the alternator is rather buried, but the battery, excellent transparent-lidded fuse box and distributor are all readily accessible.

The 633/635 range

As far as Britain is concerned there are just two big BMW coupés (the carburettor 630 CS is not available here). The 635 CSi is the flagship of the BMW range at £16,499 (without those expensive extras) and is only available with manual transmission. Since the 635 introduction the 633 CSi is now the automatic version of the range, at £15,379; you can no longer buy a manual 633.

Boot is generous but its single skin sides must be borne in mind when carrying loose sharp-cornered items. Bulge under lid contains handsome and quite comprehensive tool kit

HOW THE BMW 635 CSi PERFORMS

MAXIMUM SPEEDS

Gear	mph	kph	rpm
Top (mean)	140	225	5,950
(best)	140	225	5,950
4th	118	191	6,350
3rd	84	135	6,350
2nd	62	100	6,350
1st	40	65	6,350

TEST CONDITIONS:
Wind: 25-35 mph (0-3 mph for max speed runs)
Temperature: 12 deg C (54 deg F)
Barometer: 28.7 in. Hg (973 mbar)
Humidity: 85 per cent
Surface: dry asphalt and concrete
Test distance: 785 miles

Figures taken at 7,400 miles by our own staff at the Motor Industry Research Association proving ground at Nuneaton, and on the Continent

All Autocar test results are subject to world copyright and may not be reproduced in whole or part without the Editor's written permission

ACCELERATION

FROM REST

True mph	Time (sec)	Speedo mph
30	3.2	31
40	4.9	41
50	6.7	53
60	8.5	64
70	11.0	74
80	14.3	85
90	18.7	95
100	23.4	105
110	30.3	113
120	41.0	122

Standing ¼-mile: 16.2 sec, 85 mph
Standing km: 29.8 sec, 109 mph

IN EACH GEAR

mph	Top	4th	3rd	2nd
10-30	—	8.6	5.0	3.6
20-40	9.1	7.1	4.5	3.1
30-50	8.9	6.4	4.3	2.8
40-60	8.7	6.4	4.1	3.5
50-70	8.0	6.7	4.3	—
60-80	9.6	6.7	5.4	—
70-90	10.0	7.1	—	—
80-100	10.9	8.6	—	—
90-110	13.1	12.3	—	—
100-120	18.0	—	—	—

FUEL CONSUMPTION

Overall mpg:
17.5 (16.3 litres/100km)

Constant speed

mph	mpg	mph	mpg
30	35.5	70	29.5
40	35.3	80	27.4
50	35.3	90	25.2
60	31.4	100	22.4

(Manufacturers' figures)

Autocar formula: Hard 15.6 mpg
Driving Average 19.0 mpg
and conditions Gentle 22.5 mpg
Grade of fuel: Premium, 4-star (98 RM)
Fuel tank: 15.5 Imp galls (70 litres)
Mileage recorder: 1.6 per cent short

Official fuel consumption figures
(ECE laboratory test procedures, not necessarily related to Autocar figures)
Urban cycle 14.8 mpg
Steady 56 mph 28.3 mpg
Steady 75 mph 23.4 mpg

OIL CONSUMPTION
(SAE 20/50) 800 miles/pint

BRAKING

Fade (from 85 mph in neutral)
Pedal load for 0.5g stops in lb

	start/end		start/end
1	25-50	6	55-35
2	25-23	7	50-35
3	30-25	8	50-30
4	40-30	9	45-30
5	50-30	10	45-30

Response (from 30 mph in neutral)

Load	g	Distance
10 lb	0.18	167 ft
20 lb	0.30	100 ft
30 lb	0.55	55 ft
40 lb	0.75	40 ft
60 lb	0.95	31.7 ft
Handbrake	0.30	100 ft

Max gradient: 1 in 3

CLUTCH
Pedal 40 lb; Travel 6 in.

WEIGHT
Kerb, 30.8 cwt/3,447 lb/1,564 kg
(Distribution F/R, 56.6/43.4)
Test, 34.3 cwt/3,847 lb/1,745 kg
Max payload 794 lb/360 kg

DIMENSIONS

OVERALL LENGTH 187·25"/4756
OVERALL WIDTH 69"/1753
OVERALL HEIGHT 54·5"/1384
WHEELBASE 103"/2616
FRONT TRACK 56"/1422
REAR TRACK 58·5"/1485
GROUND CLEARANCE 5"/127
SCALE 1:35
OVERALL DIMENSIONS in/mm

Turning circles: Between kerbs, L 34ft. 1in. R 34ft. 4in.
Boot capacity: 14 cu. ft.

PRICES

Basic	£14,101.72
Special Car Tax	£1,175.14
VAT	£1,222.14
Total (in GB)	**£16,499.00**
Seat Belts	Standard
Licence, tankful petrol, delivery charge and number plates	£150.00
Total on the Road (exc insurance)	**£16,649.00**

EXTRAS (inc VAT)
*Philips 860 FM stereo/AM/cassette player with electric aerial £299.00
*Electric sunshine roof £562.00
 Manual sunshine roof £439.00
*Air conditioning £1,051.00
*Headlamp wash/wipe £161.00
*Electric l.h. door mirror £60.00
*Fitted to test car

TOTAL AS TESTED ON THE ROAD £18,782.00

SERVICE & PARTS

Change	Interval		
	5,000	10,000	20,000
Engine oil	Yes	Yes	Yes
Oil filter	Yes	Yes	Yes
Gearbox oil	—	—	Yes
Spark plugs	—	Yes	Yes
Air cleaner	—	Yes	Yes
Total cost	£12.27	£39.73	£46.85

(Assuming labour at £6.50/hour)

PARTS COST (including VAT)

Brake pads (2 wheels) — front	£16.46
Brake pads (2 wheels) — rear	£11.85
Exhaust complete	£145.37
Tyre — each (typical)	£62.51
Windscreen, laminated, tinted	£74.80
Headlamp unit	£14.54
Front wing	£221.98
Rear bumper	£168.78

WARRANTY
12 months/unlimited mileage

SPECIFICATION

ENGINE
Type: Front, rear drive
Head/block: Al. alloy head/cast iron block
Cylinders: 6, in line
Main bearings: 7
Cooling: Water
Fan: Viscous
Bore, mm (in.): 93.4 (3.68)
Stroke, mm (in.): 84 (3.31)
Capacity, cc (in³): 3,453 (210.7)
Valve gear: Ohc
Camshaft drive: Chain
Compression ratio: 9.3-to-1
Ignition: Electronic breakerless
Fuel injection: Bosch L-Jetronic
Max power: 218 bhp (DIN) at 5,200 rpm
Max torque: 224 lb ft at 4,000 rpm

TRANSMISSION
Type: Getrag five-speed
Clutch: Hydraulic, diaphragm spring

Gear	Ratio	mph/1000rpm
Top	1.0	23.55
4th	1.263	18.65
3rd	1.776	13.26
2nd	2.403	9.80
1st	3.717	6.34

Final drive gear: Hypoid bevel
Ratio: 3.07-to-1

SUSPENSION
Front—location: MacPherson strut
 springs: Coil
 dampers: Telescopic
 anti-roll bar: Yes
Rear—location: Independent, semi-trailing arm
 springs: Coil
 dampers: Telescopic
 anti-roll bar: Yes

STEERING
Type: ZF ball and nut
Power assistance: ZF hydraulic
Wheel diameter: 14.7 in.
Turns lock to lock: 3.5

BRAKES
Circuits: Twin, split front/front and rear
Front: 11 in. dia. ventilated disc
Rear: 10.7 in. dia. ventilated disc
Servo: Vacuum
Handbrake: Centre lever, rear drum within disc

WHEELS
Type: Cast al. alloy
Rim Width: 6½in.
Tyres—make: Michelin
 —type: XDX radial tubeless
 —size: 195/70VR14in.
 —pressures: F36, R36 psi

EQUIPMENT
Battery: 12V 66 Ah
Alternator: 65A
Headlamps: 110/220W
Reversing lamp: Standard
Hazard warning: Standard
Electric fuses: 19
Screen wipers: Two-speed plus intermittent
Screen washer: Electric
Interior heater: Water valve
Air conditioning: Extra
Interior trim: Leather or cloth seats. pvc head-lining
Floor covering: Carpet
Jack: Screw pillar
Jacking points: Two each side, under sills
Windscreen: Laminated, tinted
Underbody protection: Paint system, bitumastic, pvc

HOW THE BMW 635 CSi COMPARES

BMW 635 CSi £16,499

Front engine, rear drive
Capacity 3,453 c.c.
Power 218 bhp (DIN) at 5,200 rpm
Weight 3,447 lb / 1,564 kg
Autotest 6 January, 1979

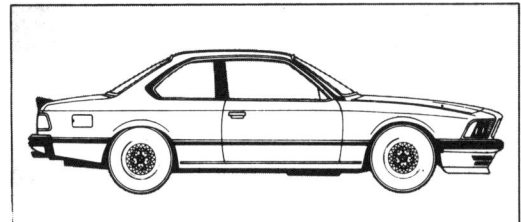

Jaguar XJ-S £15,149

Front engine, rear drive
Capacity 5,343 c.c.
Power 285 bhp (DIN) at 5,500 rpm
Weight 3,902 lb / 1,767 kg
Autotest 7 February, 1976

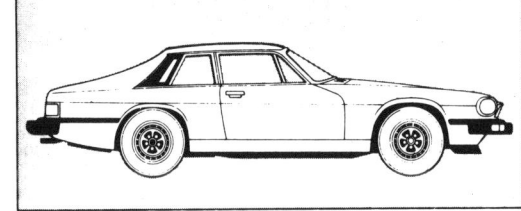

Lotus Elite 503 £12,861

Front engine, rear drive
Capacity 1,973 c.c.
Power 155 bhp (DIN) at 6,500 rpm
Weight 2,552 lb / 1,157 kg
Autotest 18 January, 1975

Maserati Kyalami £21,996

Front engine, rear drive
Capacity 4,136 c.c.
Power 270 bhp (DIN) at 6,000 rpm
Weight 3,836 lb / 1,740 kg
Autotest 8 July, 1978

Mercedes-Benz 450 SLC (A) £18,250

Front engine, rear drive
Capacity 4,520 c.c.
Power 225 bhp (DIN) at 5,000 rpm
Weight 3,685 lb / 1,673 kg
Autotest 11 October, 1975

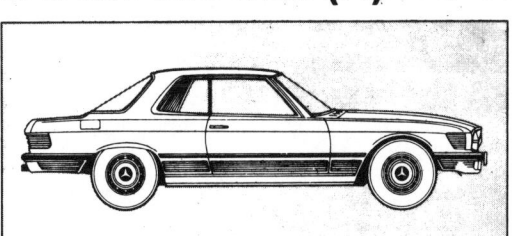

Porsche 928 £20,498

Front engine, rear drive
Capacity 4,474 c.c.
Power 240 bhp (DIN) at 5,500 rpm
Weight 3,347 lb / 1,518 kg
Autotest 30 September, 1978

MPH & MPG

Maximum speed (mph)
Jaguar XJ-S	153
Maserati Kyalami	147
Porsche 928	142
BMW 635 CSi	140
Mercedes 450 SLC (A)	136
Lotus Elite 503	124

Acceleration 0-60 (sec)
Jaguar XJ-S	6.9
Porsche 928	7.5
Maserati Kyalami	7.6
Lotus Elite 503	7.8
BMW 635 CSi	8.5
Mercedes 450 SLC (A)	9.0

Overall mpg
Lotus Elite 503	20.9
BMW 635 CSi	17.5
Porsche 928	17.1
Jaguar XJ-S	15.4
Maserati Kyalami	15.3
Mercedes 450 SLC	14.1

Clearly, on all counts but economy the Jaguar XJ-S is dominant. Its larger engine naturally puts it on top as it should, and its combination of sheer performance and flexibility and refinement is completely unrivalled anywhere in the world. It is also still relatively good value for money.

Happily there are no duds in this field. The Lotus will appeal to the purist who values efficiency, an old Chapman strongpoint, though it is the least refined of the bunch. The two most expensive cars vie closely with each other for second best performance; the BMW follows the Maserati and the Porsche, and is really closer to the Lotus than its dog-leg-first-gear limited 0-60 time would suggest. The Mercedes, because of its lack of a manual box option, falls behind on acceleration and economy.

ON THE ROAD

Top for roadholding must come, in strictly alphabetical order since it is very hard to say which is absolute topmost, the Jaguar, Lotus, and Porsche, with the Mercedes close behind. These cars all have relatively well-behaved rear suspension. The BMW 635 is much better than previous semi-trailing arm cars from Munich. The Maserati's rear end goes relatively easily too.

All steer well for all ordinary purposes, the Jaguar lacking some feel but having the best gearing for its size, needing only three turns for a 36ft lock. It rides very well and exceptionally quietly, and its brakes are well balanced. In power assisted form, the Elite's steering also lacks enough feel, and would be better with slightly higher gearing and more willing self-centring; ride is good.

The Maserati's steering, superb in its liveliness, feel and accuracy, is spoilt only by a clumsy 39ft lock; the ride is acceptable if rather firm. Feel, without road shock, characterises the Mercedes direction, plus accuracy. The 928's steering is geared nearly as high as the Jaguar's, has a little feel, is accurate except under braking when it wanders slightly but is otherwise superbly stable. The BMW could do with more feel, but does everything else in this respect just as well as the rest.

SIZE & SPACE

Legroom front / rear (in)
(seats fully back)
Maserati Kyalami	43/35
BMW 635 CSi	42½/32
Mercedes 450 SLC (A)	41/33
Jaguar XJ-S	42/30
Porsche 928	41/37
Lotus Elite 503	44/22

All of these are practical touring cars as well as high performance sporting machines, the four larger cars especially so; they in particular have the boot space of a conventional four-door saloon, but of course lack something in back seat space. The Lotus and the Porsche, both slightly smaller cars, are a little limited in luggage room, but their designs sensibly come to terms with the problems of rear seating in relatively cramped space by shaping the seats to encourage a knees-up posture.

The Maserati suffers mildly from the usual Italian car proportions of seating in front — pedals a little close, wheel a little remote — which is awkward for the longer-legged Anglo-Saxon. The Porsche and the Lotus are both cars in which one seems at first almost to sink; you are distinctly surrounded by car in the driving seat. This applies less so to the Jaguar and least of all to the two big German cars; the Mercedes and its Bavarian rival are cars in which you sit commandingly. Overall, the BMW appeals greatly in size and space, yielding nothing to the others here.

VERDICT

Undeniably, for best value for in all respects except economy and perhaps looks — and that is of course very much a matter of opinion — the Jaguar XJ-S is unapproached. Its trump card is of course its quietness and refinement; the near-silent sport cars at last.

The Lotus falls behind in refinement, which it is reasonable to assume is expected in this corner of the market. It is likely that for solid quality the Mercedes has no peer, and its refinement is good. The Kyalami will appeal to the person who wants something subtly different; it is distinguished, and also very enjoyable. The Porsche is also expensive, but an incredibly good all-rounder with no vices. The Jaguar aside, there is no clear favourite. The BMW will find favour with the man who ranks style as important as performance, and who prefers a vestige of sporting sound to total refinement; it is a worthy contender here, practical and very satisfying to drive.

UNDERSTATED ELEGANCE

If it weren't a BMW, the 633CSi would surpass our expectations

by Fred M. H. Gregory

BMW's reputation as a maker of exquisitely engineered automobiles is well deserved and secure. It rests on a succession of cars that have proven themselves both in competition and on the highway. BMWs are fast, nimble, well constructed and, that most difficult thing for high-performance cars, practical. They run like Porsches and feel like Mercedes. But the emphasis has always been on performance. Or, at least, it used to be.

In the 633CSi, the boys from Bavaria have made some concessions, evidently with an eye toward attracting the well-heeled, luxury-loving American buyer.

In the past, BMW has worked a particularly lucrative vein in the U.S. Its cars have been pricey, but not excessively so; they've been sporty, but in a

conservative way; they were cars, as someone once said, for people who like to drive fast without being obvious about it.

With the 633, BMW took a considerable step up in class—price class, that is. At about $30,000, the 633 is not the next logical move from owning a 320. But it is a short step sideways for a person contemplating a Mercedes-Benz 450SLC or a Jaguar XJ-S. And this buyer, presumably, wants some comfort and luxury for his money.

Consequently, the 633CSi is less tightly strung than its European counterpart and not as satisfying as it could be for someone who revels in precision motoring. Aside from blanching at its intimidating price, that is the worst thing you can say about the car. In nearly all other respects, it is a superior piece of machinery.

Every car manufacturer ought to force its engineers to sit behind the wheel of a 633CSi, so they could see what designing a driver's environment is all about. Nothing has been left to chance or added as an afterthought. The seat is a firm leather cup, which grips without confining. At first, it's hard on the bottom, but after a couple days of driving, you become accustomed to this. You have to spend a little time playing with the infinite number of adjustments available to dial the car to fit your physique. The seat tilts, lowers, rises, reclines and goes back and forth; and the steering wheel telescopes as well. Regardless of what position you ultimately choose, short of lying straight back, you will find all the controls within easy reach and the instruments in plain view. The latter are large, cleanly rendered, easy to read and illuminated by gentle amber light.

Visibility all around is excellent, and once you've clicked shut the unobtrusive and comfortable belts, the feeling you get is one of security and total command.

The car comes to life easily and quietly, the 3.3-liter fuel-injected straight-six engine ticking over without a shake or shudder. Its 177-horsepower takes you smoothly through the gears with no apparent flat spots of hesitation. For a high-performance car, the 633's acceleration is not terribly impressive, but it is adequate.

The transmission shifts cleanly and

positively, with short throws that leave you in no doubt about what gear you're in. An overdrive 5th gear would be nice, however.

Through the legal speed ranges, the 633CSi behaves flawlessly, passing slower traffic with ease and cruising contentedly with no wander. The steering is variable power assist. As the engine speeds up, the assist falls off, giving you more feel of the road and more positive control. The steering wheel transmits all the information you need, while damping out normal vibrations.

When you put heavy demands on the car, though, the compromises made to soften its ride become apparent. In hard cornering, it tends toward understeer and imbalance, with more body roll than one would like and not quite enough power to confidently control direction with the throttle. At high straight-line speeds, the car floats a bit and demands all your attention. Triple-digit speeds are not recommended. Were it not a BMW, this criticism of the 633 would not be necessary, but one expects more from this otherwise outstanding car.

When the 633 was first introduced, a few years ago, its understated but highly original style was lavishly praised. This apparently was warranted, since the look of the 633 has endured so well. It is not dated in the least but has become more familiar as other manufacturers have copied it. The large greenhouse, short nose and deck style are the vogue these days, and there are so many cars designed in this motif that the 633 tends to blend in with them. This causes people who are unfamiliar with the car's virtues and history to say: "It just doesn't look like 30 grand." If you want to flaunt your bucks, a 633 is not the way to go. Then again, neither are Savile Row suits, antique Persian carpets or other illustrations of subtle elegance.

After 800 intensive miles in this car, we came to grow more fond of it. Its laudable points became more obvious. The workmanship and attention to detail deserve as much praise as the car's economy. At steady cruise speeds of around 60, the 633 got nearly 22 mpg, and in on-off hard city driving, with the air conditioner cranked up, it managed 12.4. And these are real numbers, not ones generated in a laboratory.

As a GT car, the 633CSi falls short of the mark; but as a luxury coupe with no affectations, it sets a standard. SCg

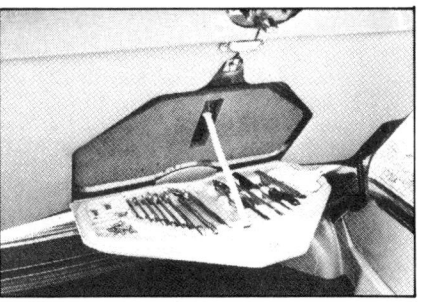

SPECIFICATIONS
BMW 633CSi

GENERAL
Vehicle type .... Front-engine, rear-drive, 2-door, 4-pass. coupe
Base price $29,265
Options on test car .... (All luxury items standard)
Price as tested $29,465

ENGINE
Type .... Inline six, water cooled
Bore & stroke .... 89 x 86 mm
Displacement .... 3210 cc/196 cu. in.
Compression ratio 8.4:1
Fuel system .... Bosch L-Jetronic fuel injection
Recommended fuel .... 87 AKI, 91 RON
Valve gear .... SOHC
Horsepower (SAE net) .... 177 at 5500 rpm
Torque (SAE net) .... 196 lb.-ft. at 4000 rpm
Power to weight ratio .... 19.4 lb./hp.

DRIVETRAIN
Transmission .... 4-speed manual
Final drive ratio 3.45:1

DIMENSIONS
Wheelbase .... 103.4 in.
Track, F/R .... 56.0/58.5 in.
Length .... 192.7 in.
Width .... 67.9 in.
Height .... 53.7 in.
Curb weight .... 3430 lb.

CAPACITIES
Fuel capacity .... 16.5 gals.
Trunk capacity .... 18.7 cu. ft.

SUSPENSION
Front .... Independent, telescoping staggered legs with helical springs & torsion bar stabilizer
Rear .... Independent, telescoping preloaded legs, rubber-mounted wishbones, torsion bar stabilizer

STEERING
Type .... ZF ball and nut, speed-related power assist
Turns lock-to-lock 3.62

BRAKES
Front .... 11.0-in. vented discs
Rear .... 10.7-in. vented discs

WHEELS AND TIRES
Wheel size .... 6½J x 14
Wheel type .... BBS/Mahle alloy
Tire make and size .... 195/70 VR 14
Tire type .... Steel radial

Road Test

BMW 628 CSi

Smaller-engined version of BMW's slingshot 635 aimed more at the well-heeled executive than sporting enthusiast. But with less power, has this £16,000 quality coupé lost its appeal?

IT HAS been over a year since BMW dropped its 3.3 litre-engined 633 CSi coupé from its UK model range leaving the larger-engined, more sporting and more expensive 635 CSi as the only remaining 6-series.

With the recent introduction of the Mercedes Benz 280 SLC coupé to Britain, however, BMW decided it needed a more closely-matched contender in this lucrative section of the market — hence the arrival of the new 2.8 litre 628 CSi coupé which we test here.

In Germany, the 628 has been available since the beginning of the year and like the UK version's, its fuel-injected 2.8 litre six cylinder engine is derived from that of the 528i saloon. But the Continental car has a four-speed gearbox while the UK 628 has a five-speed overdrive unit.

It would take a keen eye to spot the new car from its predecessor, the 633. As before, the 628 has a steel monocoque chassis, the floor pan of which is adapted from the 5-series saloon's. The front suspension is by MacPherson struts and coil springs, with power-assisted ZF worm and roller steering, and at the rear there are the pair of familiar BMW semi-trailing arms, again with coil springs. The suspension itself can be uprated by fitting gas-filled shock absorbers, stiffer springs and thicker anti-roll bars which are available as no-cost option.

Outwardly the 628 is distinguished from the more sporting 635 only by its lack of spoilers and wide-rimmed alloy wheels. Inside too, little has been

changed though the leather upholsery fitted to the test car can be replaced with velour cloth at no extra cost, as too can the front seats, which can be interchanged with the more curvaceous Recaro seats normally fitted to the 635.

Normal standards of value-for money do not pertain when you are considering cars in the 628's price range. But the current strength of the pound has allowed BMW to price the new coupé more competitively in the UK as well as offer a number of no-cost options. At £16,635, the 628 is close to £1,000 cheaper than its main rival, the Mercedes 280 SLC (£17,600) which for Britain is only available in automatic form. Other rivals include the superb Jaguar XJS (£19,187); the Porsche 911 SC (16,732) and its bigger stablemate, the 928 (£21,827); the Lotus Elite 2.2 (£16,433); and the Ferrari 308 GT4 (£17,534) which is soon to be replaced by the new Mondial. Cheaper rivals include the Opel Monza S (£12,808) which is now available with a 5-speed gearbox and the Lancia Gamma Coupé (£10,500).

We still find it hard to understand why coupé versions sell for considerably more than their saloon counterparts and in this case the 628 is no less than £4,200 more expensive than the similar-engined, yet larger-bodied 728i which has the advantage of two extra doors and a more spacious interior. Nevertheless, the 628 fills an important gap in the BMW range. It is also an extremely refined, comfortable and superbly finished 'quality' coupé with excellent ride and handling qualities together with good fuel economy. Its outright performance for a car with sporting pretentions can only be judged as fair, with its shortage of low-speed torque being the main shortcoming. It also falls down on poor rear seat legroom and mediocre heating and ventilation.

Similar to that of the 528i, the 2,788cc straight-six engine has an identical power output of 184 bhp (DIN) at 5,800 rpm with maximum torque of 177.2 lb ft (DIN) produced at 4,200 rpm.

Not surprisingly, the 0-60 mph acceleration time of 8.3 sec was identical to that of the similar heavy 528i we last tested (w/e December 30, 1977) with little to choose between the coupé's 0-100 mph time of 23.6 sec and the saloon's 23.9 sec. Although its acceleration may seem brisk, the BMW is outsprinted by all but one of our chosen rivals, which admittedly are either lighter or more powerful or both. Unfortunately we have not yet been able to test the Mercedes 280 SLC, but it should be considerably slower than

the BMW, according to the manufacturers's claims (0-60 mph in 11 sec).

If the 628's through-the-gears acceleration is to be judged as mediocre for its class, then its fourth and fifth gear pick-up can only be described as poor. In fourth, acceleration from 30 to 50 mph takes 9.8 sec while in fifth the same increment takes a yawning 13.9 sec. These figures confirm our testers' subjective opinions that the BMW feels no fireball, certainly by £16,000 standards. While the engine pulls cleanly from very low revs, its poor torque below 3,500 rpm gives the car a sluggish feel and if brisk progress is to be maintained, then you must use the gear to keep the engine spinning above this level.

We weren't able to verify the car's claimed maximum speed of 130 mph, but timed checks on MIRA's banked circuit suggested that this figure was within the 628's capabilities.

As with all applications of the excellent Bosch L-Jetronic fuel-injection system, we have sampled, starting is instant and warm-up impressively hesitation-free. As yet the Bosch-BMW 'Motronic' engine management system where a micro-computer controls both ignition timing and fuel injection is not yet available on the 628.

One of the areas the 'Motronic' system is designed to improve is fuel economy, yet even without it most owners should be satisfied with the 628's economy. We achieved 19.8 mpg overall despite our usual hard driving, a figure which most drivers should be able to stretch to around 23 mpg. Surprisingly, the 628 is less economical than the original four-speed 633 we tested (w/e October 16, 1976) which returned 20.8 mpg. But even this latest figure is competitive when compared with those of our chosen rivals (which vary from the Jaguar's 13.5 to the Porsche 911's 20.4).

Whereas the more sporting 635 CSi uses a close-ratio five-speed gearbox, where first is selected by moving the lever across to the left and back — considered awkward by many drivers — the 628 uses the five-speed Getrag gearbox from the 735i. This has a conventional 'H' gate for the lower four gears with the overdrive fifth to the right and forward.

Though the gearchange retains that slight rubbery feel that has always been a feature of big BMWs, it is precise and extremely slick and complemented by a progressive clutch and well-engineered throttle linkage. The synchromesh proved unbeatable and once mastered, the movement of the stubby mushroom-headed gearlever is one of the delights of driving the 628.

Unlike the 735i which suffered from excessive gear whine in the lower ratios, the coupé's transmission is extremely quiet. The lower ratios themselves are fairly widely spaced and at 6,500 rpm, the in-gear maxima are 36, 62 and 98 mph. With fifth gear giving a long-legged 25.8 mph/1000 rpm, cruising at 100 mph is relaxing and seemingly effortless as the engine is spinning at just under 3900 rpm.

The power-assisted ZF recirculating ball steering is sensibly weighted and every bit as confidence-inspiring as a good unassisted set-up, yet it makes light work of turning the wide Michelins at parking speeds. The steering *feels* particularly direct for the car understeers very little in the dry and

Top: facia looks rather austere but is efficiently laid out. Above: front seats are comfortable and comprehensively adjustable to ensure good driving position. Below: rear seats look accommodating but are not really suitable for adults

turns readily in response to the helm. Generally, the car adopts an impressively neutral attitude through the majority of corners, though there is sufficient power to push the tail out on tight bends if desired. Long, fast corners can be taken with confidence and on a motorway at speeds, the 628 is impressively stable.

On good surfaces, the BMW's ride is well-damped and comfortable while retaining a characteristic firmness which betrays the car's sporting background. The suspension is only caught out over broken surfaces when potholes cause the body to jar harshly, but to some extent, this sounds worse than it really is. The 628 also suffers less from fore and aft pitching on long-wave undulations, a shortcoming of the 633.

With 10.7 in ventilated discs at the front and 10.7 in non-ventilated discs at the rear, and with high-pressure hydraulic (not vacuum) assistance, the coupé's braking system is as good on the road as it is on paper. The brakes are powerful and free from fade, though the pedal does have a slightly 'mushy' feel to it when depressed slightly. Braking from high speed in the wet can occasionally result in the front wheels locking prematurely though in the dry they are superb.

In terms of accommodation, the 628 is very much a two-plus-two, despite its large overall dimensions. By saloon car standards, the rear seats are extremely cramped for adults and even if a compromise is reached with the front seat occupants, then most passengers of average height will still find their head touching the headlining. That said, there is still more rear-seat room than in the Ferrari, the Porsche and the Jaguar. The seats themselves are comfortable and well-shaped yet probably more suited to youngsters. In the front though, there is plenty of leg- and headroom and with the front seats adjustable for height, tilt, backrest rake, and a steering column that telescopes, most drivers should be comfortable behind the wheel. Inside the car there is plenty of storage space for oddments — though not on the centre console — and at 12.5 cu ft, the boot is

Comprehensive and clear main instrument display (top) is supplemented by this additional bank of warning lights (below)

enormous by coupé standards.

Comprehensive and clear instrumentation, housed in a binnacle beneath a single pane of clear plastic which is not entirely reflection-free is retained for the 628.

All-round visibility is excellent thanks to the low waistline and slim roof pillars together with the electrically-adjustable driver's and passenger's door mirrors, fitted as standard. The headlamps are powerful on main beam, but rather weak on dip, and the wipers, although excellent in their sweep, are much too slow on their fastest setting.

One of the car's other shortcomings, which has gone unchanged since the 6-series was introduced four years ago, is the heating and ventilation system. It is possible to get lots of heat but the temperature control is unprogressive and slow to respond to changes. The result is a rather stuffy interior which is not helped by the lack of throughput from the ventilation system under ram effect alone.

The generally quiet booster fan helps but unless it is switched to its fastest and noisiest setting, the flow of cold air is poor. In addition, the range of directional control afforded by the 'cheese cutter' ventilation grilles is marginal.

Although the 628 is the quietest 6-series BMW we have tested (72 dBA at 70 mph for example), it is still no match for the Jaguar XJS in this respect. The engine is pleasantly restrained even when revved hard and only coarse road surfaces cause road noise to become obtrusive. Without the optional electric sunroof (£642) fitted to the test car, wind noise would have been even lower.

As with all current BMW's we have sampled, we were considerably impressed with the high quality of finish. The test car's seats were trimmed in rich-looking dark blue leather (most of our testers said they would have preferred the optional and more supportive velour trim) with matching carpet. Although wide use is made of moulded plastic inside the car, it fits well and does not look out of place. The body panels also had a good fit and the doors shut with a satisfying 'clunk'.

The 628 is well-equipped, though by £16,000 standards, not exceptionally so. Standard fittings include electric tinted windows front and rear; central locking for the doors and boot; power steering; adjustable seats and steering wheel; alloy wheels with steel radial ply tyres; a comprehensive took kit; and the choice of either leather of velour trim, uprated suspension and Recaro seats.

Optional extras on the 628 include automatic transmission (£360); an electric sunroof (£642); headlamp wash/wipe (£211); air-conditioning (£1,212); a limited slip differential (£231); a cruise control (£342) and metallic paintwork (£337). A radio/cassette player is also an extra.

Above: engine compartment is a plumber's nightmare but all necessary items are accessible

MOTOR ROAD TEST NO 51/80 ●
BMW 628 CSi

PERFORMANCE

CONDITIONS
Weather	Wind 0-17 mph
Temperature	44-46°F
Barometer	29.7 in Hg
Surface	Dry tarmacadam

MAXIMUM SPEEDS
	mph	kph
Mean	130	209
Terminal Speeds:		
at ¼ mile	86	138
at kilometre	107	172
Speed in gears (at 6500 rpm):		
1st	36	58
2nd	62	100
3rd	98	158

ACCELERATION FROM REST
mph	sec	kph	sec
0-30	2.9	0-40	2.2
0-40	4.3	0-60	3.9
0-50	6.1	0-80	5.9
0-60	8.3	0-100	8.9
0-70	11.1	0-120	12.6
0-80	14.1	0-140	16.8
0-90	17.8	0-160	23.1
0-100	23.6		
0-110	31.4		
Stand'g ¼	16.1	Stand'g km	29.6

ACCELERATION IN TOP
mph	sec	kph	sec
20-40	13.7	40-60	8.4
30-50	14.2	60-80	8.8
40-60	14.0	80-100	8.7
50-70	13.9	100-120	10.2
60-80	15.3		
70-90	17.8		

ACCELERATION IN 4th
mph	sec	kph	sec
20-40	10.0	40-60	6.3
30-50	9.8	60-80	6.1
40-60	9.6	80-100	5.8
50-70	9.7	100-120	5.9
60-80	10.3	120-140	6.3
70-90	10.8	140-160	7.2
80-100	11.8		

FUEL CONSUMPTION
Overall	19.8 mpg
	14.3 litres/100 km
Govt. tests	16.0 mpg (urban)
	41.5 mpg (56 mph)
	30.7 mpg (75 mph)
Fuel grade	97 octane
	4 star rating
Tank capacity	18.5 galls
	84.1 litres
Max range*	420 miles
	670 km
Test distance	880 miles
	1416 km

*Calculated from an estimated touring fuel consumption of 23 mpg.

NOISE
	dBA	Motor rating*
30 mph	61	8.5
50 mph	65	11
70 mph	72	18
Max revs in 2nd	82	36
(1st for 3-speed auto)		

*A rating where 1 = 30 dBA and 100 = 96 dBA, and where double the number means double the loudness

SPEEDOMETER (mph)
Speedo	30	40	50	60	70	80	90	100
True mph	29	39	49	59	70	80	89	99

Distance recorder: 2 per cent fast

WEIGHT
	cwt	kg
Unladen weight*	28.6	1453
Weight as tested	32.3	1641

*with fuel for approx 50 miles

Performance tests carried out by Motor's staff at the Motor Industry Research Association proving ground, Lindley.

Test Data: World Copyright reserved; no unauthorised reproduction in whole or part.

GENERAL SPECIFICATION

ENGINE
Cylinders	6 in line
Capacity	2788 cc (170 cu in)
Bore/stroke	86/80 mm (3.39/3.15 in)
Cooling	Water
Block	Cast iron
Head	Light alloy
Valves	Sohc
Cam drive	Chain
Compression	9.3:1
Induction	L-Jetronic Bosch injection
Bearings	7 main
Max power	184 bhp (DIN) at 5800 rpm
Max torque	177.2 lb ft (DIN) at 4200 rpm

TRANSMISSION
Type	5-speed manual, Getrag
Clutch dia	9.4 in
Actuation	Hydraulic

Internal ratios and mph/1000 rpm
Top	0.813:1	25.8
4th	1.000:1	21.0
3rd	1.3981:1	15.0
2nd	2.202:1	9.5
1st	3.822:1	5.5
Rev	3.7051:1	
Final drive	3.45:1	

BODY/CHASSIS
Construction	Unitary all-steel
Protection	Plastic coating on underside. All internal sections sprayed with rust inhibitor

SUSPENSION
Front	Independent by MacPherson struts, coil springs, lower transverse arms, trailing arms, anti-roll bar.
Rear	Independent by semi-trailing arms, coil springs, anti-roll bar.

STEERING
Type	ZF Recirculating ball
Assistance	Yes

BRAKES
Front	10.7 in dia. ventilated discs
Rear	10.7 in dia. discs
Park	On rear
Servo	Yes
Circuit	Dual
Rear valve	Yes
Adjustment	Automatic

WHEELS/TYRES
Type	Alloy 6J × 14
Tyres	Michelin XWX 195/70 VR 14
Pressures	33/33 psi F/R (normal) 34/37 psi F/R (full load/high speed)

ELECTRICAL
Battery	66 Ah
Earth	Negative
Generator	Alternator 65 Amp
Fuses	17
Headlights type	4 × Halogen
dip	110 W total
main	220 W total

Make: BMW. **Model:** 628 CSi
Maker: Bayerische Motoren Werke AG, 8000 Munich 40, Petulring 130, West Germany
UK Concessionaires: BMW (GB) Ltd, Ellesfield Avenue, Bracknell, Berks. RG12 4TA
Price: £13,352.51 basic plus £1,112.71 Car Tax plus £2,169.78 VAT equals £16,635

The Rivals

Additional rivals to the 628 CSi include the Aston Martin V8 Saloon (£34,498), the BMW 635 CSi (£18,950), Datsun 280 ZX 2+2 (£9,596), Lotus Eclat 2.2 (£16,043), the Porsche 928 (£21,827), the Reliant Scimitar GTE (£10,324), the Lancia Gamma Coupé (£10,500), and Vauxhall Royale Coupé (£11,262).

BMW 628 CSi — £16,635

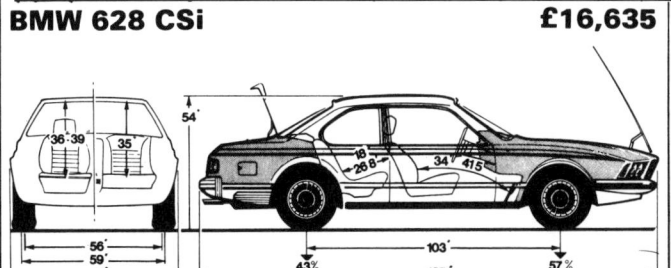

Power, bhp/rpm	184/5800
Torque, lb ft/rpm	177.2/4200
Tyres	195/70 VR 14
Weight, cwt	28.6
Max speed, mph	130†
0-60 mph, sec	8.3
30-50 mph in 4th, sec	9.8
Overall mpg	19.8
Touring mpg	—
Fuel grade, stars	4
Boot capacity, cu ft	12.5
Test Date	December 27, 1980
†estimated	

Smaller engined successor to the 633 Coupé gives brisk performance for size though lacks low speed torque. Outstanding features include excellent roadholding and handling coupled with very good fuel consumption and a light, precise gearchange. Comfortable, if somewhat hard, seating with good front seat legroom, but rear seats cramped. Large boot. Ventilation system still needs improvement. A well-finished car with an air of quality and refinement.

FERRARI 308 GT4 — £17,534

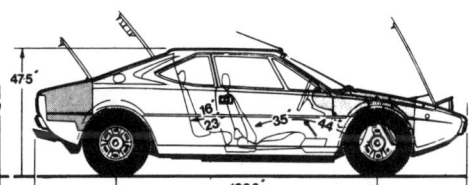

Power, bhp/rpm	255/7600
Torque, lb ft/rpm	210/5000
Tyres	205/70 VR 15
Weight, cwt	25.3
Max speed, mph	150†
0-60 mph, sec	6.4
30-50 mph in 4th, sec	5.1
Overall mpg	14.1
Touring mpg	18.7
Fuel grade, stars	4
Boot capacity, cu ft	5.0
Test Date	January 11, 1975
†estimated	

Mid-engine coupé powered by 255 bhp V8 giving outstanding performance and fair economy. Mediocre gearchange. Nominally a 2+2 but tiny rear seats not suitable for adults; boot is small. Roadholding exceptional, handling less precise and responsive than that of 246 Dino. Visibility good, heating and ventilation disappointing. Although production has ceased, the replacement Mondial has just become available but will not arrive in UK till next spring.

JAGUAR XJS — £19,187

Power, bhp/rpm	300/5400
Torque, lb/rpm	318/3900
Tyres	205/70 VR 15
Weight, cwt	34.3
Max speed, mph	145†
0-60 mph, sec	7.6
30-50 mph in kickdown, sec	2.8
Overall mpg	13.5
Touring mpg	16.1
Fuel grade, stars	4
Boot capacity, cu ft	8.4
Test Date	October 25, 1980
†estimated	

Now available only in automatic form, the XJ-S continues to provide an exceptional combination of performance, refinement and comfort at a price none of its rivals can beat. Economy improved with new digital fuel-injection, though still a thirsty car. Styling hasn't improved with familiarity, and cramped in the rear, but if you can afford the fuel bills the XJ-S remains one of the world's finest and most desirable cars.

LOTUS ELITE 2.2 — £16,433

Power, bhp/rpm	160/6500
Torque, lb ft/rpm	160/5000
Tyres	205/60 VR 14
Weight, cwt	22.9
Max speed, mph	130†
0-60 mph, sec	7.5
30-50 mph in 4th, sec	7.2
Overall mpg	20.6
Touring mpg	—
Fuel grade, stars	4
Boot capacity, cu ft	6.6
Test Date	November 1, 1980
†estimated	

Still a trend setter for exotic cars of the future. Exceptional handling and roadholding and good performance and refinement despite an engine which, even with the increase in capacity, is only 2.2-litres. Economical for such a fast car with four proper seats and a useful tailgate. Very comfortable and lavishly equipped. Visibility to the rear can be a problem but apart from this the Elite is now very difficult to fault.

OPEL MONZA S — £12,808

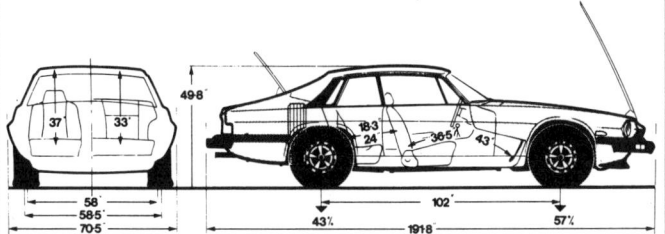

Power, bhp/rpm	180/5800
Torque, lb ft/rpm	183/4200-4800
Tyres	205/60 VR 15
Weight, cwt	27.3
Max speed, mph	130†
0-60 mph, sec	9.4
30-50 mph in kickdown, sec	3.4
Overall mpg	17.4
Touring mpg	—
Fuel grade, stars	4
Boot capacity, cu ft	10.7
Test Date	October 7, 1978
†estimated perf. figs. for auto version	

Hatchback coupé version of Opel's prestige Senator. A refined and comfortable four-seater with excellent performance for its price, fine ride and handling, combined with good accommodation for a two-door. Comprehensive specification includes a sunroof and a stereo radio/cassette player as well as power steering. Interior finish could be more plush and its engine is rather thirsty. Holds strong appeal for the sporty driver, and represents excellent value.

PORSCHE 911 SC — £16,732

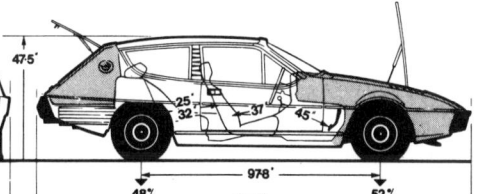

Power, bhp/rpm	204/5900
Torque, lb ft/rpm	195/4300
Tyres	185/70 Vr 15, 215/60 VR 15
Weight, cwt	22.9
Max speed, mph	148†
0-60 mph, sec	5.7
30-50 mph in 4th, sec	6.3
Overall mpg	20.4
Touring mpg	—
Fuel grade, stars	4
Boot capacity, cu ft	9.8
Test Date	November 15, 1980
†estimated	

Even faster and substantially more economical than before, the latest 911 SC is, without question, one of the world's most satisfying (as well as sensible) supercars. Superb handling and roadholding and exceptionally high standards of build and finish remain traditional facets. As does poor suppression of road noise. Tiny rear seats make it little more than a two-seater. In all, though, a true gem among supercars.

BMW 633CSi
Bavaria's flagship gets a 5-speed gearbox

PHOTOS BY DAVID GOOLEY & RON WAKEFIELD

THINGS ARE QUIET at BMW. It has been four years since a fully new model has been introduced in the regular series-production line. The 633CSi, BMW's luxury-sports coupe and top of the current American model lineup, was introduced in early 1976 in Europe; a 3.0-liter version of it (the 630CSi) came to America in 1977 and the 3.2-liter edition now sold here has been on the North American market since 1978.

Each year, however, the big BMW coupe gets detail improvements. In the 1980 *Guide* a significant improvement in fuel economy and driveability was noted, thanks to last year's change from thermal reactors to a 3-way catalytic converter with oxygen-sensor feedback for emission control. For 1981 the mechanical changes are limited to adoption of the 5-speed gearbox that went into the 528i in 1980; a ZF 3-speed automatic remains optional, and with it cruise control can be ordered. There are also some minor interior and equipment changes such as a digital quartz clock replacing the "old-fashioned" dial type, revision of the central locking system to provide locking and unlocking from the right door or trunk lock as well as the driver's door, and adding a little light to the ignition key that points the way to the ignition switch when a button on the key's side is pressed.

As time goes on and the general downsizing process continues, the BMW coupe—already a step up in size over its predecessor, the 3.0CS—looks bigger and bigger. Nor is its weight modest by current standards. But its proportions and styling are outstanding, and the recent mechanical changes have helped keep its fuel efficiency high enough that the clientele for a $40,000 car seems to find it quite acceptable. Perhaps more to the point, the 633CSi is the fuel-economy leader in its class, despite its ample dimensions. Credit goes to its relatively modest engine size and its standard 5-speed gearbox, a device not available on most of the competition.

All of its direct competitors—Porsche 928, Mercedes-Benz 380SL/SLC, Jaguar XJ-S—have bigger engines and more cylinders. In the past few years, indeed, BMW has experimented with more complex engines, building a 3.8-liter V-8 and a 4.5-liter V-12 as production proposals before deciding against them. Now the Bavarian maker has publicly decreed that six cylinders will be its upper limit, a decision influenced by both the long-term energy picture and BMW's expertise in the inline 6-cylinder configuration.

As an example of what can be done with the engine type, the 633CSi's "big six" lends credence to BMW's position. Part of the company's pitch for six cylinders is that this count provides almost as much refinement of operation as eight or 12, and this engine's smoothness supports the pitch. You know by its sound

BMW will build no cars with more than six cylinders. *The BMW cockpit: luxury and clever ergonomics.*

AT A GLANCE

	BMW 633CSi	Jaguar XJ-S	Porsche 928
Curb weight, lb	3400	3890	3370
Engine	inline-6	V-12	V-8
Transmission	5-sp M	3-sp A	3-sp A
0–60 mph, sec	8.4	7.8	8.1
Standing ¼ mi, sec	16.8	15.9	16.2
Speed at end of ¼ mi, mph	84.5	90.5	89.5
Stopping distance from 60 mph, ft	160	157	138
Interior noise at 50 mph, dBA	65	64	72
Lateral acceleration, g	0.754	0.726	0.811
Slalom speed, mph	55.5	56.6	59.7
Fuel economy, mpg	20.0	15.0	est 18.0

that it's a six, but the sonorities are interesting and virile and the delivery of power is butter-smooth right up to the point where the ignition cutout punctuates the exercise (at 6200 rpm, down from the previous 6400). Performance remains unchanged and strong; the 633CSi handily breaks the 8.5-second mark in reaching 60 mph from a standstill.

In general, the emission control by 3-way catalyst and oxygen-sensor feedback allows the BMW six to run well and respond spontaneously to the accelerator pedal. But this test car did have one hitch, in that its engine ran rather unpredictably after a cold start. Idle speed fluctuated widely, and as the engine warmed up the speed fluctuation moderated into mere roughness and finally gave way to a normal, smooth idle. Even at that, though, the speed (1100 rpm) seemed too high, and a heat shield under the catalytic converter rattled as the engine idled.

Aside from these problems, the 633CSi was as pleasant as ever to drive—indeed, thanks to the new 5th gear, more pleasant than ever. In the first four gears, except for an insignificant change in the 1st-gear ratio, overall gearing is the same as before. Likewise for shifting, which is wonderfully smooth and precise—including getting into 5th. The overdrive 5th reduces engine speed by 19 percent, increasing highway and suburban mileage (our test mileage was up from 19.0 to 20.0 mpg this year). But it has little effect on the CSi's already moderate noise level. In fact, at certain speeds it can even increase engine noise coming into the car. At 30 and 70 mph, for example, we got readings 1 decibel higher in 5th than in 4th, a difference attributable to the engine's exhaust note, which at some speeds intensifies noticeably with a little extra throttle opening. (To maintain the same road speed in 5th gear, more throttle opening is required—that's why an overdrive ratio gives better fuel economy.)

In urban driving the coupe's large size sometimes makes it a little difficult to find a suitable parking space, though its huge glass areas do make it relatively easy to maneuver into a tight space. But it is on the open road where this GT car truly shines. If we in America were allowed to drive fast, the BMW could strut its stuff as it does in its homeland; here we must make do with its relaxed mechanical sounds and good directional stability at American-style freeway speeds and the competence it demonstrates on winding rural highways.

This competence is provided by BMW's traditional attention to suspension, brakes, steering and body integrity. Like other "normal" BMWs (that is, excluding the M1), the 6-Series coupe has MacPherson struts at the front and semi-trailing-arm independent suspension at the rear. For the North American market only, BMW fits BBS-Mahle alloy wheels with 6½-in. rims; they are shod with Michelin XVS tires in place of the XDX formerly used. The change in tires—the new ones are less expensive and not capable of sustaining extreme cruising speeds—was made to save weight, according to BMW, and we did not find any marked difference in the way they handled.

In normal-to-brisk driving on winding roads, then, the 633CSi remains basically the same: responsive, stable and capable of high cornering speeds. BMW's variable-assist power steering, with decreasing assist at high engine speeds, contributes to the good impression by retaining plenty of road feel at all speeds. But as one approaches the cornering limits, BMW's chassis philosophy requires increasing expertise at the wheel to cope with the final oversteer. In tight, low-speed corners, this characteristic manifests itself as the lifting of the inside rear wheel, a bothersome trait that can be dispensed with only if one does not drive so energetically or does order the optional limited-slip differential.

Most people—in particular, most Americans—will never encounter any of this. Nor will they ever experience the CSi's braking limits, which are also high. Its stopping distances in "panic" situations aren't the world's shortest, nor are its brakes utterly free of fade, but they do a good all-around job—as they should, with their large-area ventilated discs all around. Stopping distances could probably be improved by the use of wider tires, but BMW has not followed the trend to the sportier 60- or 55-series tires used by some makers of sporting machinery.

The CSi's cockpit typifies the contemporary BMW approach to the driver's working environment and passengers' accommodation. BMW puts special emphasis on driver ergonomics, curving the instrument panel in plan view so that everything is within easy reach. Instruments, both the large main ones and the smaller dials, are easily read, and all the controls are logically arranged and pleasantly precise in operation.

Most of the coupe's interior materials are plastic of one kind or another. In its contours as well as its functional aspects, the padded dash is a remarkable example of what modern manufacturing processes can accomplish. The door and side panels, though beautifully thought-out and flawlessly assembled, tend more to the austere than the lavish. And the seats don't convey any of the boudoir effect one finds in domestic luxury coupes of similar size, but their leather-upholstered contours were designed with comfort and good support in mind.

Standard equipment in the 633CSi is extensive: air conditioning, electric window lifts, a 2-way electric sunroof (that lifts at the rear or slides open), electric outside mirror, central locking, power-assisted steering, metallic paint and a comprehensive radio-cassette stereo system are all on the long list of items included in the basic price, with automatic transmission (a 3-speed ZF device) and the limited-slip differential the only optional items. One thing, however, causes BMW owners some grief. Because the radio is of high quality and easily adaptable to other cars, the theft rate for the stereo system is high.

But so is the general satisfaction of 633CSi owners with their luxury GT coupes. Our only reservations concern its large size, small fuel tank and (despite the improvements) a rate of fuel consumption that can only be justified by its high performance and considerable luxury. Of course its price is high too, but the dollar's dramatic comeback over the past year should at least give it price stability for a while.

ROAD TEST
BMW 633CSi

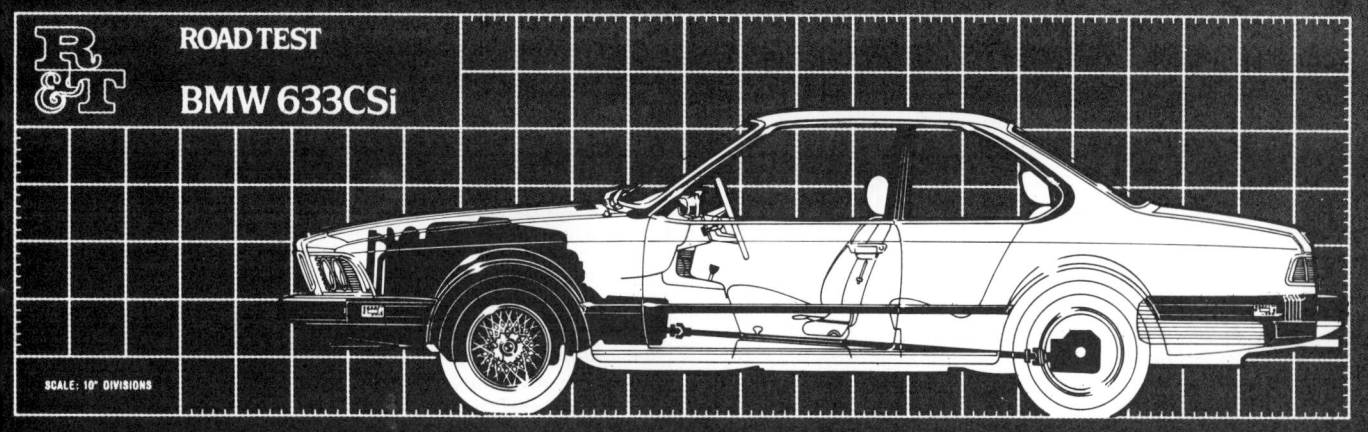

SCALE: 10" DIVISIONS

PRICE
List price, all POE	$35,910
Price as tested	$35,910

Price as tested includes standard equipment (power strg, air cond, elect. window lifts, elect. sunroof, central locking, AM/FM cassette/stereo, etc)

IMPORTER
BMW of North America, Inc, Montvale, N.J. 07654

GENERAL
Curb weight, lb/kg	3400	1542
Test weight	3600	1633
Weight dist (with driver), f/r, %		55/45
Wheelbase, in./mm	103.4	2626
Track, front/rear	56.0/58.8	1422/1494
Length	192.7	4895
Width	67.9	1725
Height	53.7	1364
Ground clearance	3.7	94
Overhang, f/r	42.3/47.0	1074/1194
Trunk space, cu ft/liters	15.7	445
Fuel capacity, U.S. gal./liters	16.5	62

INSTRUMENTATION
Instruments: 85-mph speedo, 7000-rpm tach, 999,999 odo, 999.9 trip odo, coolant temp, fuel level, clock
Warning lights: oil press., brake system, alternator, coolant temp, low fuel, oxygen sensor, brake-pad wear, rear-window heat, check control panel, seatbelts, hazard, high beam, directionals

ENGINE
Type	sohc inline 6
Bore x stroke, in./mm 3.50 x 3.39	89.0 x 86.1
Displacement, cu in./cc 196	3210
Compression ratio	8.4:1
Bhp @ rpm, SAE net/kW	174/233 @ 5200
Equivalent mph / km/h	133/214
Torque @ rpm, lb-ft/Nm	188/255 @ 4200
Equivalent mph / km/h	107/172
Fuel injection	Bosch L-Jetronic
Fuel requirement	unleaded, 91-oct

Exhaust-emission control equipment: 3-way catalyst

DRIVETRAIN
Transmission	5-sp manual
Gear ratios: 5th (0.81)	2.79:1
4th (1.00)	3.45:1
3rd (1.40)	4.83:1
2nd (2.20)	7.59:1
1st (3.82)	13.18:1
Final drive ratio	3.45:1

ACCOMMODATION
Seating capacity, persons	4
Head room, f/r, in./mm 37.0/35.0	940/889
Seat width, f/r 2 x 21.5/2 x 20.5	2 x 546/2 x 521
Seat back adjustment, deg	70

MAINTENANCE
Service intervals, mi:
Oil/filter change	7500
Chassis lube	none
Tuneup	15,000
Warranty, mo/mi	12/unlimited

CHASSIS & BODY
Layout	front engine/rear drive
Body/frame	unit steel
Brake system	11.0-in (280-mm) vented discs front, 10.7-in. (272-mm) vented discs rear; vacuum assisted
Swept area, sq in./sq cm 547	3528
Wheels	cast alloy, 14 x 6½J
Tires	Michelin XVS, 195/70HR-14
Steering type	recirc ball, power assisted
Overall ratio	16.9:1
Turns, lock-to-lock	3.6
Turning circle, ft/m 37.4	11.4

Front suspension: MacPherson struts, lower lateral links, compliance struts, coil springs, tube shocks, anti-roll bar
Rear suspension: semi-trailing arms, coil springs, tube shocks, anti-roll bar

CALCULATED DATA
Lb/bhp (test weight)	20.6
Mph/1000 rpm (5th gear)	25.5
Engine revs/mi (60 mph)	2350
Piston travel, ft/mi	1328
R&T steering index	1.35
Brake swept area, sq in./ton	303

RELIABILITY
Owners of earlier-model BMWs reported 9 problem areas and 3 disabling reliability areas compared to overall Owner Survey averages of 11/6. So we expect the overall reliability of the BMW 633CSi to be better than average.

ROAD TEST RESULTS

ACCELERATION
Time to distance, sec:
0-100 ft	3.4
0-500 ft	9.1
0-1320 ft (¼ mi)	16.8
Speed at end of ¼ mi, mph	84.5

Time to speed, sec:
0-30 mph	2.6
0-60 mph	8.4
0-80 mph	14.9

SPEEDS IN GEARS
5th gear (4700 rpm)	120
4th (5800)	124
3rd (6200)	91
2nd (6200)	59
1st (6200)	34

FUEL ECONOMY
Normal driving, mpg	20.0
Cruising range, mi (1-gal. res)	310

HANDLING
Lateral accel, 100-ft radius, g	0.754
Speed thru 700-ft slalom, mph	55.5

BRAKES
Minimum stopping distances, ft:
From 60 mph	160
From 80 mph	275
Control in panic stop	very good
Pedal effort for 0.5g stop, lb	25
Fade: percent increase in pedal effort to maintain 0.5g deceleration in 6 stops from 60 mph	32
Parking: hold 30% grade?	yes
Overall brake rating	very good

INTERIOR NOISE
Idle in neutral, dBA	57
Maximum, 1st gear	80
Constant 30 mph	63
50 mph	65
70 mph	72
90 mph	77

SPEEDOMETER ERROR
30 mph indicated is actually	32.0
60 mph	60.0
80 mph	78.0

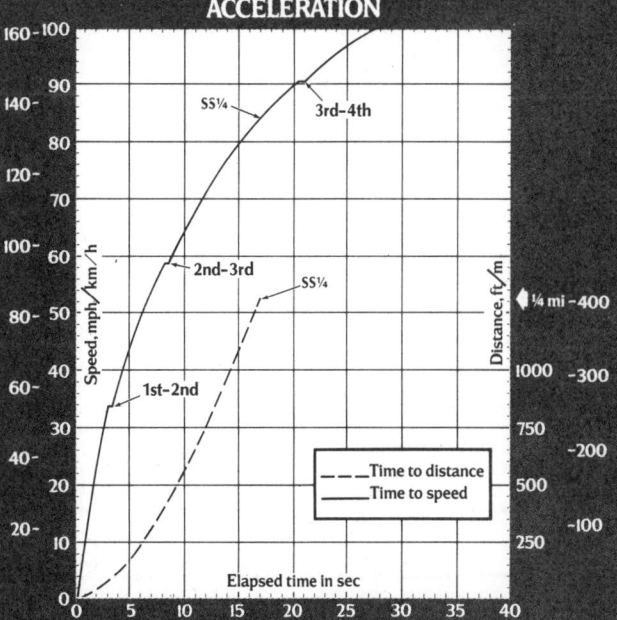

FRONTLINE
ANY OTHER BUSINESS

ACCORDING TO BMW, THEIR 628CSi coupe does more than 30mpg at motorway cruising speeds. The fact that when I tried the car it averaged 23mpg suggests either that BMW people do not cruise very fast on the motorway (though I myself never exceeded 128mph) or that my driving of the car on lesser roads was rather hectic. Admittedly I was pushing the car rather, on the occasion in question, feeling that I had a long way to go in a short time and that this was not a genuinely fast car.

To criticize a car capable of 130mph for not being fast may sound rather silly; it is in fact meant as a compliment to the serenity with which this BMW covers the ground at any speed within its range. There have been other BMW models that I have criticised for feeling uncomfortable and even unstable at almost any speed within their capabilities (I was never happy driving hard in a Five Series car) but this coupe seems to have been endowed with much firmer and tauter suspension than those unfortunate saloons. One hears tales of BMW proffering 'sports' or 'touring' suspension to suit the tastes of the individual customer or of a particular national market (you would be astonished at how soggy a car built for the USA can be) but if you ask them about it in this country they have a tendency to pretend not to know the answer, or to insist that all BMW suspension is *right*, and that if I cannot tell what I have got when I am driving it then I do not deserve the privilege of trying it. Not in so many words do they say this, being very polite people, but there is no limit to what can be conveyed by a rise of the eyebrow.

Anyway, I believe that all the coupes now come to Britain on sports suspension, the main feature of which is probably a stiffening of the springs so as to prevent undue wheel travel leading to excessive camber change, especially at the rear – stiffer in roll as well as in bump. To support my belief, I can report that never in the course of a drive from Bracknell to Torquay and back did the car ever give me a moment of displeasure in its roadholding, handling, steering, or any other of the compartmental aspects that determine whether a car is nice to drive. The CSi, even in its low-powered 2.8litre form with only 184bhp under its bonnet and no fancy aerodynamic appendices under its nose or over its tail, is a very nice car indeed.

My wish had been to try the quick one, the 635CSi, the coupe with knobs on. In highly developed form, that car has had an amazing succession of victories in the sort of racing represented by our Tourist Trophy, racing for the European Touring Car Championship. I long ago abandoned any supposition that performance in this kind of professional racing necessarily bore any relationship to the performance of the production car on the public road; but it seemed to me that the BMW coupe was the right sort of car to win that sort of championship, and that the production version of it probably came nearer to the idea of a superb high-speed road car than all the extravagantly modified and wildly contorted family saloons that usually seek publicity on those circuits. You may remember that I had been tremendously impressed by the M535i saloon, which behaved magnificently but was typical of the irrelevance of racing to a family saloon (what did I say earlier about the ordinary Five Series?); the 635CSi promised to be even better and a lot more plausible, less a boy racer.

Unfortunately, the 635 on the BMW press fleet was in dire need of tender loving care on the day appointed for me to borrow it. I gather that it is a very popular car with all those who have access to the fleet, and finds itself going out for testing whenever a mechanic or a staff member has a spare moment. That is probably one of the best recommendations the car could have, but it did not really help me. Instead, help came in the form of the 628 which could be spared for a couple of days and would get me out of my immediate difficulties.

These were that I had to be in Torquay within the next four hours and that I had no transport suitable for the three passengers expecting to accompany me. Faced with that sort of pressure, the 628 coupe is just the right machine to restore that equanimity without which it is dangerous to embark on a journey of serious length.

It has quite enough performance to make a long, high-speed journey feasible, without being such a muscle-bound brawler as to make the driver feel that he had better go very carefully lest he go too far. Moreover, it is so quiet and comfortable, so admirably

> **The BMW has enough performance to make a high-speed trip possible without being such a brawler as to make the driver go carefully.**

adaptable to the curious shapes and sizes in which people are made, so luxuriously supportive and functionally positive that the relaxation which it imparts increases with every passing mile, until that happy state is reached when brain and hands and feet are alert and active but all other tensions have disappeared. Beyond that point no further relaxation is desirable, but the rhythm of pressures on the steering wheel and the gearlever (both of them pleasant to use) imposes a natural limit of relaxation that is automatically appropriate to the urgency of one's driving.

That urgency is further limited by some of the car's other characteristics. The Getrag five-speed gearbox is not the delightful close-ratio device that I trust is still available to special order, but a wide-ratio affair in which fifth is suitable only for the maintenance of a road speed already reached in a lower ratio, while bottom gear is a screamer that may do wonders for the 0-60 time (9.0sec, which is not particularly quick for a car of this calibre) but forces the intervening ratios to be separated by gaps that are wide enough to be sometimes uncomfortable, especially when overtaking. The ZF servo-assisted steering is also a little low-geared, and I could not help thinking that the 195/70VR14 tyres could with advantage have been broader. Looking at the specification, I see that one of the available options is 'competition' suspension, which is presumably even stiffer and shorter in travel than the sports variety, and may permit the fitting of wider wheels with correspondingly more expansive tyres; but if one were to ask for all that, one would be inclined to ask also for the 3.5litre engine and all the other nice things which go with it.

The price one is prepared to pay must obviously enter the debate. The 635CSi, when I last enquired, was listed at £19,329, for which one presumably gets (and I can do no more than presume, until BMW's administration of tender loving care is complete and I can try the car) a machine capable of 140mph, of sprinting from 0 to 60 in 7.5sec, and of responding even more crisply to instructions than the 628. The lesser car was priced at £16,968, so the relatively little extra would be well worth spending if one were keen on driving hard and very fast. On the other hand, if the performance of the 628 were more than adequate for your needs (as I suspect is the case for the majority of interested customers), all that marvellous style and luxury, stability and solidity and general assurance of quality with reassurance of safety, could be yours with the price of a 2CV6 to spare. Do you, though, if you are of such stuff as BMW buyers are made of, know what a 2CV6 is? I am reminded of the observation of the late Mr Nubar Gulbenkian when talking about his town carriage, which was a London taxi refinished and refurnished in the manner appropriate to a multi-millionaire's station: *It will turn on a sixpence – whatever that is.*

The coming of decimalisation has denatured the steering of subsequent cars. There may be the sound of greater precision in saying that a car will turn on a five pence, but the coin is actually larger. In fact the steering of the best cars turns on banknotes of very large denominations; but back in the days when only those possessing them could even consider owning a car, Lanchester gave instruction in their drivers manual for turning in a narrow street without using reverse gear. What they recommended amounted to a powered skid turn; goodness knows that the politicians, busy-bodies and holier-than-thou (I beg thy pardon, holier-than-I) fraternity would make of it.

LJK SETRIGHT

Better BMW

Latest BMW 635 CSi features big changes under the skin

By Sam Brown

Above, right: good economy is a big feature of the new Coupé. External giveaways are big front spoiler and TRX wheels and tyres

ONLY the deeper, wrap-round chin spoiler and the distinctive squat Michelin TRX wheels and tyres give a clue that the BMW 635 CSi that makes its debut in Britain later this month is any different from the extra muscle version of the 6-series car announced in July 1978.

But beneath the smooth-flowing and still attractive bodywork lies an extensive series of modifications covering both power unit and suspension, that combine to provide performance with rare economy together with handling that at last approaches the standard the car should have had in the first place.

An ABS braking system as standard is offered on the Coupé for the first time and this, together with a whole package of electronics controlling engine management and servicing, combines to boost the price by £3,621 on that of the current car to £22,950, moving it up firmly to challenge the 928 Porsche. The 628 CSi is also improved but does not have the ABS and TRX wheel and tryre packages so that, at £17,895, it is only £927 more expensive than the old car.

The eagle-eyed will spot that the 635 engine has been slightly reduced in capacity to 3,430 c.c., by reducing the stroke from 93.4 to 92mm and enlarging the bore from 84 to 86mm. Main reason for this is to enable the standardisation of cylinder blocks with the Eta engine. Harmonising the bore size enables the block to be used on this engine and also to be stroked down for the 2.7 Eta unit now on sale in the United States. The European Eta is due early next year, possibly at about the same time as the diesel.

Another less obvious change is in weight saving. The new Coupé benefits from revised pressings and castings in certain areas of the car, including the engine, all of which add up to a reduction in kerb weight of 156lb on the previous model.

Remarkable economy for a 3.4 litre engine is achieved by a range of changes and modifications but most significantly by the application of the third generation of the Bosch/BMW fuel injection system.

The new system does not have the cold start valve included in the old L-Jetronic unit. Instead there is a longer injection opening period, an arrangement claimed not only to save fuel but which is simpler, doing away with the temperature time switch, wiring for the cold start valve and the associated circuitry. Some time ago, BMW followed Jaguar in arranging for the injection to cut out on over-run, switching on again automatically to preserve engine idling at 1,200rpm; on the new system, it cuts in 200rpm lower. Other minor alterations, including lower idling rpm add up to a claimed 9 per cent fuel saving. There is a choice of three gearboxes within the standard specification, covering automatic and five speed overdrive or five speed close ratio manual gearbox.

Certainly the 650 mile plus drive on the UK launch exercise from Dover to Marseilles produced some extremely good figures. Over the first stage between fuel stops our car returned a corrected 27.6mpg, averaging (according to the standard in-car computer) 61.9mph. The slightly shorter second stage, which included longer sustained periods at over 100mph, reduced the overall figure to 26.2mpg. Going by the computer's constant consumption read-out facility, 70mph indicated speed gave 36.6mpg and at 100mph 15.1mpg, suggesting a 400 mile range from the still modestly sized 15½ gallon tank.

One of the major criticisms of the previous Coupé was of the somewhat twitchy ultimate handling performance. BMW have tried to meet these objections with a fairly drastic revision of the suspension.

At the front the new Coupé the 5-series arrangement with a double link at the base of the strut to reduce steering bias if one side runs into different surface conditions. At the back the layout of the radius arms has been revised with the semi-trailing arms set at 13 degrees and the inclusion of camber compensating links – as already used on the 528i.

For the first time on a Coupé, the ABS anti brake locking system is featured as standard – as are TRX wheel and tyre combinations, previously only available as an option and then after some bodywork modifications. The 628 has what appears to be a TRX package but it is not.

Inside there have been changes to the seating and trim and the previous four spoke steering wheel is replaced by a three spoke version.

Instruments and minor controls are presented in a revised layout with speedometer and rev. counter in two large dials under a binnacle before the driver. The rev. counter features an "economy" gauge and between the dials lie the fuel gauge and the electronic service interval indicator. The indicator is linked to the electronic speedometer and incorporates the system test light to supplement the already comprehensive test panel set to the right of the facia. The angled centre console includes a new heating and ventilation control system of dials and slides.

On the road

The Coupés have always been smooth and this latest 635 is the smoothest yet, delivering really effortless performance. 100mph is an absurdly easy cruising speed, and it doesn't feel hard-worked even close to the 142mph maximum of the five speed manual version we drove. The overdrive fifth provides for very relaxed cruising although the level of road noise intrusion was higher than might be expected, with some bump-thump.

The large door mirrors seem to be responsible for the bulk of the wind noise but with the standard electric sunroof open wind roar is very intrusive.

Suspension improvements have produced some obvious gains and although we will have to wait to explore the ultimate handling behaviour there appear to be immediate benefits in smoothing out the rather joggly ride at slow speeds and reducing the inclination to sudden rear end breakaway.

The 7-series-derived variable power steering is well weighted with plenty of feel and adds to the impression of a confident and competent long distance cruiser further improved. □

Below: revised dash layout is cleaner. Three spoke wheel is also new

Bottom: new engine has latest BMW/Bosch fuel injection for economy plus performance

SPECIFICATION

	635CSi	628CSi
ENGINE		
Head/block	Front, rear drive	
	Al. alloy/cast iron	
Cylinders	6 in line	
Main bearings	7	
Cooling	Water	
Fan	Viscous	
Bore, mm (in.)	92 (3.62)	86 (3.38)
Stroke, mm (in.)	86 (3.38)	80 (3.15)
Capacity, cc (in.)	3,430 (209.3)	2,788 (170.1)
Valve gear	single Ohc	
Camshaft drive	Chain	
Compression ratio	10.0 to 1	9.3 to 1
Ignition	Digital electronic	
Injection	Bosch BMW system	
Max Power	218 at 5,200 rpm	184 at 5,800 rpm
Max Torque	224 lb ft	174 lb ft
SUSPENSION		
Front – location	Independent, MacPherson struts, lower links	
– springs	Coil	
– dampers	Telescopic	
– anti-roll bar	Yes	
Rear – location	Independent, semi-trailing arms	
– springs	Coil	
– dampers	Telescopic	
– anti-roll bar	No	
STEERING		
Type	ZF recirculating ball	
Power assistance	Hydraulic (variable rate)	
SERVICE DATA		
Fuel Tank: (Imp gallons/litres)	14.5/65.8 (reserve 1.8/8.2)	
PERFORMANCE (Mfrs figures)	Manual/Automatic	
Maximum speed (approx.):	142/137	132/127
0-60 mph (sec)	7.1/8.7	8.8/10.6
ECE/Government Official Fuel Consumption figures		
Urban (manual – mph)	18.0	19.8
Steady 56 mph (90 khp)	39.2	41.5
Steady 75 mph (120 kph)	32.1	32.1

Road Test

The executive saloon with the sporting appearance helps to keep the BMW 635CSi at the top of the ladder.

The sporting life

Unbeatable BMW is a well used phrase relating to the Bavarian marque but when it comes to their latest version of the 635CSi at a cool £22,950 MARCUS PYE is able to relate that the tag still applies to BMW's flagship.

The prevailing British economic climate over the past few years has failed to suppress sales at the loftier end of BMW's extensive range of prestige motor cars. Indeed, the British arm of the German company expects continuing growth of their market share during the next 12 months. The Bavarian manufacturer's laudable policy of introducing revisions to its models only when significant benefits can be reaped — never change for change's sake — has long been one of its hallmarks, and one which, coupled to the technical excellence of their engineering and craftsmanship of the highest order, guarantees the success of the Munich machines on all markets and at every level.

The flagship of the BMW fleet for Britain is the 635CSi coupé, introduced some years ago. Yet now selling in unprecedented numbers, albeit in much revised form since last June. Although the latest version of the 635 is outwardly barely distinguishable from its predecessors, it is a very different beast beneath the skin. A new engine, revised suspension and standardised ABS braking system are features of the 1982 car, while extensive use of advanced electronics in the ignition and fuel metering departments endow the sizeable machine with remarkable petrol consumption — perhaps the most important criterion of the redesign.

State of the art electronics also play a major role in the cockpit, where a multi-purpose, programmable trip computer, the service interval indicator and a comprehensive safety check panel keep the driver abreast of his vehicle's performance. The new heater is electronically regulated too, with characteristic BMW efficiency

Elegant lines for a classic car.

and completes the list of basic driver aids.

Despite the designers' accent on fuel economy when planning the revisions, the 635's outstanding performance has not suffered as a result. The cylinder bore of the lusty straight-six engine has been reduced from 93.4mm to 92mm and the stroke unchanged, giving a capacity of 3430cc (rather than 3453cc in its original configuration), although the compression ratio has been raised from 9.3:1 to 10:1. Power output is apparently unchanged at 218bhp, while the coupé is now 156lbs lighter than earlier versions.

Much of the interest in the new car lies under the long, graceful, bonnet where the engine sits purposefully far back in the shell. A great deal of work, both on the test bench and in prototype installations, has been put into finding the optimum ignition and fuel injection settings to compromise performance with economy. The Bosch L-Jetronic ignition and injection equipment retains these settings by means of a programmed electronic pulse which allows for every variation of speed, engine and ambient temperature and barometric pressure likely to be encountered. Additionally the fuel supply to the engine is automatically discontinued when the throttle is closed above 1200rpm, with a corresponding swing of the economy gauge needle (located at the bottom of the tachometer dial) back to zero fuel consumption.

The engine rasps into life at the first turn of the key seemingly irrespective of conditions. It pulls urgently and smoothly to the 6000rpm red line in the first four ratios (fourth being direct) and in effortless silence once top is engaged. Only above 5000rpm does the unit's wail become audible in the cabin — until then the only discernable noise emanates from the heater fan (fairly obtrusive if powerful at the top end of its variable speed range) and from the wind which whistles round the panoramic electrically-adjustable wing mirrors and the large but deflectorless sliding sun roof.

Owners may specify the five-speed 'overdrive' gearbox or a three-speed automatic transmission (at no extra cost) although a close-ratio 'sports' five-speeder was available at one time to those intent on diminishing the car's fuel economy. The actual change on 'our' overdrive unit was typically BMW, precise, notchy and of longish throw. The box takes a little getting used to, particularly as the clutch on ours took up the drive high in its travel. The pedal is pleasantly weighted, nonetheless, gearchanging becoming slick and rewarding after a few miles.

On starting the sleek machine the 'check control' panel runs through its cycle leaving only an amber warning triangle flashing on the dashboard if all is well. This is extinguished by a dab on the brake pedal and the Bee-Em is ready to go. The safety check facility (which monitors brake, tail, dip beam and number plate lights, engine oil and coolant levels plus the condition of the washer bottles) also has an instant test button.

Once on the move, the 635 accelerates very quickly for a car weighing just over 28cwt, and its handling, under normal conditions, is exemplary for a machine with the capacity and style of a limousine, and the power and aerodynamics of a sports car. Power-assisted ZF recirculating ball steering is beautifully light for manoeuvring yet assumes ideal weight and feel when the 635 is travelling at speed. The attractive alloy road wheels, shod with ultra low profile (220/55 aspect) Michelin TRX radials on our example, contribute to the car's mean, aggressive stance.

Suspension geometry has been revised and the components uprated on the new coupés. A double-joint linkage at the lower end of the front struts helps to overcome steering pull if one corner runs into a deep puddle or softer ground. Semi-trailing arms at the rear, set at 13deg incidence, and camber compensating links improve road-holding dramatically. No longer does the 635 take on an oversteering attitude if the driver lifts off suddenly mid-corner. Indeed, the coupé's wide tyres have a high limit of adhesion on dry roads, where the car corners most neutrally at the sort of speeds which the average BMW driver will encounter. Pushed harder, the softly sprung 635 rolls increasingly into understeer and anyone who intends to drive really rapidly would be well-advised to invest in one of the uprated front suspension kits marketed by Tom Walkinshaw Racing. Revised spring/damper rates and a stouter anti-roll bar would make a world of difference . . .

The standard ABS anti-lock braking system, which BMW pioneered with Bosch and first presented 10 years ago, is a spectacular success. Valued at £872 when it was merely an extra, the system has to be worth every penny if only for the peace-of-mind it instills in the driver. Once acclimatised to the initial sensation of varying feedback through the brake pedal as each wheel is individually retarded, one basically forgets that the ABS is installed — until wet or slippery conditions are encountered, that is. Because the wheels cannot lock, regardless of pressure on the pedal, braking distances are most dramatically reduced in adverse conditions. The dual circuit discs provide excellent stopping power anyway and were not prone to fade, even during 'press-on' motoring. We also tried an emergency stop from high speed in pouring rain and were amazed by the speed of deceleration and the car's controllability.

Accommodation within the BMW's cockpit is luxuriously roomy for up to four adults although the front seats were not as comfortable as many we have tried. The seats are adjustable for reach, rake and height but are both unnecessarily hard and lacking in lateral support. I did not find the back rest quite long enough over an arduous cross country journey, consequently was difficult to set the integral head restraint in a relaxing position. Individually-contoured rear seats (with their own inertia reel belts) are fine to travel in — but the occasional necessity of carrying a fifth passenger in the centre is made awkward by the belt anchorage points and the raised squab between

Top: The layout of the control area is as superb as we have come to expect. Above: Individually contoured rear seats provide luxurious comfort. Below: The built in Heyco tool kit to be found in the boot.

rear seat wells.

Seats apart, the driving position is commanding, with good visibility from the effectively pillarless coupé. The steering column and associated controls are adjustable and the restyled instrument binnacle is absolutely superb — more like the cockpit of an aircraft than a car. Wash/wipe and indicators are operated by stalks on the column, while lighting is controlled by BMW's familiar dash-mounted knob with twist rheostat facility for the binnacle illumination. Auxiliary driving lamps and rear high-intensity lights run off the same switch while the heated rear screen and hazard flashers are centrally mounted above the efficient heater ventilator.

Both electric front windows proved recalcitrant (one failing to operate throughout and the other not closing properly on occasions) but the steel sunroof, which opens normally or tilts according to which way the switch is pressed, was a marvellous and reliable asset. Not so good is the front ashtray beside the cigar lighter. It opens towards the driver with the result that ash can easily escape during stop-start runs, littering the map pocket. A large glove box opens downwards in the usual way, while a cavernous boot carries not only a quality Heyco tool kit but also an EEC warning triangle.

All in all though, the BMW 635CSi is a wonderful driver's car. It is simplicity itself to operate, fast yet frugal. The knowledge that more than adequate acceleration is on tap when needed — helped by a sympathetic torque curve — is reassuring but at the same time vast mileages can be covered at high average speeds with miserly fuel consumption. The sight of the fuel consumption indicator ebbing back towards minimum figures is a challenge in itself, indeed it becomes a topic of major interest as to how little petrol one can use.

We managed an incredible 27.6mpg over a relatively short distance on the BMW's launch in France recently but an overall figure of 23.4mpg over the 1200 miles of our more extensive test speaks for itself. The car was driven over a mixture of motorway, A and B roads and around town with little conscious effort to record artificially high figures. Most owners could improve upon this with ease. Our consumption was calculated on a full tank to full tank basis as the trip computer's average readout was, we thought, a fraction optimistic. Better was its range assessment — with the 15 gallon fuel tank the big Bee-Em can travel over 350 miles.

The computer, which sits beside a *Blaupunkt* radio/cassette player, is programmed to deliver a plethora of facts, although an evening with the instruction manual is recommended if you are to master its intricacies. Air temperature, mpg, projected arrival times, average speed, fuel range and a stop watch are among the facilities in addition to the digital clock.

BMW claim a 0-60mph time of 7.1sec and a maximum of 142mph from the revised 635, the former figure over a second quicker than that of its predecessor, presumably mainly due to its lighter weight. The widely spaced lower ratios carry the car to 40mph, nearly 70mph and over 'the ton' while the direct fourth should be good for around 135mph.

For the relatively high price of £22,950 you do get a great deal of motor car. The BMW may be more expensive than comparable models from Porsche and Jaguar but its rivals offer similar levels of performance — at double the British legal limit — with the penalty of less agreeable fuel consumption figures. Certainly if you can afford any of the options, then paying your £1.75 per gallon will not come hard but the quality of finish, the strength and reliability of the BMW cannot be overlooked. With better front seats and tauter front suspension, it really could be unbeatable at the price. ■

BMW 635 CSi £22,950

Specification

Cylinders/capacity	6 in line/3430cc
Bore x stroke	92mm x 86mm
Valve gear	ohc
Compression ratio	10.0:1
Fuel system	Bosch L-Jetronic
Power/rpm	218bhp at 5200rpm
Torque/rpm	224ft lbs at 4000rpm
Gear ratios	3.822; 2.223; 1.398; 1.0 and 0.813:1
	(Automatic 2.478; 1.478 and 1.0:1 optional)
Final drive	Hypoid 3.07:1
Steering	Power-assisted ZF recirculating ball and nut
Brakes	BMW/ABS anti-lock system, discs front and rear
Wheels	light alloy, 6½J x 14ins, 220/55 VR14 tyres
Suspension (F)	Independent, MacPherson struts, anti-roll bar
Suspension (R)	Independent, semi-trailing arms, spring struts anti-roll bar

Dimensions

Wheelbase	103.4ins
Track (F/R)	56.2ins/58.5ins
Length	197.2ins
Width	67.9ins
Kerb weight	3152lbs

Fuel

Urban/56mph/75mph	19.8/41.5/32.1mpg
Testing	23.4mpg
Tank	14.5galls plus 1.8gall reserve

Century puts more super in the Bavarian supercoupe

Sumptuous, yes. Seductive, *jawohl*, as few others. Civilized as Dom Perignon. Sure and silent as an assassin's knife. But at this office the first question asked concerning a turbocar—*any* turbocar—is invariably, "whaddaya spose that rig will *do*?" So let us dispense with blood-red sunsets and secret valleys flashing past in the misty dawn to get right down to standing starts. What'll this tricked-up supercruiser do? Well, it'll just run heads-up with any XJ-S Jaguar in town, is all. Just *dust* Porsche 928s. Just *frazzle* Ferrari 308s. In fact, this car will show a tidy set of heels to anything in its class, and is even in the ballpark with somewhat smaller quickies like the everlasting Porsche 911SC and Datsun 280ZX Turbo. Heady company indeed for a leather-lined luxo coupe that scales in at 3400 lb. "Golly, officer, it didn't *feel* like 110. Yes, I know it was only four blocks. Guess I just sort of got carried away."

The Century/BAE BMW 633 CSi Turbo is one of those rare cars that put Des Moines-to-Omaha luncheon runs into the realm of plausibility. We make the distinction between the plausible and the practical here, you understand, a distinction you would very likely have to flesh in with considerable detail before some local magistrate were you to actually attempt such a feat. But the point is that performance

---

If you're looking for a 633 CSi that'll run with the other exotics, this is the only way to go

---

such as this has not been a conspicuous part of the 633's character heretofore. Since its introduction (as the 3-liter 630) in 1976, this baron of Bimmers has been super in every sense but the one that counts most with hard-core enthusiasts: horsepower. A bump to 3.2 liters in 1978 had lit-

by Tony Swan
PHOTOGRAPHY BY PAUL CHAMBERLAIN

tle impact on the car's relative sluggishness, and ultimately it was left to our friends at Century BMW in Alhambra, California, to cue up more quick.

Century is North America's biggest BMW outlet, and promises to become even bigger with the recent development of its own in-house performance branch, Century Turbo Performance Center. Century Turbo operates in cooperation with BAE, the Torrance turbo specialists. BAE has developed turbo kits for every car BMW offers in North America (except for the 528e). The thing that makes the BAE effort noteworthy in this regard is that the various kits have all been emissions certified, which, of course, is the only way a factory-franchised dealer like Century could ever afford to get involved in such a project.

The 633 installation is standard BAE fare, which is to say highly professional. The cosmetics are appealing and, more important, so are the results. The BAE kit employs a Garrett AiResearch T04 turbocharger blowing through the 633's Bosch L-Jetronic injection system at 6 psi max. This marriage has proven itself to be particularly well-suited to turbocharging in

BMW 633 CSi Turbo

BMW 633 Turbo

other applications, and so it is with BMW's big six: smooth, quiet, and potent. The system is beautifully tuned to minimize the low-end sluggishness that still plagues some turbocars, which makes plenty of sense since off-the-line torpor is one of the gripes against the stock 633. However, the system also delivers plenty of the high-end omigawd here-we-go rush that makes turbocars so much fun to drive. At its peak—achieved at 5500 rpm—the 633's Century setup delivers 260 hp at 5500 rpm, which is 80 more than stock.

As the acceleration numbers indicate, this is enough to stir the blitzen Bimmer, 3-speed ZF automatic and all, along at a pretty respectable rate—a startling rate, in fact, at least to some of the other big-buck GT bravos we encountered out there in the streets.

As you will perhaps have noted from the pictures, this BMW differs from standard in more ways than the $3500 worth of alterations that have been made under the

hood. The car rides on fat Pirelli P7s mated to handsome 3-piece Epsilon modular wheels. And rides low, thanks to the substitution of shortened coil springs for the stock units, front and rear. The new springs have substantially higher rates than stock, and are matched to Bilstein gas dampers all around—strut inserts up front, height adjustable shock absorbers at the rear. The suspension package also includes heavier anti-roll bars front (28 mm) and rear (19 mm), both of them adjustable. It all adds up to a suspension setup that's a good deal stiffer than stock, but not beyond the realm of civilized sports motoring. The Century 633's suspension, with help from the P7s, did seem to have a capacity for transmitting lots of little lumps and bumps into the driver's ears and fingertips, but no one on the staff seemed to consider this a real source of irritation. Rather, it figured as a worthwhile tradeoff for the new level of athletic prowess the big coupe displayed on Mulholland Drive and
Continued on page 73

ROAD TEST DATA

BMW 633 CSi Turbo

◢ SPECIFICATIONS

GENERAL
- Vehicle manufacturer .... Bayerische Motoren Werke AG, Munich, Fed. Rep. of Germany
- Vehicle importer .... BMW of North America, Inc, Montvale, NJ 07654
- Body type .... 4-pass., 2-door coupe
- Drive system .... Front engine, rear drive
- Base price $37,402
- Major options on test car ..... Century Turbo conversion, suspension package, Epsilon wheels, Pirelli P7 tires
- Price as tested $46,453

ENGINE
- Type .... Inline 6, water cooled, cast block, aluminum head
- Displacement .... 196 cu in. (3210 cc)
- Bore & stroke .... 3.50 x 3.99 in. (89.0 x 86.1 mm)
- Compression ratio 8.4:1
- Induction system .... Bosch L-Jetronic fuel injection, turbocharged
- Valvetrain .... SOHC
- Crankshaft .... Forged, 7 main bearings
- Max. engine speed .... 6600 rpm
- Max. power (SAE net) .... 260 @ 6700 rpm
- Max. torque (SAE net) .... 196 @ 4000 rpm
- Emission control .... 3-way catalyst, oxygen sensor, EGR
- Recommended fuel .... Premium unleaded

DRIVETRAIN
- Transmission .... 3-sp. auto.
- Transmission ratios (1st) 2.48:1
- (2nd) 1.478:1
- (3rd) 1.1
- Axle ratio 3.25:1
- Final drive ratio 3.25:1

CAPACITIES
- Crankcase .... 5.3 qt
- Cooling system .... 12.7 qt
- Fuel tank .... 16.5 gal
- Luggage .... 15.7 cu ft

SUSPENSION
- Front .... MacPherson struts w/gas pressure strut inserts, coil springs, anti-roll bar
- Rear .... Semi-trailing arms, coil springs, gas pressure shocks, anti-roll bar

STEERING
- Type .... Recirculating ball, power assist
- Ratio 16.9:1
- Turns, lock-to-lock 3.6
- Turning circle, curb-to-curb .... 37.4 ft

BRAKES
- Front .... 11.02 in. vented discs, power assist
- Rear .... 10.7 in. vented discs, power assist
- Swept area .... 547 sq in.

WHEELS AND TIRES
- Wheel size .... 7 x 15 in.
- Wheel type .... Epsilon 3-piece modular
- Tire size 225/50VR15
- Tire mfr. & model .... Pirelli P7
- Tire construction .... Steel-belted radial

DIMENSIONS
- Curb weight .... 3400 lb
- Weight distribution (%), F/R 55/45
- Wheelbase .... 103.4 in.
- Overall length .... 192.7 in.
- Overall width .... 67.9 in.
- Overall height .... 53.7 in.
- Track, F/R .... 56.0/58.8 in.

CALCULATED DATA
- Power-to-weight ratio .... 13.08 hp/lb
- Brake swept area to weight ratio .... 303 sq in./ton
- Top speed .... 140 (est.)

SKID PAD
- Lateral acceleration .... 0.754 g

◢ TEST RESULTS

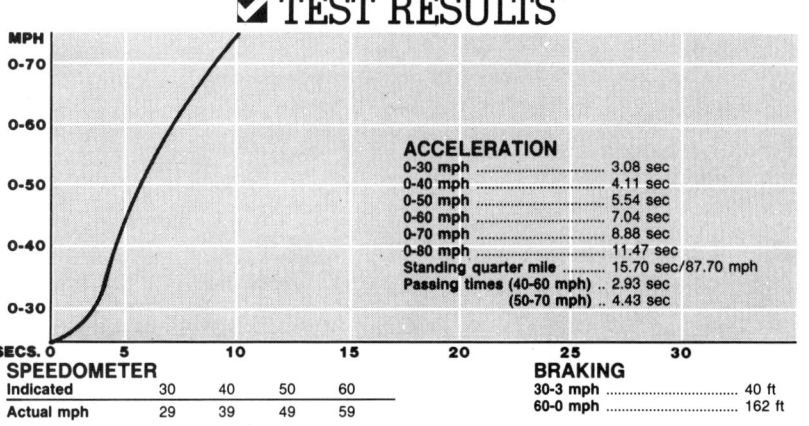

ACCELERATION
- 0-30 mph .... 3.08 sec
- 0-40 mph .... 4.11 sec
- 0-50 mph .... 5.54 sec
- 0-60 mph .... 7.04 sec
- 0-70 mph .... 8.88 sec
- 0-80 mph .... 11.47 sec
- Standing quarter mile .... 15.70 sec/87.70 mph
- Passing times (40-60 mph) .. 2.93 sec
- (50-70 mph) .. 4.43 sec

SPEEDOMETER
Indicated	30	40	50	60
Actual mph	29	39	49	59

BRAKING
- 30-3 mph .... 40 ft
- 60-0 mph .... 162 ft

Driving Impression:
BMW 635CSi

The lads from Munich perform a brainlift
BY INNES IRELAND

They're a busy bunch of lads down there at the BMW plant in Munich, always beavering away at something new. While the research and development chaps are spending long hours in their efforts to make their 1.5-liter turbocharged inline-4 engine last a little bit longer or be a bit more reliable in the Brabham BT50 Formula 1 car, others strive to maintain the company's policy of continual research to improve their bread-and-butter motorcars. Bread-and-butter is perhaps not an apt description for the subject of the latest developments, the 635CSi, but quite a lot of bread and a great deal of butter are required to be able to afford one. In the UK market it retails at around the equivalent of $46,000!

With the modifications that have been introduced in the past year to the 3-, 5- and 7-Series cars, it is logical for similar moves to be made with the 6-Series cars. Whereas subtle changes were made to the exterior of the odd-numbered models, those fortunate enough to own the current 6 model will be happy to learn that no such changes have been made to this series, and in that respect their precious cars will not be outdated. As Dr Walter Hasselkus, Managing Director of BMW (GB) Ltd, put it, "This is not just a facelift; it is a brainlift," the improvements being primarily confined to technical developments.

Interestingly, for the first time BMW showed a new model to a small select group of British journalists before showing it to the Continental press, for Great Britain, even in these days of recession, remains the second largest market for this particular model after the home market is considered. I can't quite think why I was invited to be a member of this small select group, but the feather in my cap was obvious although the timing, coming between the Detroit and Canadian Grands Prix, was awkward.

By twisting the R&T Editor's arm severely up his back in an attempt to get the holders of the *Road & Track* sporran to unleash the strings and let the moths out, it was agreed that I should make the flight back to Europe between races. But only after I explained that I could drink a prodigious amount of whiskey if left lying about in some hotel for five days!

Because of the delay caused by the restart of the Detroit GP, the police escorted me to the airport where a tolerant British Airways held the flight, enabling me to leave Detroit on the night of the GP and arrive in London the following morning. The trip to join my car in Marseille was more luxurious, for this was completed in a private Fan Jet Falcon, lunch and drinks served en route, on Tuesday.

There were six of us and with three cars awaiting our arrival, it was two per car. After being handed a route card, we were asked to estimate our fuel consumption for the 160-mile trip, which included *autoroutes*, town work and a good deal of zooming about in the mountains. Having been surprised by the relative economy of other models, I deliberated a long time before writing 27.6 mpg (Imperial gallons, that is, equivalent to 23.0 mpg, U.S. style). The other chaps were plumping for 22 to 24 mpg.

The most significant new features in the 635CSi are in the engine and suspension. The former has had its capacity reduced from 3453 to 3430 cc but the compression ratio has been raised from 9.3:1 to 10.0:1 so that it still produces 218 bhp at 5200 rpm with maximum torque of 224 lb-ft being reached at 4000 rpm. The figures for bore and stroke are now 92.0 and 86.0 mm respectively. New digital engine electronics ensure that the amount of fuel being delivered and the ignition timing are best suited to the conditions of air temperature, barometric pressure and engine temperature balanced against the performance demanded of the engine at any given time.

Front suspension is by BMW's double-jointed MacPherson-strut system having a small kingpin offset, negative camber and an anti-roll bar. The double-jointed linkage, as in current 5- and 7-Series cars, overcomes the "pull" felt through the steering when one wheel goes through rain puddles or soft ground. At the rear, semi-trailing arms are set at 13 degrees, the spring struts having a linkage that compensates for camber changes. Unfortunately I can't make a direct comparison, but it is claimed that this revised geometry improves the adhesion provided by the rear suspension.

Taking time to familiarize myself with the various controls, I was last to leave the airport. When passing one of my colleagues on the *autoroute*, I was amazed to find that the economy gauge set into the lower segment of the rev counter was indicating 28 mpg though we were doing 100 mph. At this speed, engine and wind noise are very low and there is no significant increase even at 135 mph, when I found the 635CSi's directional stability to be of a high order. The quoted maximum is 142 mph and I'm sure this is not an exaggeration.

I didn't time myself accelerating away from the *péage* but 0–60 mph is given as 7.1 seconds, which isn't exactly hanging about for a 3150-lb car, some 155 lb lighter than its predecessor, incidentally. What I did find was that the rev limiter cut in at 6000 rpm whereas the redline sector doesn't start until 6100 revs are reached. The engine is so smooth all the way up that cutting in at this level could be an embarrassment when making a quick overtaking move in, say, 3rd gear, and knowing this engine would run happily to 6500 rpm without any danger, I would like to see the rev limiter cut in at this more realistic figure.

Autoroute bends, even the tighter ones in the mountains, can be taken easily at 120-plus mph, the car exactly following the line it is given with little body roll. But it is on the narrow twisting mountain pass roads that the car becomes a delight. Stability is easily controlled by dropping down to the appropriate gear on the approach, then aiming at the apex. The precision and feel of the steering allowed me to place the car to within an inch of my chosen clipping point, the 205/70VR-14 tires fitted on the attractive 14 x 6½J alloy rims producing tremendous grip even when I threw the car through S-bends with a very rapid change of direction. Handling characteristics are neutral, and I could really lean on the suspension on downhill sections using a higher gear than one would normally choose without any sign of either front- or rear-end breakaway.

An antibrake-locking system (ABS) is fitted that gives this car quite remarkable stopping capabilities. A light pedal pressure is all that is required in most circumstances, the pedal having good sensitivity and progression. But to try to fool the ABS, I took the car along a dusty, gravel-strewn unmade road at about 45 mph and got hard on the brakes. Without a trace of locking a wheel, it pulled up in an incredibly short space—and dead straight at that. Where this system would really come into its own would be in emergency applications, particularly in the wet. It would allow the driver to steer the car accurately away from danger while still braking hard, a thing he would be unable to do were he to have the front wheels locked.

There are lots of nice features built into the new 635CSi as standard fittings such as the electrically controlled sunroof, the panel of warning lights that will indicate any one of many faults or low-liquid levels and now a fuel cut-off at revs over 1200 where no fuel is supplied to the engine when the throttle is closed. This is one of the ways that BMW has proved its search for greater efficiency has not been in vain. The official figures show that fuel consumption is improved by 21 percent in urban use and 7 percent at a constant 75 mph, this last figure being a remarkable 32.1 mpg. Even so, for a car of this nature the 14.5-Imperial gal. tank gives a realistic range of slightly more than 350 miles which, to my mind, is 100 miles too short considering the car's performance, comfort and ease of driving.

What pleased me most was that I arrived first at the "drinks" stage and, finally, when the technician punched the computer at the end of my run I was right on the money; it read 27.6 mpg. And I didn't have to freewheel down the mountains to get it up, nor blast up and down in 1st gear in front of the 12th century abbey where we stayed to get it down!

Sadly, next day, after another session with the car, it was back to London to catch the evening flight to Montreal and the Canadian Grand Prix.

What'll she do at the cash register? The answer is still plenty

Continued from page 71

other similar stretches of choice twisties. Rolling in league with it fatter footprints (P225/50s versus P195/70s stock), the stiffened 633 truly thrived in the tricky world of decreasing radius and high-speed sweepers, with scarcely a trace of body roll. Although we were unfortunately unable to get the car on our skidpad, we feel certain the suspension modifications and P7s add up to much more than the standard 633's .75 lateral g capability. It is a surefooted setup that gives its driver the feeling he would have work very hard to get himself into trouble.

The P7s didn't have quite so salutory an effect on the 633's braking performance. The stopping distances were certainly acceptable, and control good, but we would expect the P7's excellent sticking qualities and the 633's 4-wheel disc brakes to haul the car down from 60 mph in something less than 150 ft.

About the only other modifications included in this package were inside the cockpit—an excellent Momo steering wheel with a Turbo logo horn button (important, right?) and a VDO boost gauge. Which brings us to the next performance question: What'll she do at the cash register? And the answer, as you might have guessed, is

It'll run heads-up with any XJ-S, dust 928s, and frazzle Ferrari 308s

still plenty. The base car, a 1981 model, stickered at $37,402. The turbo kit, installed, adds $3500 to this total, and the Century Turbo Performance Center will look after the setup for five years or 50,000 miles for an additional $895. Considering the vagaries of turbomotors, this service contract figures as a real bargain.

Going for the suspension package, which was handled here by Suspension Techniques of Rosemead, California, adds another $1493, with the wheels ($344 each) and tires ($364 each) running the total up another $2832. While the various suspension elements may be separated from one another, we feel certain the improvement in handling wouldn't be nearly as dramatic without the big P7s, and would opt for the whole package. In fact, the handling improvements would be welcome even without the extra punch of the turbomotor.

Add in the Momo wheel ($205.90) and the boost gauge ($125), and you have $9050.90 worth of additions to a $46,453 BMW 633 CSi. Even for senior Bimmer buyers, that's a hefty premium—almost 20%. But if you're looking for a 633 CSi that'll run with the other exotic iron, it's the only way to go.

For details on the Century Turbo Performance Center program contact Century Motor Sales, 1811 W. Main St., Alhambra, CA 91801, (213) 570-8444. BAE is located at 3032 Kashiwa St., Torrance, CA 90505, (213) 530-4743.

  # BMW 6 and 7 series

IT IS many years since BMW made any sudden or radical changes to the type of passenger cars they build, for one range usually follows on logically from the last. Thus, the famous four-cylinder overhead camshaft engine was introduced over 20 years ago, and its six-cylinder derivative came along in 1968, yet both types are still made today.

In the late 1970s, existing ranges of big four-door saloons, and 2+2 seater coupés, all with six-cylinder engines, were replaced by — respectively — the current 7-Series saloons, and the 6-Series coupés. All these new cars shared essentially the same chassis engineering, suspensions, and running gear, so we can certainly survey them as *Buying Secondhand* prospects in a single group.

All of these cars, which are still being made, were fast, all were originally expensive, and most have come to suffer considerable depreciation. Even so, careful choice can turn up a very desirable machine. They are detailed, and styled, in a typically German way, and have few obvious rivals. Here in Britain, however, they have had to sell in the rarefied Jaguar XJ/Daimler, Mercedes-Benz and Rover bracket where competition is fierce. It is against such rivals that we judge 6-Series and 7-Series cars today.

Defining the pedigree

Although the first 6-Series Coupé was unveiled in February 1976, and the original 7-Series saloon not until May 1977, the design of the two types of car was commonized wherever possible. Although the coupé had a 103.4in. wheelbase, and the saloon 110in., they shared the same basic overhead-cam slant-six engines, Getrag manual or ZF automatic transmissions, the same type of semi-trailing link independent rear suspension, power-assisted ZF steering gear, and MacPherson strut independent front suspension (though the 6-Series did not inherit the saloon's "double-joint" layout until mid-1978). All the cars had fat 14in. high-speed radial ply tyres, four wheel disc brakes, and a great deal of mechanical and electronic sophistication. Because the styling included the famous BMW "kidney" grille at the front, and was a more modern version of what had gone before, the cars were quite unmistakeably BMW. As before the Coupé bodies were built by Karmann, and the saloon shells by BMW, with final assembly at the Dingolfing factory in Bavaria.

Above all, all these BMWs were massively strong (even the Coupés weighed more than 3,300 lb, ready for the road), powerful, very fast, and nimble if a touch ponderous. The "image" offered, too, was of great importance — if you bought a Volvo, it seemed, you were wedded to safety: if you bought a Mercedes-Benz, it was safety allied to technical excellence. But if you bought a BMW, it was to enjoy German quality construction with fast, sporting, motoring.

Engines

All the cars shared the same type of seven-bearing, overhead camshaft, straight six cylinder engine, but there were considerable detail differences between models. In seven years, we can identify the use of five different engine sizes, four ranges of peak power outputs, and two distinct types of fuel supply, though in no case was the engine ever set up to be an "economy" unit. It has always provided high specific outputs, sophisticated features, and a silky ability to rev, and produce real broad-chested torque.

The smallest engines have been 2,788 c.c., with a complex four-choke carburettor on the original 728 saloon, and Bosch L-Jetronic fuel injection on all other types. For two years, the 730 saloon was supplied with a 2,985 c.c. engine, also fitted with the four-choke carburettor, though this was not fitted to a British-market coupé.

The most persistent engine size in use has been the longer-stroke 3,210 c.c. unit, always fitted with Bosch injection, and found in 633CSi Coupés, and the 733i/732i saloons, whose differences are described below.

The top of the range is provided by two slightly different "3.5-litre" sizes — the 3,453 c.c. on cars used until mid-1982, and the slightly smaller 3,430 c.c. which replaced it. Both types have fuel injection, and both have the same (218 bhp) quoted peak output. The size reduction came by changing the bore and stroke from 93.4×84.0mm to 93.0×86.0mm, to provide more commonization with other derivatives in the range.

Physically, all the engines are the same size — and all provide a real bonnetful, and one usually identifies the cars simply by reading the chrome script on the boot lids!

ENGINE AND BODY AVAILABILITY

Engine	2,788 c.c. Carb	2,788 c.c. Inj	2,985 c.c. Carb	3,210 c.c. Inj	3,210 c.c. Inj	3,430 c.c. Inj	3,453 c.c. Inj
DIN bhp	170	184	184	197	200	218	218
628CSi Coupé	–	1980-83	–	–	–	–	–
633CSi Coupé	–	–	–	–	1976-78	–	–
635CSi Coupé	–	–	–	–	–	1982-83	1978-82
728 Saloon	1977-79	–	–	–	–	–	–
728i Saloon	–	1980-83	–	–	–	–	–
730 Saloon	–	–	1977-79	–	–	–	–
732i Saloon	–	–	–	1980-83	–	–	–
733i Saloon	–	–	–	–	1977-79	–	–
735i Saloon	–	–	–	–	–	1983	1980-82
735SEi Saloon	–	–	–	–	–	1983	1981-82

Note: All cars had the same basic straight-six cylinder engine, with single overhead camshaft valve gear.

Transmission choice

All cars were made available with manual or automatic transmissions, but there has been a great deal of development of types in seven years. The original cars had four-speed, all-synchromesh manual gearboxes, but an "overdrive" five-speed gearbox was standardized on the 635CSi, and optional from the start up of "Mk 2" 7-Series saloon production in July 1979. All 633s, 728s, 730s and 733s, however, had four-speed boxes. For 1981, all 7-Series manual transmission saloons had the five-speed transmission, as did the 628CSi coupé. The 635CSi later got a choice of overdrive or direct-top sports five-speed transmission.

Three-speed ZF automatic transmission was optional on all cars but standard on the 735i Special Equipment saloon. However, from June 1982 (6-Series) and October 1982 (7-Series), this transmission was replaced by the new four-speed ZF automatic, in which lock-up top ratio was an overdrive, to encourage better economy. A large proportion of all big BMWs sold in the UK have automatic transmission, even the fastest and most sporting coupés.

Body and model choice

All 6-Series cars are two-door coupés, with conventional separate boot compartments, and generous 2+2 seating, while all 7-Series cars are conventional four-door, five-seater, saloons. In the main, their model names denote exactly what engine is fitted — a 732i, for instance, is a 7-Series saloon with 3.2-litre fuel injection engine. However, at first, there was slight confusion, for 633CSi and 733i models both had 3.2-litre engines, not 3.3-litre units as one might infer.

All are extremely well equipped, as one would expect from this type and class of motor car, but the largest-engined varieties had even more standard fittings than the (relatively) basic cars. From time to time, the updating of models has led to optional equipment either being standardized, or made available on variants lower down the range. Purely as an example, the 728 range inherited fuel injection at the end of 1979, took on central locking soon afterwards, got the on-board computer and ABS braking options for 1981, and the four-speed automatic gearbox option for 1983.

The only version to have obvious aerodynamic aids is the 635CSi, which not only has a deep front spoiler, but a rubberized spoiler across the tail.

Mechanical development

The most significant of all has been the provision of ABS anti-lock braking, optional from the summer of 1979, and standard on the 635CSi from mid-1982, while another has been the adoption as standard of the 13-degree trailing arm rear suspension on 1982½ Coupés and 1983 model saloons.

Another significant improvement (6-Series Coupés only) was that the double-joint type of MacPherson strut suspension, always fitted to the saloons, was standardized on 635CSi Coupés from introduction, and on all Coupés from mid-1979, while Michelin TRX tyres (and their unique, metric-measure, wheel diameters) also became available on Coupés in due course, and standard from mid-1982.

In 1982, the accent was on add-

Both the 6- and 7-Series were well-equipped and luxurious, with comprehensive instrumentation. This 7-Series model featured a functions check panel (right of steering wheel)

APPROXIMATE SELLING PRICES

Price Range	628CSi Coupé	633CSi Coupé	635CSi Coupé	728/728i Saloon	730 Saloon	733i/732i Saloon	735i Saloon	735iSE Saloon
£4,600-£5,000				1978	1978			
£4,900-£5,300						1978		
£5,400-£5,800		1977						
£5,900-£6,300				1979				
£6,300-£6,700					1979	1979		
£6,700-£7,100		1978						
£7,600-£8,000				1980				
£8,400-£8,800						1980		
£9,500-£9,900				1981		1981	1980	
£10,300-£10,800			1979					
£11,600-£12,100				1982		1982	1981	
£11,900-£12,400								
£12,900-£13,400	1981		1980					
£14,200-£14,700						1982		
£15,400-£15,900			1981				1981	
£18,200-£18,700	1982		1982					1982
£20,000-£20,500			1982/3					

Note: *Automatic transmission was available on all models, and adds between £500 (early models) to £1,000 (1982 models) to values. Other desirable extras like TRX tyres, and ABS braking also add to values, which are not strictly quantifiable.*

Opposite page: Top, 1980-model BMW 735i; bottom, a 635CSi coupé from the year it was introduced in Britain, 1978

All cars shared the same type of seven-bearing, straight six engine and physically all engines were the same size, well filling the underbonnet space. This injection engine is in a 7-Series car

SPARES PRICES

	628 CSi Coupé	635 CSi Coupé	728i Saloon	735i Saloon
Engine assembly-bare (exchange)	£1,100.22	£1,621.59	£1,166.22	£1,166.22
Gearbox assembly (exchange)	£651.10	£722.74	£624.91	£722.74
Clutch pressure plate (exchange)	£51.12	£58.95	£51.12	£58.95
Clutch driven plate (exchange)	£42.37	£42.37	£42.37	£42.37
Automatic gearbox with convertor (exchange)	£779.30	£779.30	£749.98	£779.30
Propellor shaft universal joint repair kit (exch)	£24.28	£24.28	£24.28	£24.28
Final drive assembly (exchange)	£456.32	£499.30	£477.81	£477.81
Brake pads – front (set, new)	£26.00	£26.00	£26.00	£26.00
Brake pads – rear (set, new)	£14.10	£14.10	£19.21	£19.21
Suspension struts – front (pair)	£90.57	£90.57	£71.53	£71.53
Suspension struts – front (pair)	£80.14	£99.16	£63.61	£63.41
Radiator assembly (new)	£152.63	£116.46	£99.71	£99.71
Alternator (exchange)	£214.87	£214.87	£212.04	£212.04
Starter motor (exchange)	£84.44	£116.01	£84.44	£116.01
Front wing panel	£356.20	£356.20	£131.24	£131.24
Bumper, front (new)	£418.61	£424.71	£100.97	£100.97
Bumper, rear (new)	£212.96	£212.96	£147.52	£147.52
Windscreen, laminated	£136.94	£136.94	£116.91	£116.91
Exhaust system complete	£160.85	£395.09	£315.02	£376.76

All the above prices include VAT at 15 per cent.

SPECIFICATION AND PERFORMANCE

6-Series coupés	633CSi	635CSi	635CSi
Tested in *Autocar* of:	16 Oct 1976	6 Jan 1979	7 Aug 1982
Specification:			
Engine size (c.c.)	3,210	3,453	3,430
Engine power (DIN bhp)	200	218	218
Car length		15ft 7.2in	
width		5ft 9in	
height		4ft 6.5in	
Boot capacity (cu.ft.)		14.5	
Turning circle (kerbs)		34ft 6in approx	
Unladen weight (lb)	3,280	3,447	3,175
Max. payload (lb)	740	794	794
Performance:			
Mean maximum speed (mph)	131	140	139
Acceleration (sec):			
0-30 mph	2.9	3.2	2.7
0-40 mph	4.4	4.9	4.3
0-50 mph	6.1	6.7	5.6
0-60 mph	8.1	8.5	7.3
0-70 mph	10.7	11.0	9.8
0-80 mph	13.7	14.3	12.4
0-90 mph	17.4	18.7	15.4
0-100 mph	21.9	23.4	19.4
0-110 mph	27.6	30.3	25.8
0-120 mph	39.4	41.0	34.2
Standing ¼-mile	14.9	16.2	15.6
Consumption:			
Overall mpg	20.6	17.5	21.8
Typical mpg – easy driving	27	23	28
average	23	19	24
hard driving	18	16	20
Mpg at steady 70 mph	28.0	29.5	n/a
Fuel grade	4-star	4-star	4-star
Oil consumption (mpp)	1,000	800	Nil

SPECIFICATION AND PERFORMANCE

7-Series saloons	733i 4-speed	732i 4-speed	735i 5-speed	735i Automatic
Tested in *Autocar* of:	6 Aug 1977	17 Nov 1979	27 Dec 1980	27 Jan 1983
Specification:				
Engine size (c.c.)	3,210	3,210	3,453	3,430
Engine power (DIN bhp)	197	197	218	218
Car length			15ft 11.3in	
width			5ft 10.9in	
height			4ft 8.3in	
Boot capacity (cu.ft.)			16.9	
Turning circle (kerbs)			35ft 6in approx	
Unladen weight (lb)	3,585	3,471	3,472	3,506
Max. payload (lb)	993	1,036	1,036	1,124
Performance:				
Mean maximum speed (mph)	122	127	129	130
Acceleration (sec):				
0-30 mph	3.1	2.8	2.8	3.0
0-40 mph	5.0	4.4	4.2	4.4
0-50 mph	6.8	6.0	5.7	5.8
0-60 mph	8.9	8.0	7.5	7.8
0-70 mph	12.2	11.4	10.4	10.0
0-80 mph	15.5	14.4	13.3	12.7
0-90 mph	20.7	18.6	16.8	16.3
0-100 mph	28.0	24.6	22.9	20.7
0-110 mph	38.8	33.1	30.7	27.1
0-120 mph	–	46.4	–	37.7
Standing ¼-mile	16.7	16.4	15.9	15.9
Consumption:				
Overall mpg	19.4	19.8	17.3	19.6
Typical mpg – easy driving	25	26	23	26
average	21	22	19	22
hard driving	17	18	16	18
Mpg at steady 70 mph	25.8	n/a	n/a	n/a
Fuel grade	4-star	4-star	4-star	4-star
Oil consumption (mpp)	2,000	Nil	900	Nil

CHASSIS IDENTIFICATION

6-Series Coupés:

October 1976: 6-Series introduced to the UK. This and all subsequent cars with 2+2 fixed-head coupé body style, and six-cylinder engine. Original imports were of 633CSi only, with 3,210 c.c. From:

		Chassis No.
633CSi Manual	4380	4380001
633CSiA Automatic	4390	4390001

October/December 1978: 633CSi dropped, in favour of larger-engined (3,453 c.c.) 635CSi. Changeover at:

Last 633CSi	–	4381133
First 635CSi	5550	5550001

September 1980: Introduction of smaller-engined 628CSi (with 2,788 c.c. engine), and changes to 635CSi. 628CSi had no spoilers. 635CSi now with leather seating, central locking, and DME (digital motor electronics) engine management. From:

First 628CSi Manual	5595	5595002
First 638CSi Automatic	5595	5595001
635CSi changes from:		
635CSi Manual	5550	5550548
635CSi Automatic	5585	5585508

June 1982: Revisions for 1983 and subsequent models. No style differences, except bigger spoiler. 3,430 c.c. engine on 635CSi, much mechanical change. — —

7-Series saloons:

June 1977: 7-Series saloons, with four doors and straight six-cylinder engines, introduced to UK. Original range comprised 728, 730 and 733 models (2.8, 3.0, 3.2-litre engines). Opening imports:

728 Manual	5720001
728 Automatic	5725001
730 Manual	3235001
730 Automatic	3255001
733i Manual	5750001
733I Automatic	5755001

Autumn 1979: Re-alignment of 7-Series range. 730 and 733i derivatives dropped, and replaced by 732i (3,210 c.c.) and 735i (3,453 c.c.) models, without major changes

Last 730 model	no records
Last 733i Manual	5750513
Last 733i Automatic	5757244

First imports of 732i and 735i models at:

732i Manual	5335001
732i Automatic	5345001
735i Manual	7420000
735i Automatic	7420500

At the same time, 728 became 728i, with fuel injection engine, as:

728i Manual	4470001
728i Automatic	4475001

April 1980: Development changes, including central locking windows and tinted glass (all), plus electric window lifts (732i) and headlamp wiper/wash (735i), but no mechanical changes. From: No records

September 1980: Changes for 1981 models, include larger fuel tank, five-speed gearbox standard, and optional ABS anti-lock braking (all), and digital motor electronics (732i and 735i). Changes effective at:

728i Manual	4470420
728i Automatic	4476260
732i Manual	5335143
732i Automatic	5345510
735i Manual	7420182
735i Automatic	7425953

September 1981: All 1982 model-year cars equipped with electric window lifts, service interval indicator, and 5-speed gearboxes. —

Autumn 1982: Changes for 1982, including 4-speed ZF automatic transmission option, slight style improvement at nose.

ing economy to speed and style, for a significant amount of weight (about 100lb) was taken out of the cars, and the engine settings (allied to the new bore/stroke ratio) helped to improve steady-speed economy figures as well.

Availability and choice

The 7-Series saloons have not sold as well in this country as BMW (GB) might have hoped, for they have always been expensive (by Jaguar standards, at least) seemingly under-engined. The Coupés, on the other hand, have done well, and are flagships to make most BMW dealers proud.

The rarest models — those which you will find difficult to locate as secondhand models — include the original carburetted 728 and 730 saloons, and the more recent 628CSi Coupés. On the other hand, the most numerous, in BMW terms, will be the 733i/732i saloons, and the 635CSi coupés. In the last three years, about 55,000 BMWs have been registered in this country, of which no more than 10 per cent fall into the category we survey here.

As we have already emphasized, all these cars are fast, heavy, and relatively uneconomical. We have never tested 2.8-litre versions, but would not expect them to be any more economical than the larger-engined versions. Expect about 20 mpg from most six-cylinder BMWs, in everyday use, and performance roughly in line with engine sizes — but there is no doubt that the overdrive five-speed manual and four-speed automatic transmission versions offer the best fuel economy/relaxed driving combination.

The cars are still in current production, and there are 145 BMW dealers in the UK, so the supply of parts and service expertise, should pose no problems. Service of the later versions, in any case, should not be too frequent a chore, as BMW have introduced their "service indicator" facia display on 1982/1983 models to let you plan ahead.

What to Look For

These were expensive cars when new, and though they may seem to offer remarkable value on the secondhand market, do not ever forget that parts and service costs are high. To get an immediate idea of condition, ask to see the car's service record (1), remembering that BMW intervals are now variable, not at rigid mileages, and try to find out about the previous owner or owners (2). Most new 6-Series and 7-Series cars were originally bought by companies, and have usually been faithfully and impeccably serviced. Most first owners seem to keep them for up to three years.

The mileage at which you buy a secondhand BMW may be critical (3), for sizeable expense seems to occur at about 50,000/60,000 miles, when exhaust systems, tyres (second time around), and other work all seem to coincide.

Mechanical

The 6-Series and 7-Series cars share the same basic chassis and running-gear. They are built to last, and seem to do so, but after heavy use some problems do emerge.

On the big six-cylinder engine, look for signs of top end camshaft and rocker heel wear (4), which may manifest itself as extra noise, poor fuel consumption, and ragged running. A rebuild will be expensive, but it should not be necessary for some years. Cars used in cities seem to suffer more wear quickly than high-mileage motorway cars.

Another problem, rare but nevertheless present, may be timing chain wear at high mileage (5). Although there should be no problems with the Bosch fuel injection system, the carburettor cars may occasionally show temperament (6) — the unit used is complex, and should not be tackled by a DIY man. One possible problem on injection engines is internal collapse of the air box in the manifolding (7), which will affect the engine idle.

Be sure there are no signs of cooling system water loss (8), of contaminated cooling water (9), or of poor radiator circulation (10), as these alloy-headed en-

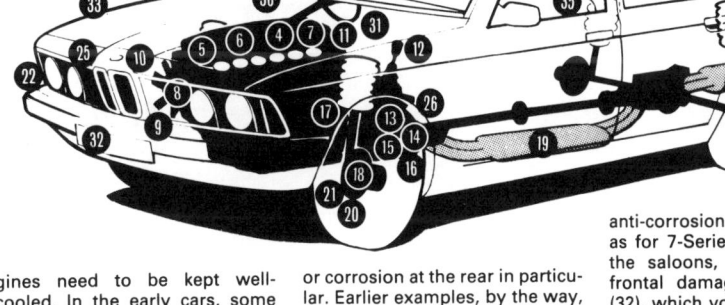

gines need to be kept well-cooled. In the early cars, some head cracking was noted, but this should have been rectified by now on all surviving examples (11). The engines should never be in need of bottom-end attention.

The majority of all cars have automatic transmission, which should always be smooth and silky in operation. Be suspicious of a jerky change (12), or excessive slip (13), both of which indicate wear, which may have been caused by persistent oil loss. Is there any sign of a leaky transmission? (14).

The manual transmissions all have fine reputations — but we recommend a five-speed gearbox model, if you can find one (15); these were not always available, or even optional, on earlier examples. There should never be a rear axle, or drive shaft problem, just so long as the lubricants have not leaked away (16).

Dampers and suspension in general are robust, and long-lived, but look for some front suspension bush wear (17) which may have affected the steering. If you see a car with grubby wheels, this is usually due to brake dust, for pads rarely last for more than 12,000 miles (18), as they are intentionally soft. The wheels usually clean up well, if you use a mild detergent and elbow grease. Tyres are very expensive (especially TRX), but last for at least 25,000 miles in most cases.

Other chassis points to watch include exhaust systems (19), good for about 50,000 miles, cracking of brake cylinder castings (20), and perhaps some signs of brake disc scoring (21), or corrosion at the rear in particular. Earlier examples, by the way, sometimes suffered from front brake locking, which may explain the incidence of accident repair damage around the nose (22)!

Body and trim

Dealers we talked to showed us four and five-year-old cars, coupés and saloons showing no signs of rust, but exhibiting high mileages. Certainly the shells are very solidly built, and serious corrosion should not be present, even on early examples (23). There is comprehensive underbody and sill protection, and the front spoilers are plastic. More recent cars have a six-year body warranty (ask about it — 24).

On 7-Series saloons, look for some paint chipping around the nose (25), and some spotting around wheel arches (26), plus corroded boot lid edges (near the base of the rear window — 27) on older cars. Also on these cars, there may have been evidence of condensation and consequent deterioration in the boot compartment (28). We think that mudflaps are a desirable fitting (29), but these are not standard. Most 7-Series cars, by the way, have a sun-roof (30), and comprehensive radio/cassette equipment (31), but not all. As a guide, look for a car loaded with extras. Many are so equipped, and though more costly they offer better value.

At first 6-Series Coupés did not sell well in the UK, and some of the first Karmann-built bodies suffered early corrosion, but the standard is now high. The same anti-corrosion warranties apply, as for 7-Series saloons. As with the saloons, look for signs of frontal damage accident repair (32), which you might be able to pick up by an examination on a ramp, and possibly for signs of rust on the front wings (33), though there should be few signs of general rot, except for stone chipping damage.

Inside the car, some of the more expensive versions have leather seating, which we recommend (34), though most, with cloth trim, clean up well, even after heavy use (35). The standard of equipment is high, though do not grieve the lack of an on-board computer (36). On our long-term cars, we have found it easy to manage without consulting the complex little electronic box of tricks.

General

Most secondhand examples tend to be found at BMW dealerships, who claim 90 per cent repeat orders, and most are offered with the BMW Approved Car Warranty, which costs £125 a year for unlimited mileage and — subject to conditions, and regular service — can be extended for future years (37). We think this is a very valuable gamble. In general, there are far more 7-Series saloons than 6-Series Coupés available, with the 730 saloon and the 628CSi Coupe being the rarest (i.e., least popular) of all.

Remember, above all, that the BMWs, however reliable, need to be maintained in the manner to which they have been accustomed. You cannot buy at a bargain price, and expect to spend nothing on upkeep. In this respect the big BMWs are rather like the big Jaguars.

CAR TEST

BMW 635 CSi COUPÉ, AUTOMATIC

A fantasy car at a fabulous price—yet it's a sporting and down-to-earth Bee-Em on the road!

There is another world in motoring which few of us are ever privileged to experience — a fantasy world of ultra-expensive motor vehicles which are intended only for the very wealthy.

BMW, of Germany, is one of the motor manufacturers which features strongly in this world of ultra-motoring, with some of the most costly cars available to South African motorists. Its top Seven Series models, for instance, are now priced well over the R40 000 mark, and even these will be eclipsed by the specially-developed BMW 745i which was announced recently.

Yet unchallenged leadership in the top BMW range must go to the exclusive BMW 635 CSi — a stately and sumptuous coupé model which is available here only to special order, at a price of R65 000 (including import duty, but without General Sales Tax).

CONNOISSEUR'S CAR

The 635 CSi is a beautifully-styled, slim-pillar hardtop coupé, a tasteful blend of ultra-luxury car and blue-blooded sports car. It shares much of BMW's electronic wizardry as featured in top Five Series and Seven Series sedan models, with a standard of comfort,

roadability and performance which is all its own.

To enhance its aerodynamics, it has a slender front airdam and a trunk lid spoiler — though it is available without these, as an option. It has dual, high-penetration headlights with a two-stage dipped beam reflector to provide improved low-beam intensity and spread, plus foglights front and rear. Even the bumpers are styled for minimum drag, and wrap round for full protection.

Extra-low-profile high-speed radials are fitted on light-alloy rims.

An electrically-operated sunroof is an option (fitted on the test car) which is quiet-operating and efficient, and the twin door mirrors have a demisting heater in addition to electrical adjustment.

DETAILED ELECTRONICS

The instruments and controls are quite straightforward: tachometer, speedometer, temperature and fuel gauges in the binnacle in front of the driver, a thick-rimmed, leather-bound steering wheel operating the well-rated power steering, and (in the case of the automatic-transmission test car) a T-handle transmission selector.

We will not go into fine detail on the multiple electronic information systems

KEY FIGURES
Maximum speed	211,1 km/h
1 km sprint	29,4 seconds
Terminal speed	178,0 km/h
Fuel tank capacity	70 litres
Litres/100 km at 80 (ECE)	7,85
Optimum fuel range at 80	891 km
*Fuel Index	10,21
Engine revs per km	1 560
National list price	R65 000
(*Consumption at 80, plus 30%)	

incorporated in the fascia, except to mention that these include a strip-light service interval indicator, the multi-function BMW on-board computer with LED digital clock, the check control unit for an instantaneous monitoring of operating functions such as engine oil and coolant levels, fuel consumption indicator, automatically-regulated air-conditioning and heating systems, stereo radio/tape deck with balance control, and pilot lights covering such functions as brake fluid level, engine oil pressure and ABS anti-lock braking system. There is also a very positive-acting cruise control — one of the best we have experienced.

One small failing which we recorded was that on the right-hand-drive model, the stalk switches on the steering columns interfere with access to the ignition switch: we found that the driver's arm tended to knock the windscreen wipers and washers on when reaching for the ignition switch.

DRIVING COMFORT

This is a most comfortable vehicle to sit in and to drive. The driving seat has a wide range of adjustment to suit the individual, and the steering column is adjustable for reach — which is much more important than rake adjustment.

There are such convenience features as twin dome lights, an illuminated vanity mirror, and power windows all round.

The coupé is tailored as a four-seater, with head restraints at all four seats. The front seat backrests fall well forward to allow clear access to the rear seats through the wide, coupé-style doors. While there is a good standard of headroom in the rear seats, legroom is limited — particularly if the occupants are long-legged adults.

The doors are large, but not heavy, and we found them easy to work with and to close.

SOPHISTICATED ENGINE

The engine is one of the world's most sophisticated six-cylinder units — a 3,5-litre overhead-cam motor with the DES (digital electronic) fuel injection system, which applies an automatic cut-off of fuel supply when coasting, extending down to 1 000 revs.

This engine has 12 balance weights for turbine-smooth operation, and we found on the test car that it runs easily up to 5 900 revs, at which point a safety governor limits its exuberance.

The automatic transmission is a ZF three-speed with a high level of efficiency — now being joined in production by a new 4-speed unit. It is teamed with a 3,077 to 1 final drive ratio to give reasonably long-legged gearing (37,8 km/h per 1 000 revs) without detracting from pulling power.

ROAD PERFORMANCE

This is a fairly big car, nearly 4,8 metres long and with a mass of just over 1,4 tons. Yet in spite of this it makes a deceptively-fast start from rest: there is no wheelspin, but the transmission reacts firmly and gets the car to 60 km/h in 4,4 seconds. Once moving, it whirls away to 80 in 6,1 seconds and 100 in nine seconds, covering the first kilometre inside 30 seconds.

There is abundant power available in road use. For instance, the coupé will accelerate from 60 to 80 km/h inside two seconds, just by flooring the pedal, and it will climb a 1-in-7 gradient even at cruising speeds.

We did a full measurement of maximum speed both ways on a level road, registering an average of 211,1 km/h (131,2 m-p-h) with the speedometer reading 223. This was confirmed by the accurate tachometer, which read just over 5 650 at the maximum and showed a minimal slip loss in the transmission. With air-conditioning operating, incidentally, maximum speed drops by about four km/h.

ECONOMY, SOUND, BRAKING

We have based our fuel consumption tables and graphs on the official ECE figures which are mandatory in Europe, and these show remarkable capability for a big-engined automatic:

The 635 CSi retains a distinct BMW family appearance, particularly from the front or rear, with spoilers marking the biggest difference.

a steady 7,85 litres/100 km at 80, and 9,01 at 100. As with any powerful automatic, though, there is a penalty for heavy-footed driving, and the ECE lists 14,9 litres/100 km as the urban-cycle result for this car.

The 635 CSi is magnificently-quiet on a good road surface, registering only 74,5 decibels inside the car at a steady 100. This rises to 81,5 with a window open, and 83,0 on a coarse road surface — which is rather high for a top-quality car.

The brakes are superb: responsive, well-balanced and consistent, stopping the car repeatedly from a true 100 km/h (108 indicated) in between 3,3 and 3,4 seconds.

SMOOTH RIDE

BMW suspensions are among the best in the world, and the CSi has a full measure of this blue-blooded, all-independent system which makes light of any road surface. This is not the sort of car you would normally throw around on rough roads, but you could do so in full comfort and safety — it's built to take it.

Under more normal road conditions it is gentle and refined, handling easily and with a fair standard of inherent sportiness and load-capability. It is particularly steady at cruising speeds, and not deflected by gusting side winds.

There is a mild tendency to develop tyre whine on a smooth road surface — though this does not detract from a road-hugging grip.

TEST SUMMARY

This very special Bee-Em really belongs with castles and mansions and country estates. With its considerable price inflated by import duty in South Africa, it is well outside the reach of the ordinary motorist.

Yet as a road car, it has that common touch of outstanding comfort, drivability and good manners which usually characterise BMW cars, and it is sure to retain its status as a collector's item as the years go by.

SPECIFICATIONS

ENGINE:
- Cylinders . . . . 6 in line
- Fuel supply . . . . . Bosch L-Jetronic injection
- Bore/stroke . . . . . . 82,0/86,0 mm
- Cubic capacity. . . . . . 3 430 cm³
- Compression ratio. . . . . . 10,0 to 1
- Valve gear . . . . o-h-v, single o-h-c
- Ignition. . . . . digital electronic system
- Main bearings . . . . seven
- Fuel requirement . . 98-octane Coast, 93-octane Reef
- Cooling. . . . . . . 9-bladed fan with viscous coupling; electric auxiliary fan; oil cooler

ENGINE OUTPUT:
- Max. power I.S.O. (kW) 160
- Power peak (r/min) 5 200
- Max. usable r/min 6 000
- Max. torque (N.m) 310
- Torque peak (r/min) 4 000

TRANSMISSION:
- Forward speeds . . three, ZF 3-HP-22 automatic
- Selector . . . . console T-handle
- Low gear . . . . 2,478 to 1
- 2nd gear . . . . 1,478 to 1
- Top gear . . . . direct
- Reverse gear . . . . 2,090 to 1
- Final drive . . . . 3,077 to 1
- Drive wheels . . . . rear

WHEELS AND TYRES:
- Road wheels . . . . forged alloy
- Rim width . . . . 6,5 inches
- Tyres . . . . . 220/55 VR 14 steel belt radials

- Tyre pressures (front) . . . . . 180 to 220 kPa
- Tyre pressures (rear) . . . . . . 200 to 250 kPa

BRAKES:
- Front . . . . ventilated discs
- Rear . . . . discs
- Pressure regulation . . . dual circuits
- Boosting . . . . vacuum servo
- Handbrake position . . between front seats

STEERING:
- Type . . . . ZF ball and nut, power-assisted
- Lock to lock . . . . 3,5 turns
- Turning circle . . . . 10,1 metres

MEASUREMENTS:
- Length overall . . . . 4,755 m
- Width overall . . . . 1,725 m
- Height overall . . . . 1,365 m
- Wheelbase . . . . 2,630 m
- Front track . . . . 1,430 m
- Rear track . . . . 1,460 m
- Ground clearance . . . . . . 0,095 m
- Licensing mass. . . . . . . 1 430 kg

SUSPENSION:
- Front . . . . independent
- Type . . . . . . anti-dive coil struts, stabiliser bar
- Rear . . . . independent
- Type . coil struts, semi-trailing arms, stabiliser bar, dual-jointed half-axles

CAPACITIES:
- Seating 4/5
- Fuel tank . . . . 70 litres
- Luggage trunk . . . . 410 dm³

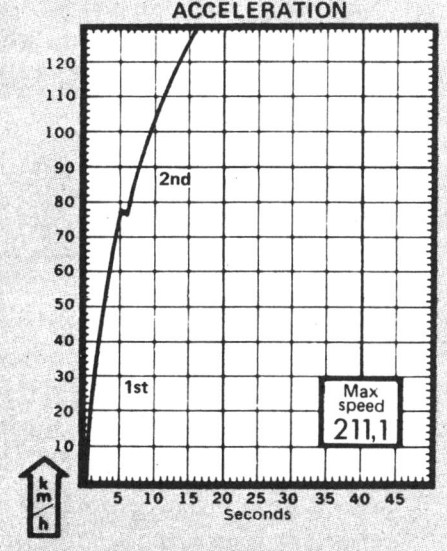

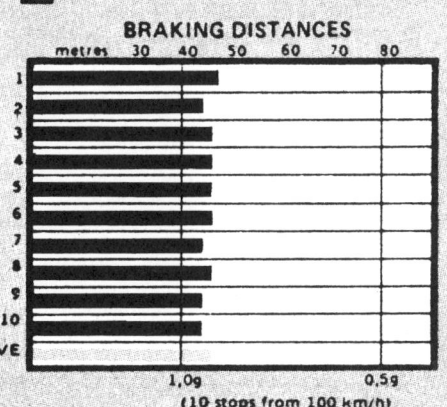

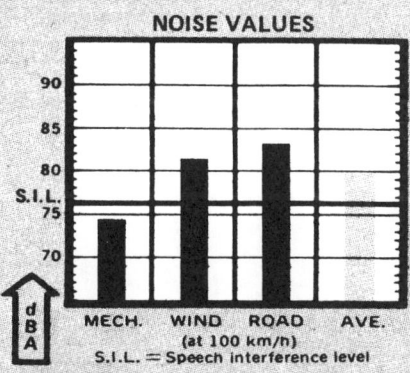

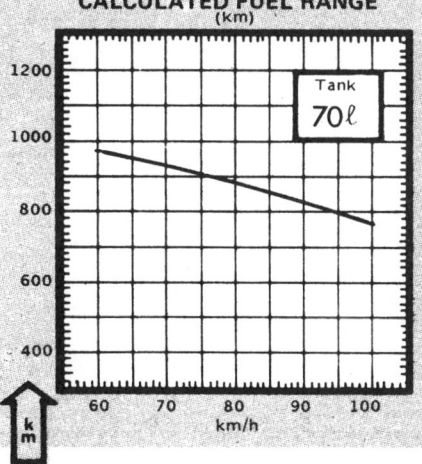

test — BMW 635 CSi automatic

PERFORMANCE

MAKE AND MODEL:
- Make . . . . BMW
- Model . . . . 635 CSi automatic

PERFORMANCE FACTORS:
- Power/mass (W/kg) net 111,9
- Frontal area (m²) 2,35
- km/h per 1 000 r/min (top) . . . 37,8

INTERIOR NOISE LEVELS:

	Mech.	Wind	Road
Idling	61,0	–	–
60	68,5	–	–
80	72,0	77,5	79,5
100	74,5	81,5	83,0
Average dBA at 100			79,7

ACCELERATION (seconds):
- 0-60 4,4
- 0-80 6,1
- 0-100 9,0
- 1 km sprint 29,4

OVERTAKING ACCELERATION (A/T):
- 40-60 1,8
- 60-80 1,8
- 80-100 2,9

MAXIMUM SPEED (km/h):
- True speed 211,1
- Speedometer reading 223
- Calibration:
  - Indicated: 60 70 80 90 100
  - True speed: 52 62 72 82 92

FUEL CONSUMPTION (litres/100 km; based on official ECE figures):
- 60 7,16
- 70 7,46
- 80 7,85
- 90 8,40
- 100 9,01

BRAKING TEST:
From 100 km/h
- Best stop 3,3
- Worst stop 3,5
- Average 3,36

GRADIENTS IN GEARS:
- Low gear . . . . 1 in 2,7
- 2nd gear . . . . 1 in 4,4
- Top gear . . . . 1 in 6,5

GEARED SPEEDS (km/h):
- Low gear 79,2
- 2nd gear 132,8
- Top gear 196,3
- (Calculated at engine power peak – 5 200 r/min).

TEST CONDITIONS:
- Altitude . . . . at sea level
- Weather . . . . fine and hot
- Fuel used . . . . 98 octane
- Test car's odometer . . . . 5 743 km

WARRANTY:
12 months, unlimited distance.

TEST CAR FROM:
BMW South Africa, Rosslyn, Transvaal.

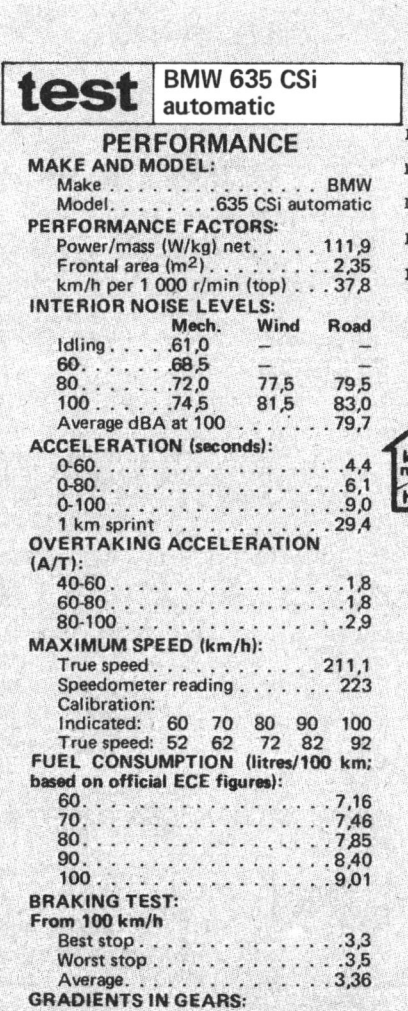

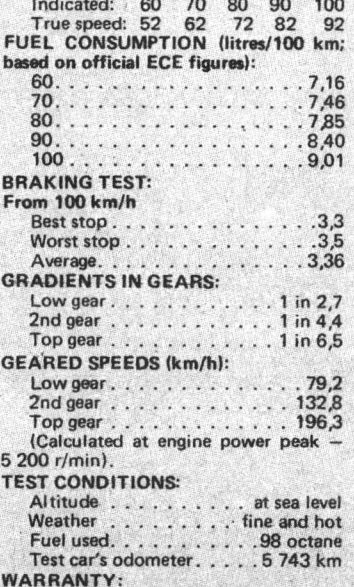

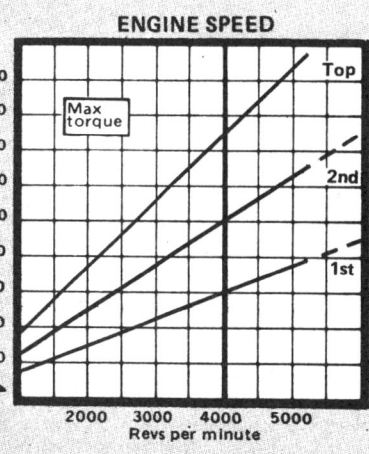

IMPERIAL DATA
- **ACCELERATION (seconds):**
  - 0-60 m-p-h 8,5
- **MAXIMUM SPEED (m-p-h):**
  - True speed 131,2
- **FUEL ECONOMY (m-p-g) (ECE):**
  - 50 m-p-h 35,8
  - 60 m-p-h 32,2

CRUISING AT 100
- Mech. noise level . . . . 74,5 dBA
- 0-100 through gears . . . . . 9,1 seconds
- Litres/100 km at 100 (ECE) . . . 9,01
- Optimum fuel range at 100 . . 777 km
- Braking from 100 . . . . . . 3,36 seconds
- Maximum gradient (top) . . . 1 in 6,9
- Speedometer error . . . . 8% over
- Speedo at true 100 108
- Tachometer error . . . . negligible
- Odometer error . . . . 2,6% over
- Engine r/min at 100 . . . . 2 649

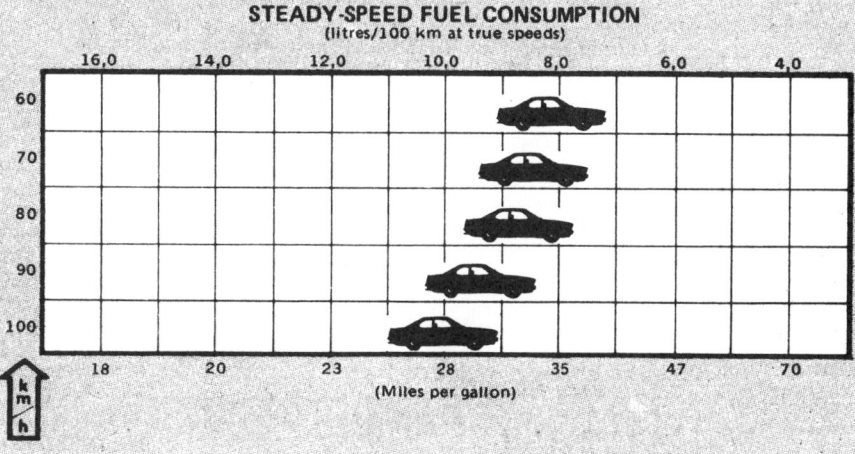

TRACK TEST

The Frank Sytner Grace International Racing BMW 635CSi, supported by a multitude of sponsors, with Tiff at the wheel.

Graceful racer

Rovers dominated the 1983 Trimoco championship, although a couple of BMWs upset the ranks. TIFF NEEDELL tried out one of the Rover challengers, Frank Sytner's 635CSi.

The 1983 Trimoco British Saloon Car Championship has been dominated by Tom Walkinshaw's Rover Vitesses. Ready to go at the beginning of the year, the 'Yellow and Blue' formation proceeded to win every round of the championship. However, as the year progressed, the challengers emerged, the strongest being a pair of BMW 635CSis, the Sytner and Cheylesmore cars. Recently I had the opportunity to visit the Silverstone Grand Prix circuit and sample the Sytner machine.

Frank Sytner has been a well known name in British racing circles for almost 10 years, competing in FF1600, FF2000, Sports 2000 and BMW 323i County Championship cars, but last year was the first time he moved up and tried 'big' cars, a TWR 3.5 Rover. Then, Frank, the car and the team somehow failed to get it all together and that simply doubled Frank's renowned determination to come back and win.

As Nottingham's famous BMW dealer and newly appointed Great Britain Alpina Concessionaire, Frank obviously chose one of his own products to challenge the car that almost beat him, and was happy to forge a partnership with Grace International Racing. For 1983 saw the return of one of saloon car racing's most famous partnerships when Malcolm Gartland officially returned from retirement and rejoined Ted Grace at Grace International Racing.

History lesson

In the late '60s and early '70s, Malcolm and Ted had been responsible for the Falcons, Camaros, Capris and BMWs that saw so much success in the hands of Brian Muir and Frank Gardner. With the demise of Group 2 in 1974, Malcolm called it a day, while Ted continued preparing the Camaros of, firstly, Frank Gardner, and then Stuart Graham, staying on with Stuart to prepare his Capris. He then spent a couple of relatively quiet years before this BMW project came up.

It was during 1982 that Ted was asked, by a major sponsor, to look at the Group A scene and come up with a proposition for 1983. In England, the British 'Group 1½' series was dominated by Rovers and Capris, the rules not really suiting any BMW options, but the Europeans were having their first year of Group A. BMW took the title with a 528i that scored on reliability but couldn't match the Jaguar for speed.

However, when BMW upped the 635 production to the magical 5000 mark this seemed the obvious choice for Ted and the dozens of European privateers. With the deal still not settled, Ted went ahead and ordered a shell and all the bits from BMW Motorsport in Munich and started to build the car last December. Sadly, the original sponsor fell through, but, by February, Frank Sytner had been in touch and they struck a deal to do a limited season of testing and racing with the real target being the 1984 championship.

Frank arrived with a deal for engines from his Alpina contacts in Buchloë, Germany, and backing from Chartered Trust Finance PLC (a division of the Standard Chartered Bank), Linneys Colour Printers, Radio Trent, Mobil, Woodland Generators and gsi Computer Software. You can see why Frank had to move up to a big saloon so as to provide enough bodywork for his sponsors! In April they began their testing at a time when the Rovers had already won four of the 11 Trimoco rounds. On May 1 they made their race debut at the Donington European championship round. It was not a fairy tale beginning, however, for a new wheel worked its way off and dumped Frank into one of uncle Tom's sandpits on only the 12th lap.

Six weeks later, and the team's Trimoco debut saw Frank sitting on the front row with two Rovers beside him and three behind. A switch from Dunlop to Pirelli rubber had provided a much more satisfactory tyre deal and Frank fought a splendid battle with Peter Lovett for second place only to miss the chicane with two laps to go and finish a penalised third. Two weeks later he made second stick at Donington and at the British Grand Prix he came out fourth, the winner of a hectic dice with Hans Stuck in the Cheylesmore BMW.

After that, things went off-song slightly — punted by Stuck at Donington, a tyre troubled 10th at the Tourist Trophy, and a broken gear linkage at the Silverstone finals day. It obviously wasn't the perfect season, but Frank knows that he and his BMW could match either Stuck or Jonathan Palmer in the Cheylesmore twin and, with further testing scheduled, he is confident of getting ahead of the Rovers in 1984.

Track impressions

Out on the track the BMW feels already well-sorted, being smooth and confidence inspiring. The Rover (AUTOSPORT, March 17, 1983) had given me a definite feeling of uneasiness turning in to corners and a slight lack of confidence under braking — admittedly in a very new and unsorted machine — but the BeeEm felt docile and forgiving. I had made the mistake of putting in ear plugs before I clambered over the horizontal 'side-bar' of the crash cage and into the seat, failing to notice the

Above: The Alpina straight six 3.5 BMW engine. Below: The team with, from left to right, Malcolm Gartlan, Frank Sytner, Ted Grace and one of the mechanics.

TRACK TEST

exhaust pipes poking out of the right hand side of this left hand drive machine. The consequent blissful silence must have added to the sense of smoothness, but it also adds a dangerous lack of sound feedback. I 'felt' the rear brakes lock for the chicane on one lap, but only when I blipped the throttle on the down change did I realise that the engine had stalled and the inside rear was *still* locked up!

Just like a road car...

The controls of the 635 were as simple as a road car. There was an adjustable seat, a good average 'racing' weight of steering giving enough feedback and the positive gear change of the Getrag five speed 'box. Sitting on the left with the right hand gear shift and its 'racing' pattern (having first back on a dog leg with second to fifth in the H-pattern) made me feel quite at home — I'm accustomed to just such a layout in single-seaters and Group C cars — although having so much car to my right did cause a few missed apexes.

Handling characteristics

The handling was basic 'understeer in, oversteer out' but with its own little 'pitching' characteristic. Having turned in on the optimum racing line and then fed in the power before the apex, the 635 would set itself into an understeer-oversteer-understeer-oversteer pitching motion. I felt this could be due to the radial tyres but Frank was quick to point out that it happened on the Dunlops as well and that it was a characteristic of the car that he has just learned to drive around. Apparently, the use of the optimum line and standard 'brake-turn-power' technique provokes this pitching. Frank turns in very early, waits for the front to bite and then progressively adds the power. If the pitching sets in then Frank knows it's a slow lap, for it obviously scrubs off that vital edge of speed.

Even before I spoke to Frank I had found myself automatically driving round the problem without knowing it, sometimes trying to turn in to a corner too fast, which caused excessive understeer and forced me to delay any application of power until the front had regained some grip. By that time I was well into the corner, and only then would the 635 take the power and proceed in an orderly fashion.

As usual, the adverse camber of Copse exaggerated the turn-in understeer and was its usual frustrating self. Beckett's could be attacked with an induced turn-in oversteer but that encouraged the pitching, while a smooth understeer turn-in usually led to a close encounter with the grass as I ran out of road cresting the little hump encountered just at the exit of the corner. Stowe and Club were both a pleasure in the 635, although the early morning dampness made the ribbed concrete run-offs very slippery indeed, exaggerating the oversteer on the exit on more than a few occasions! Club was possibly the worst place to have to sit on the left because the very fast late apex is so crucial to a quick lap at Silverstone and every inch of the circuit is vital. That, plus Club's very large kerb, means it must be approached with caution.

"Out on the track the BMW feels already well sorted being smooth and confidence inspiring."

Up to the chicane and the brakes were very impressive for a big saloon. Relatively small master cylinders gave the pedal a reasonable amount of travel, but only a little 'foot power' was needed to clamp the pads to the ventilated discs and haul down the speed. This, however, did encourage over-confidence and led to my long rear wheel lock-up. Nevertheless, the brakes were surprisingly good and had I had next year's cockpit adjustable balancer installed I could have braked to even greater depths. The chicane itself saw the 635 change direction with little fuss and even a misjudgement of that right hand wheel position did little to stir the BeeEm from its path, even with only two wheels in contact with the tarmac!

The two, straight six, 3.5-litre, 12-valve, single overhead cam engines used by the team are both Alpina creations, while the team itself carries out tuning and rebuilds. A figure of 280bhp is quoted at 6800rpm with a 7200 limit. Interestingly, these figures are virtually identical to those of the 3.5 Rover V8. The injection and 'engine management' is by Bosch and the Getrag gearbox delivers the power to one of a choice of three homologated ZF differentials. The Pirellis are 11ins all round mounted on 16ins BBS rims. The car needs ballast to stay above the 1184kgs limit and, although retaining the standard 635 layout, has homologated front uprights, fabricated swinging arms at the rear and centre lock wheels all round. Fully adjustable anti-roll bars replace the standard 'fixed' units and Bilstein shock absorbers help to smooth out the bumps.

A civilised racer

All in all, the 635 BMW provides a very civilised way to enter international motor sport. The basics are provided by München, but there is plenty of room for Grace International Racing or anyone else to tune and improve their individual steeds. Although a very smooth and forgiving car to drive, I feel there is much to come yet from the BeeEms for that very reason. If it is smooth you can usually go quicker by making it nervous and twitchy, perhaps losing some of that soft nicety in exchange for hard grip. However, all that's a matter for testing over the next four months before Frank sets out to prove that his BMW lives up to his dealership's 'unbeatable' motto. ■

Needell (left) chats to Sytner.

Mean and purposeful, the Sytner BMW might well prove 'unbeatable' next year.

RACE TEST

For 71 laps Jeremy Shaw was able to enjoy the pleasures of driving a Group A BMW 635CSi at the Tourist Trophy.

No ordinary day

When JEREMY SHAW heard of the opportunity to drive a BMW635CSi in the Tourist Trophy, he wasted no time in offering his services.

It had been a fairly ordinary Thursday morning. I had been busy in the office sorting out a few of the general matters that assist in co-ordinating a weekly magazine such as our own, answering various telephone enquiries and tapping out a couple of news stories for the following week's edition.

The 'phone rang again. "Hello, is that Jeremy Shaw?" asked a voice at the other end. It was a female voice, so perhaps I showed a touch more interest than usual. I confirmed that, yes, in fact this was he, and asked whether I could be of assistance.

"Well," continued the voice, "we are running a car in a race at Silverstone next week — the Tourist Trophy, do you know the one I mean? (I muttered again in the affirmative) — and we are looking for a couple of drivers . . ."

Silence. It was my turn to speak but nothing really came to mind. My brain was still confused from the previous question. "P-Pardon," I stammered. "You'd like me to, er, suggest a couple of drivers?" This was no longer any normal Thursday.

Calls to inform us that so-and-so is driving a car to be sponsored by such-and-such at a faraway race meeting are reasonably common at AUTOSPORT, but not so callers who seem to have any difficulty in arranging a *pilote*.

After expressing pleasantries and a few details of the drive and the lady's requirements, I noted that her company were based in West London, not too far away from our offices. With little further ado, I had hopped into my car and quickly made my way towards our agreed rendezvous, "not far from Shepherd's Bush", she had said. The significance of the telephone call sunk in on the way. She was looking for the names of a couple of drivers who wouldn't mind sharing the 'chores' of driving pukka Group A BMW 635CSi in one of this country's most prestigious motor racing events, the RAC Tourist Trophy. Perhaps, I noted unbelievingly, 'He' had taken notice of me after all!

On arrival at the modern, functional office/

Jeremy Shaw — journalist turned Group A racer.

workshop building which served as the headquarters of Autosport & Design — a company where business is in the distribution of a vast selection of go-faster goodies for up-market motor cars — I was introduced to one of the directors, John Cannon, and to Eve Wheeler, the lady who had telephoned me less than an hour before.

The situation was explained to me. They, Autosport & Design (the similarity between their name and ours is purely co-incidental) imported a plethora of motor accessories and tuning kits, including a range of the same for BMW cars from the German firm, Hartge Tuning. They had recently become sole Hartge importers for the UK and as such wanted to take advantage of the appearance of Hartge-prepared Group A cars at the TT. It was a good opportunity to cement their relationship and thoroughly prove the articles they were marketing.

A deal had been struck to insert a couple of drivers of their choice into one of four Hartge-prepared cars that would be in the race. Which is where I came in. Could I help? How could I not?

After extolling the virtues of one Jeremy Shaw as a racing driver-cum-motoring writer, informing them of my experience of a couple of 24-Hour races (I didn't say that they had been at Snetterton, nor that I had crashed in one of them, even if it hadn't been totally my fault) plus my international forays (I had driven a Clubmans car at Zandvoort only a few weeks earlier), I said that I would be only too happy to offer my services. "Why not?" they said, "that sounds fine."

The intervention at that point of a secretary with a cup of tea could not have been more opportune. It gave me the chance to reflect upon my apparent good fortune and the fact that I had for many years hankered after the possibility of driving a big saloon in a race such as the TT. Since spectating at the classic Spa 24 Hours in 1972, watching Hans Stuck, Gerry Birrell and Alex Soler-Roig stage a 'dead-heat' 1-2-3 finish for the works Ford Capri team, this kind of racing has always held a special fascination for me. Now I was to have the chance of joining the modern-day saloon 'aces' on equal footing. It took a bit of believing.

First impressions

It was on the Friday before the race that I first got to see and drive the car, taking part in an all-to-short unofficial session. After being introduced to brothers Herbert and Rolf Hartge, who supervise preparation of the car and carry out various developments of their own, and to Dominique Fornage, a French garage proprietor who actually owned the car that I was to drive, I was strapped into the bare cockpit.

The slim but sturdy Recaro seat gave a very upright driving position but after adjustment of the seat belts to suit my tall frame, I found the driving position to be very comfortable. I wasn't unduly concerned about the left-hand driving position — after all, I had driven plenty of times on the road in such cars and my single-seater racing experience meant that the right-hand gear lever should be no problem — and so I spent the next couple of minutes familiarising myself with the controls, which included a variable speed windscreen wiper, heater/demister and, honestly, electric windows! There's no chance of getting bored on the Hanger Straight, I thought.

Talking of the longest straight on the 2.932-mile GP circuit, I suddenly remembered that I had never actually driven a car round 'in anger' before, my experience at the Northamptonshire venue having been exclusively gleaned on the Club circuit. Oh well, in for a penny . . .

As I ventured out of the pits for a first few exploratory laps — the track was still damp from some earlier rain but slick tyres were deemed in order — I soon noticed that the engine pulled strongly and cleanly and that, despite the weight of the car and the width of the slick tyres, the steering was not as heavy as I would have anticipated.

The car tended to 'wander' around the track, following the smallest ripples in the tarmac, until the radial Pirelli tyres had reached working temperature but then the car gave plenty of 'feel' and seemed particularly safe under braking.

This session did not allow me to gain full assessment of the car's (or my) capabilities, but it did enable me to work out gearing for each of the corners. Copse, the first right-hander after the pits was taken in third gear, although a brisk, cold breeze tended to rather upset the car here, strong enough to blow the car wide on the exit and inducing understeer. It was then up through fourth to top — flat through Maggott's — to Beckett's, which was posing my biggest problem; where to brake? The corner, anyhow, was negotiated in third gear, then fourth before the flat-out Chapel Curve and fifth soon afterwards for the long blind towards Stowe.

A terminal speed of something approaching 140mph was reached before the very quick right-hander, and this provided the most exhilarating cornering of the entire lap. A dab of the brakes a little before the 100 yard board, and a flick into fourth gear was all that was required before standing on the throttle again and hugging the inside of the tarmac. The slight ridge of kerbing towards the end of the corner, still on the inside, served to maintain the

RACE TEST

slightly sideways attitude of the car and, with minor corrections of the steering, kept it drifting wide to the clipping point on the outside. Fantastic!

Fifth gear was required before reaching Club Corner, another very fast right-hander which has a rather higher kerbing on the inside and which, unlike at Stowe, one definitely did not want to clip. Abbey Curve was easily flat out in top gear, after which maximum revs of around 6400 were reached before it was time to stand on the anchors for the Woodcote chicane — another new corner to me and one at which I found difficult to judge a correct turn-in point — which was taken in third gear. Fourth was selected soon after crossing the start-finish line, holding that ratio past the pits to Copse to complete a lap.

The event

The blustery, damp conditions which were to prevail throughout the weekend were particularly evident during the two practice sessions on the Saturday, although wet tyres were never actually fitted to the car. My determination to ensure that the car stayed in one piece led to my being reasonably cautious during qualifying, although I was reasonably pleased to note that it was my time rather than that of Dominique, who was also at the circuit for the first time, which was to stand for our grid placing. This was, admittedly, rather a long way down the field, 29th out of 44 to be exact.

Race-day dawned even wetter than before, but Fornage had kindly insisted that I do the first stint in the car, so it was I who ventured out for the warm-up

Conditions at the start of the race were appalling, the spray making it almost impossible to see anything ahead.

Handling someone else's £20,000 car induced caution...

session, the Bee-Emm on Pirelli wets for the first time.

Here I had another admission to make. My power over the weather gods, or whatever, had determined that I had never driven a racing car in wet conditions before. Here was another new experience! It was one that I soon began to enjoy. The initial horror of seeing so many cars on the circuit at the same time soon passed as I concentrated on acclimatising myself... even if it was for only three laps before the chequered flag came out! Nevertheless, my time of 1m57.50s stood as 20th best from the session and only a gnat's slower than the admittedly troubled Walkinshaw/Nicholson Jaguar.

As the start time approached, there was no sign of the clouds lifting. This was going to be a wet race. And then came my initiation to wet *racing*. It's a damn sight different to practising, I can tell you. For a kick off, as the field gathered speed on the final pace lap and we approached Woodcote and the green flag, I honestly couldn't see a thing. The spray was completely opaque. And here was I in someone else's £20,000 motor car...

The first lap was as near to frightening as I have ever been in a racing car, and that includes having been in a couple of accidents. Visibility was virtually zero when following another car, even with the wipers working overtime. Soon, however, the situation eased. Cars became a little more spaced out and by nipping out of the slipstream it was possible to see a way past.

Already, I had lost several positions, but the controllability of the car and the grip afforded by the tyres, plus the incredible braking capability soon led to some rapid progress up the field. I was up into 17th place by lap 20 and feeling completely at ease, thoroughly enjoying the drive. Next man ahead of me was Andy Rouse, at that time leading class B. Unfortunately, after I had passed him at Stowe and then been re-taken on the exit, we then went for the same piece of tarmac at the flat-out Abbey Curve. The two cars made contact and we both spun in echelon, slithering for what seemed like an age across the sodden grass.

Soon after rejoining, none too happy (although

... but mixed wets and slicks made handling unpredictable.

neither, I should say was Andy, his class lead now down the drain), I was signalled into the pits for a change to dry rubber, the rain having abated and left a fast-expanding dry line. Another problem. The team obviously had trouble in changing one of the right-side tyres and it was after what seemed like a fortnight that I was eventually given the all-clear to resume.

For the first lap or so, the car felt diabolical. What on earth could be the matter, I pondered? It was weaving all over the place, particularly under braking. Perhaps a wheel is loose, I thought.

Not wanting to lose further time, I persevered for a couple of laps, at greatly reduced pace. Now the car was beginning to feel a bit better. Perhaps it was just the effect of cold tyres on a cold track?

It was only after I was eventually signalled in for our routine driver change, after completing 71 unforgettable laps, that I was made aware of the problem. "We couldn't get one of the tyres off on that side," said one of the team, "so we left the old rubber on..." It took a couple of seconds to sink in. "You mean, I was out there for 40-odd laps with slicks on one side and wets on the other?" I gasped. I thought it had felt a little odd!

The original plan was for team owner Fornage to relieve me at the wheel, although Dominique had been so enthusiastic about the performance of European Formula 2 Champion Jonathan Palmer at the wheel of his sister car — running in second place — that he decided to stand down. Thus, Jonathan took over for the final 30-odd laps, our car eventually being classified 20th overall, despite the earlier delays.

The race, of course, had been immensely satisfying, my lap times in the mid-1m44s bracket not being among the quickest of the day but then again reasonably respectable, given that I wasn't out when the circuit was at its driest, nor was I equipped with entirely optimum tyre wear! For fun value, though, the day was most memorable, for which my thanks go to Autosport & Design, Hartge Tuning, Dominique Fornage and his untiring mechanics. *A bientôt!*

The Hartge Tuning 635CSi of Vojtech/Enge.

The Hartge of the matter

The engine in the car I drove, as well as in those driven to second and third places in the Canon Tourist Trophy by Jonathan Palmer/James Weaver and Zdenek Vojtech/Bretislav Enge, was prepared by Hartge Tuning, run from a small town near Saarbrucken, West Germany, by two brothers, Herbert and Rolf Hartge.

Both men have long held an interest in motor racing, Herbert first racing a Formula Vee car in 1970. This was soon sold when they established a BMW agency business during the following year, but Herbert returned to the circuits within a couple of years in the first of a succession of BMW cars.

Many class wins were gained in Group 2 racing, mainly within Germany, although Hartge preferred to race as and when he liked — "I was only interested in doing the big races at the best circuits" — rather than concentrating on any championship.

In 1979 came the establishment of Hartge Tuning. "One day," recalls Herbert, "we thought how much we had spent on motor racing and how much knowledge we had gained. Then we sat down to think how we could commercialise that knowledge."

They began by producing a tuning kit for the old 323 model, machining and balancing the engine until it produced some 180bhp. There followed a succession of suspension kits, spoilers, instrument panels and other accessories for the entire range of BMW cars, most of them developed by Rolf, the engineer of the two, and marketed by Herbert.

One of the most significant of their innovations is to develop an electronic fuel injection and ignition system for the 635CSi engine, one which has apparently resulted in an extra 20bhp for the Group A engine when compared with the 'standard' Alpina unit. The electronic system has been produced in conjunction with AFT and Schrick, two other German companies, and has already drawn plenty of interest from rival tuners and teams.

ROAD CAR

Although handling is commendably neutral, the M635CSi's tail can be hung way out if desired, thanks to huge reserves of power.

M-azing!

BMW Motorsport's M1 engine transplant takes the 635CSi into the supercar league. MARCUS PYE brings road impressions from Germany.

A tranquil autumn afternoon in Southern Bavaria, the serene beauty of the undulating greenery to the left in stark contrast with the harsher landscape on the opposite side. The whole vista, bathed in lukewarm sunshine, is bisected by a pleasantly rolling, double-barrelled snake in the road, paving the way to the popular ski resort of Garmisch-Partenkirchen, and Austria not far beyond. Up ahead, the foothills give rise to rugged mountains at a remarkable rate on this day. A cursory glance at the crystal clear tachometer goes a long way towards completing the picture . . .

The chunky gear knob has long nestled far right and forward, ever since modest progress through the suburbs of Munich. The overdrive-top ratio can be put to maximum use thanks to a prodigious torque output from beneath the long, elegant bonnet. As the white needle begins to threaten the yellow segment at 6500rpm (red being a further 400rpm away), the canted 'M' symbol, with its flashes of blue, purple and red, presents the telltale clue to the Bayerische Motoren Werke's latest wondercar.

Right foot nailed to the floor and revs approaching their peak — switching attention momentarily to the tacho's partner at left. It reads 255kph for a brief burst until we throttle back, point proven — especially to the fellow in a big Mercedes whom I had waved by a couple of minutes earlier, then chased hard on the two-lane autobahn until my sleek flame red charger would go no faster. Two-five-five! "What's that *in English?*" you ask. Why, an indicated 158.44mph if the calculator serves me correctly. Nearly 160mph in a BMW 635 — extraordinary.

That stylised 'M', needless to say, is ultimately the key to the matter. Repeated, unobtrusively, on the radiator grille and boot lid of the purposeful machine, it is the hallmark of BMW's Motorsport department, headed by the genius Paul Rosche. Through the hallowed doors of his engineering shop have emerged the long line of BMW competition cars and engines and also, in recent years, the stunning M1 sports car (perhaps better known in G4/Procar racing trim in this country) and the dramatic M535i sports saloon derivative of the mid-range 5-series car.

The latest project, designated M635CSi, cleverly combines all the *gung-ho* of the M1's 24-valve, 3453cc development of the BMW potent straight-six with the subtlety and grace of the 635CSi, in its current form no slouch with a hefty 218bhp on tap. The Motorsport engine in the coupé is managed by Bosch's 'state-of-the art' Motronic fuel injection system, coupled with digital engine electronics. Canted over at 30deg for this installation, the alloy-headed powerhouse punches out a whopping 286bhp, revised inlet tracts helping to account for the 9bhp increase over M1-spec units. On a compression ratio of 10:1, peak output is achieved at 6500rpm — for which read 286bhp in a car weighing in at 1500kgs. Maximum torque is available at 4500rpm.

Attention to detail has always been high on the list of criteria in BMW's design philosophy and, of course, technical head Rosche's team could not be content 'merely' with their engine transplant. Not a bit of it. Stiffer springing was called for amid the race-bred suspension geometry, together with beefier gas-filled dampers, a new limited slip differential and, most importantly, bigger, uprated disc brakes, safeguarded with the respected ABS anti-lock arrangement.

The result is a car which handles like a dream, beautifully responsive and balanced: tenacious roadholding from the fat Michelin TRX VR radials supplementing inherently neutral cornering right up to 'normal' limits in the dry. The tail can, naturally, be provoked out of line as desired with forceful jabs of the accelerator, but sudden lift-off while on the turn produces no evil traits due to fine chassis tuning. The M635CSi can be turned-in to corners at high speed without drama, aided by the solid, positive feel of the power steering.

Behind the wheel for the first time, one appreciates

Controls are well laid out and functional.

the quality of the revised seats and, once strapped in, a panoramic view is afforded from the airy, pillarless coupé. The comfortably proportioned leather-rimmed wheel, like the driving position, is ideal for the job.

The engine is delightfully free-revving, supremely fuss-free and docile in traffic but ever-eager to thrust the projectile forward with its full — and exhilarating — potential. For all its phenomenal power (and once the needle swings past 3000rpm it feels like a jet aircraft as its brakes are released under full power to commence take off), the M635CSi is perfectly tractable and equally enjoyable to drive at all speeds.

Rev the engine at standstill, and the distinctive shrill bark of the tough six-cylinder engine tells you it's from thoroughbred stock. But, at high speed, its song is muted to a large degree, still pleasantly perceptible yet blended with surprisingly light swishing from the tyres and wind deflection over the roof and wing mirrors.

At colossal speeds, even, the ride and stability of the BMW are beyond reproach, handling being precise all the while and inspiring confidence throughout our all-too-brief acquaintance. The smartly faired-in front spoiler/air dam is a little deeper than that fitted to the 'standard' (and that is really doing it a disservice!) CSi, and is balanced by a small lip on the boot. The 'M' version's squat stance is accentuated by the eyecatching wide alloy wheels which carry the low-profile Michelins, hooded by swathed arch extensions.

As is traditional with BMWs, finish of the article is flawless and the quality of construction second-to-none. It is already a marvellous 2+2 touring limousine, yet the Motorsport tweaks add considerably to the prestige of this very desirable motor car. With a claimed maximim speed of almost 160mph, 0-60mph acceleration in a shade over 6secs and overall fuel consumption of better than 16mpg, (20mpg should easily be on with restraint) the M635CSi is perhaps the ultimate Q-car of the era — and an ideal works development exercise towards furthering the Bavarian marque's enviable reputation in the important-for-image touring car racing championships of Europe.

The car is scheduled for introduction to the German market next spring, when it will carry an approximate price tag of DM89,000. By the time the M635CSi arrives in Britain, in limited quantities, hopefully towards the end of 1984, about £30,000 should secure one. Catch it if you can . . . ■

The heart of the matter; 286bhp straight-six.

BMW M635CSi
£30,000 (approx)

Specification

Cylinders/capacity	six in-line/3453cc
Bore x stroke	93.4 × 84mm
Valve gear	dohc per cylinder bank
Compression ratio	10.5:1
Fuel system	Bosch Motronic
Power/rpm	286bhp/6500rpm
Torque/rpm	251 lbs ft/4500rpm
Gear ratios	3.5/2.08/1.35/1.00/0.81:1
Final drive	3.73:1
Steering	power assisted ball and nut
Brakes	dual circuit, servo assisted discs all round with ABS anti-lock system
Wheels	BMW alloy 15in
Tyres	Michelin TRX 240/45 VR4 15
Suspension (F)	Independent double jointed struts with torsion bar stabiliser
Suspension (R)	Independent struts with coil springs on semi-trailing arms

Dimensions

Wheelbase	109.3ins
Track (F/R)	59.5/61.0ins
Length	198.12ins
Width	71.8ins
Height	56.4ins

Performance

Maximum speed	158mph
0-60mph	6.3s

Flying the coupé

WHEN BMW launched the revitalised 635CSi, last June, the main aim was to uprate the super coupé's innards without unnecessary external cosmetic treatment. 'This is not just a face-lift, it's a brain-lift', commented the managing director of BMW GB, Dr Walter Hasselkus... but it wasn't just a brain that the surgeons transplanted.

Changes included a new suspension system to give more-predictable cornering and handling, a new, weight-saving engine that displayed a relatively teetotal approach to fuel, the stunningly effective anti-lock brake system (ABS), and more electronics than a computer factory.

The 635CSi doesn't have just one brain. Digital engine electronics, for example, provide the engine with a programmed ignition and fuel-injection system that is always perfectly tuned. The 'brain' can take in variables such as speed, engine temperature and even barometric pressure to ensure that 'pinking', for example, is a disease which afflicts only lesser lumps. And when the throttle is released at engine speeds above 1000rpm, the fuel is completely shut off and the fuel-consumption indicator goes off the high end of the dial.

Then there's the service interval indicator, or SII as we will soon be calling them. You don't service a BMW every 10,000 miles any more — you do as you are told by the computer. For example, the 'brain' counts the number of times you apply the brake pedal and change gear, and when it all adds up, a service is called for by the green lights turning to orange and then red.

Warning lamps cover every conceivable function, such as low coolant level, a failed bulb, low oil, brake fluid or screen-wash levels and, would you believe, even a warning lamp to tell you that one of the warning lamps isn't working?

The dashboard computer is another little executive toy which, at the touch of a button, provides you with information on average speed, mpg, tank range, estimated time of arrival and the possibility of ice on the road. Surely this flagship of the BMW fleet represents the state of the electronic art. To talk of perfection would be churlish, however, for we'd be saying that there was no improvement to be made. One day, soon, you won't have to push buttons at all and — maybe on a BMW before the rest of the herd — the car will simply respond to voice commands.

DRIVE called a 635CSi to heel and headed for the test track.

How it goes

America demanded a more powerful and more fuel-efficient big BMW engine, and the 635's unit is the first step in this direction. To the driver the new 3.4-litre engine remains indistinguishable from the old one — and it's none the worse for that.

The unit is very smooth indeed and will pull happily from 500rpm all the way to the electronic rev limiter at 6200rpm. Actually, it's slightly smaller than the old engine, but at 218bhp the power output is the same as before. It's a little lighter so, in theory, the car ought to be a shade quicker, but even with a stopwatch the improvement is hard to find. The power has been maintained by raising the compression from 9.3:1 to 10:1, which should help fuel economy. It is made possible, of course, by those clever digital electronics.

Power goes to the rear wheels via a long-legged, five-speed gearbox which allows the car to sit at 100mph all day without the slightest trace of exertion, while the rev counter needle hovers a little over 3000rpm. However, if economy is not your priority — and if we ever have £23,000 to spend it won't be ours — there's a no-cost option of a five-speed, close-ratio box to make this even more of a driver's car. For the less energetic — and we admit that the clutch requires a strong left leg — there is also the option of a four-speed automatic gearbox. A unique lock-up device on this electronic marvel ensures that the

QUICK TEST

automatic versions are more economical than the manual boxes right through the official government test cycles. That, so far as we know, is a first for any company in the world.

It is surely an area in which BMW can now make the Jaguar XJS look rather old hat. The British company long since stopped offering any kind of manual gearbox to customers and, in the process, made their undoubtedly fine product less of an enthusiast's car. The comparison with the XJS simply has to be made at some stage and, whatever else we thought, all our drivers voted the BMW more fun to drive.

For the first time, but not we trust the last, DRIVE went to Vauxhall's superb proving ground at Millbrook to take the BMW's performance figures. The track has the over-riding advantage that you can lap constantly, and safely, at more than 150mph with your right foot pressed into the carpet — and this we intended to try.

Basically, the Millbrook bowl is a steeply banked circle with a circumference of just more than two miles — accurately measured to within 1mm. It is designed so that cars can be driven 'hands off' the wheel at a constant 100mph around the top of the 'wall'. This means that at 150mph the steering effort is the same as if the car was doing a mere 50mph.

With the BMW, we had three ways of measuring speed — some ways being better than others. First there was the speedometer, which proved to be a rather optimistic device — high on the banking it read a steady 150mph in fourth gear. Next we had the computer, which was a little more accurate but still proved to be optimistic at maximum speed.

Thirdly, and finally we had a digital watch with a stopwatch facility capable of measuring to the nearest 1/100th of a second — and that's the way to do it. In ideal conditions we lapped the banking at an easily repeatable 53.1sec time and time again. If our £5 calculator is as good as the £10 watch, that's a speed of 136mph. Snag is that BMW assures us the 635CSi will eat 142 autobahn miles every hour, and it isn't like the Germans to boast, is it?

Fifth is an overdrive gear, so moving up a cog improves fuel consumption rather than top speed. Blasting a Metro aside for the umpteenth time, the stop watch came up with a lap time of 54.1sec — a mere 132mph. Meanwhile, the speedometer was showing 145mph, and fuel consumption on the computer display screen read 11mpg. If that was true it was a very good figure at that sort of speed.

On the open road the 635 is always fast enough, but we'd have to say that supercar rivals — particularly the Jaguar XJS — are frequently a lot faster. Were we scrubbing off speed on the banking? Maybe a little, but not 6mph. Were we simply overloaded? Well, we always test two-up and the weight could have made a slight difference, but these are questions we can't answer. Suffice to say, however, that a 635CSi flat out is still an awesome car to drive, and we were not too disappointed about the missing miles per hour.

DRIVE's usual method of timing Quick Test acceleration runs is to take the times from the car's own speedometer — after calculating the error; in this case we had problems. Some cars can suffer from 'speedo-lag', which is testers' talk for a car that goes faster than the speedometer needle. With the BMW, the problem is just the opposite.

Dump the clutch at 5500rpm and the speedometer needle shoots round to 40mph and hangs there. Before you actually reach 40mph it's time to change into second — otherwise the rev limiter cuts in to ruin your run. Meanwhile, clouds of blue smoke pour off the fat, VR rated Michelin tyres as the car lays down shrieks of expensive rubber on the road. It is a problem we solved in the past by drawing graphs, so we have to admit that our 0-30mph time is a calculated one. Nonetheless, 2.9sec must be very close to the truth.

Reaching 60mph in 7.1sec makes the car slower to this point than a humble Caterham 7 — we proved this point by running the two cars side by side at the proving ground. Beyond 60mph, however, the Caterham is left for dead as the big BMW soars on

QUICK TEST

majestically to, and beyond, 100mph.

In fourth gear the phenomenal torque ensures that the 20mph increments are rarely more than 8sec. Even in fifth the car pulls more than lustily from as little as 30mph.

The handling of BMWs has often been criticised in the past — though not by this writer. The company assumes that purchasers can tell their opposite locks from their elbows, and we have never recommended the cars to anyone who can't catch a gentle tail slide. However, it was certainly true to say that if a driver lifted off the accelerator when halfway round a corner, the back end *would* pass the front as surely as night follows day. And, it has to be admitted that driving ability and a healthy bank balance do not always enjoy a symbiotic relationship . . .

First came the 5-Series suspension modifications, which put an end to the 'lift-off oversteer', and the rest of the BMWs were to follow suit, leaving, to date, only the 316 with this 'fun' characteristic. Now, the 635's handling reveals gentle understeer at first, but if you keep pouring on the power the back end simply has to slide. Frankly, we wouldn't have it any other way — but be warned: this big 'un does not suffer fools gladly.

We know from experience that BMW's power-steering system represents the state of the art: although there are more-precise non-powered systems, the trade-off in this system's quick reaction is a fair exchange. Certainly, on Millbrook's ride-and-handling circuit we discovered that the wheel gave considerable feedback — top marks. But straight-line stability is the 635's forté. Flat out, the car feels like an arrow on wheels.

How comfortable
Frankly, the seats failed to convince our team of nitpicking

testers. Admittedly, they didn't actually promote backache, but for £23,000 one expects something extra-special. We'd certainly say that the 635's seats are comfortable but they don't have the attractions of those in, say, a Saab. We always seem to be left with the feeling that big German seats are designed for big German bottoms. On the plus side the seats can be adjusted every which-way by a system of levers, and just available is the all-singing, all-dancing BMW electrically-operated pew where even the headrests are motor-operated.

Heating and ventilation are, of course, computerised, too Just dial the temperature you want, in degrees Centigrade, and forget about it. Two sliders control where the air goes and a third controls the volume, which can be boosted by a rheostat-controlled fan. To make a good system even better, BMW includes an air conditioner as standard. We know from experience after a summertime run to the South of France that the system is powerful and effective, but on our car the air-conditioning's control button

failed soon after the vehicle arrived. Pity, really, because air conditioning is still useful in December — especially on damp, misty mornings.

Some drivers criticised the car's wind noise — pillarless designs are prone to such problems — but a glance at the speedometer usually puts an end to the complaints. You expect a bit of a whistle at . . . well . . . um, high speed. However, there were no complaints at all about the ride, which is remarkably compliant and comfortable at any speed — rather reminiscent of the XJS, and that's the best there

is. Some surfaces can make themselves felt as well as heard, but it is all rather distant and wholly acceptable to luxury lovers.

Instrumentation is simple but an object lesson to the also-rans. There's a 160mph speedometer and on the right of that is the rev counter — red-lined at 6200rpm and backed up by an electronic cut-out. Set into this dial (5-Series style) is the econometer, which on our car fluctuated happily between 0 and 70mpg in a rather useless fashion. When driving cross country we'd expect to get about 25mpg. Most owners will be delighted with that sort of economy but by carefully nursing the beasty along it's possible to send the needle into the low 30s.

However, saving the best bit until last, the exciting feature that could save your life is undoubtedly the ABS braking system. With this, any fool can execute a perfect emergency stop without ever locking a wheel or skidding a single inch. The system gives you cadence braking electronically: this means that, as a wheel is about to lock and skid, it is released to grip and then instantly braked, and so on. At the moment it is available on only BMW and Mercedes-Benz and as ALB on Honda Prelude models, and we can assure readers that the cost is recouped with the very first accident you don't have. The clever part is that, while you are obtaining maximum braking, the car can be steered in a normal fashion.

Verdict
There are other supercars which are faster than the 635 and there are cars which are more refined, silent and luxurious. There are also cars which come close to matching the BMW's performance for a mere fraction of the price. But few supercars can combine the BMW's practicality with its raw excitement when being hurried, and we suspect that none can surpass the company's meticulous quality control.

Little niceties abound. For instance, the glovebox holds a neat rechargeable torch, and when you clean the windscreen, there is a moment's pause between the water hitting the glass and the wipers moving into action, so avoiding a scratched screen.

Whether the 635 is the best supercar in the world is certainly open to question, and we suspect that speed-freaks would rule it out simply for lack of 'horses'. But we are sure that the 635CSi is one of the best — and that is something to be proud of. If you want a clincher to the argument, then try ABS brakes on a slippery road and ponder. The next time you need them could also be the last.

PERFORMANCE
Maximum speed 138mph in 4th
135mph in 5th
Acceleration from rest (sec)

mph	time	indicated mph
30	2.9	33
40	4.6	43
50	5.7	54
60	7.1	65
70	10.2	76
80	12.9	87
90	15.8	98
100	19.9	109
110	26.5	120

Acceleration in gear (sec)

mph	5th	4th
30-50	10.4	7.9
40-60	11.2	8.0
50-70	11.9	7.9
60-80	14.2	8.1
70-90	13.7	7.5
80-100	15.5	8.9
90-110	—	10.3

Standing-start ¼ mile 15.4sec
Terminal speed 92mph
Max G 0.74 (on damp concrete)

ENGINE
Type and size front-mounted straight six, 92mm bore x 86mm stroke = 3430cc, 7 main bearings with cast iron block and light alloy head
Compression ratio 10.0:1
Valve gear single overhead camshaft, chain driven
Fuel system continuous Bosch Motronic digital electronic injection, fed from 16.3 gallon tank
Max power 218bhp (DIN) at 5200rpm
Max torque 224 lb ft (DIN) at 4000rpm
TRANSMISSION
Clutch 9.45in diameter, hydraulic operation
Gearbox five-speed manual. Ratios: first 3.822:1; second 2.223:1; third 1.398:1; fourth 1.000:1; top 0.813:1
Final drive 3.07:1 to rear wheels
Mph per 1000rpm 28.9mph in top gear
Rpm at 70mph 2422rpm

CHASSIS
Suspension Front: double-joint, spring-strut axle, small kingpin offset, negative caster, ideal turning axis with two guide joints, torsion stabiliser. Rear: independent wheel suspension on semi-trailing arms, control angle 13 degrees, spring struts with auxiliary control arm
Steering ZF ball and nut with 3¼ turns between full locks, power assisted. Turning circles average 34½ft between kerbs with 51½ft in response to one turn of the wheel
Wheels light alloy with 220/55VR390 (Michelin TRX on test car)
Brakes disc brakes on all wheels (ventilated on front wheels) with hydraulic servo and ABS anti-lock system

All figures taken with the kind assistance of Vauxhall/GM's Millbrook Proving Ground

SHORT TAKE

BMW M635CSi

The spirit of the M1 lives on.

• James Bond would feel right at home in a BMW M635CSi. Emergencies in the Balkans would be in easy reach of the motorway-gobbling big coupe. Those Alpine-switchback games of tag with the sinister forces of SPECTRE would be easy sport for the BMW's acceleration and handling, even with a full complement of Q's death-dealing devices in the ample trunk. More critical yet to 007 than mere life and death, the Bavarian coupe has the proper blend of racy lines and understated elegance for whisking luscious double agents away from the baccarat tables of Monte Carlo.

Such images seem incongruent with the BMW coupes we know in America. Our 6-series cars never got their full share of the sporting legacy left by their 2800 and 3.0CS forebears. They had the misfortune to arrive in our market just when most European manufacturers were inclined to neglect performance in the wake of American safety and emissions regulations. Recent examples are much improved, but the early image has been hard to shed. In Europe, however, with such versions as the bespoilered, 218-bhp, 140-mph 635CSi on the scene, the situation is quite different. BMW works hard to maintain a sporting image in the home market, its latest move being the introduction of the M635CSi at last fall's Frankfurt show.

The "M" stands for "Motorsport," the branch of BMW that spawned the legendary M1. This new M-coupe perpetuates the bloodline with an updated version of the mighty six-cylinder M1 engine, complete with a twin-cam, four-valve, pent-roof-combustion-chambered head; a big-bore, short-stroke version of the 3.5-liter block; a tuned intake system with six individual

throttles; and a sensuous bundle of six intertwined exhaust headers. The mechanical fuel injection of the M1's big six has been replaced by a Bosch Motronic system, which controls the spark timing and the electronic injectors with far greater precision. This has allowed the compression ratio to be bumped from 9.0 to 10.5:1. As a result, the new engine develops 286 bhp at 6500 rpm, up by 9 bhp from the original, and 251 pounds-feet of torque at 4000 rpm, a peak 12 pounds-feet higher and 1000 rpm earlier than the original. Fuel economy is about fifteen percent poorer than the two-valve 635CSi's, but most of

Vehicle type: front-engine, rear-wheel-drive, 4-passenger, 2-door sedan
Price as tested (Germany): $31,459 (estimated)
Engine type: 6-in-line, iron block and aluminum head, Bosch Motronic fuel injection
Displacement .... 211 cu in, 3453cc
Power (DIN) .... 286 bhp @ 6500 rpm
Transmission 5-speed
Wheelbase .... 103.3 in
Length .... 187.2 in
Curb weight .... 3300 lbs
Manufacturer's performance ratings:
Zero to 62 mph .... 6.4 sec
Top speed .... 158 mph
Fuel economy, European city cycle .... 14 mpg
    steady 56 mph .... 29 mpg
    steady 75 mph .... 23 mpg

the difference is due to much shorter gearing (3.73 versus 3.07 final-drive ratio).

In addition to the new powertrain, the M-coupe gets low-profile 220/55VR-390 Michelin TRX tires on new modular aluminum wheels, a firmer suspension, and beefier front disc brakes. The only interior alterations are a pair of firm and supportive sport seats and a Motorsport steering wheel. Outside, the M-coupe is distinguished by leading and trailing M-badges and a slightly deeper front spoiler.

BMW claims a top speed of 158 mph and a six-second zero-to-sixty time for the M-coupe—figures we weren't able to verify with proper testing, since we drove the car only in Europe. We can say, though, that the big coupe easily kept up with a 231-bhp Porsche 911 Carrera on the straights at Hockenheim raceway. Another convincing demonstration took place on the Côte d'Azur, where we blew off a Honda 750 motorcycle in the serious local stoplight grand prix.

BMW hasn't forgotten low-speed manners in its quest for performance. The race-derived engine can claw for its 7000-rpm redline one moment, then murmur along contentedly at 1000 rpm in top gear the next. In either case, the engine is refined in sound and feel. One can sense the motor's efforts, but never any strain.

Such broad sophistication is not offered by the M-car's suspension. It's definitely locked into the high-speed mode, becoming smooth and supple only when kilometers are being devoured. Low speeds don't really generate any serious harshness, but neither is the suspension particularly absorbent. On the other hand, there's never any bobbing, weaving, or instability at any speed. The limit handling is unusually forgiving for a BMW, with slight understeer controllably giving way to oversteer as the throttle is lifted.

Sad to say, but BMW will probably never bring this M-coupe to America. Its combination of impeccable breeding, faultless manners, and upper-crust demeanor, concealing a core of sinewy strength and barely controlled energy, has a more narrow appeal in America than in Europe. Americans tend to be more single-purpose in their automotive desires, preferring to leave James Bond cars at the movies.

—*Csaba Csere*

BMW

Big Black Bimmer

Storming out of his Castle at Offa's Dyke for a dawn cross-country dash in a chariot of silken punch... Philip Llewellin cuts loose in the classic BMW 628CSi

I HAVE MET PEOPLE who think Shropshire exists only in the novels of Charles Dickens and P G Wodehouse. Others see the county as a latterday Ultima Thule, far north of Watford in the pagan wilds of Rent-a-Husky land where the natives still daub themselves in woad, and wolves stalk the forests. It may not be the most fashionable base for a freelance writer whose work takes him all over the world, but after living here on the Welsh Marches for most of my 43 years I wouldn't move for all the whisky in Scotland and to hell with China's stock of tea.

But living within a few spear lengths of Offa's Dyke is just a little inconvenient every now and then. Like when you have a meeting some 220 miles away in Essex at 9.30 in the morning. Getting there involves either going through London, which gives you a shrewd idea of what eternity will be like, or slipping off the M1 in the hope of bashing briskly through Beds and hacking hopefully across Herts before finally filtering into FoMoCo territory. Even the latter route demands a *very* early start if, like me, you believe it's better to be half an hour early than half a minute late.

It's not so bad in summer. You leave in daylight, free from worries about black ice and confident that the A5 section of the trip will pose no traffic problems. In fact the scenario looks almost attractive — despite the 4.30 alarm clock — when the brain reminds you that this time there's a BMW parked in the drive. A black 628CSi "koopay", whose superbly efficient four-speed automatic and leave-it-to-me cruise control promise to make the going that much easier than usual. A lot more petrol than the family hack? Of course. But maybe not as much as you might expect. It had already been checked out at 18.4 mpg over 150 miles of stop-start driving punctuated by a couple of brief but brisk charges along quiet roads in the hills.

The better part of 20 years had passed since I first sampled BMW engineering during a test day at Mallory Park. Early memories also include a week with an 1800Ti — priced at just under £1500 in 1967 — and several delectable days with a delightfully compact, wickedly fast 2002. It averaged the thick end of 80 mph from home to Marble Arch early one summer morning . . . in the days when I was crazy and radar was only a little more common than machine guns in medieval warfare.

It may lack the ultimate status of the 218 bhp 635CSi now providing

628CSi

exemplary transport for one of my well-heeled friends, but the smaller-engined version is an acceptable compromise for paupers whose budget can only be stretched to £18,710 instead of £23,995. The black beauty shipped up to Shropshire by BMW also had a few extras costing almost as much as a new Mini City. The ABS braking system adds £939, you pay £754 for the electric sunroof, £499 for the on-board computer, £394 for cruise control and a relatively modest £311 for a limited-slip differential. All of which pushed the price of TRX 899Y up to within sniffing distance of £22,000.

In return for that you get a very classy two-door whose styling can probably be labelled as classical.

BMW tend to mutter into their steins of Bavarian beer when asked about things like drag coefficients, and the 628CSi's rakish nose appears to be more aesthetic than aerodynamic. In fact even at Britain's legal motorway speeds there was more wind noise than expected – enough to make me wonder what levels would be reached during really quick *autobahn* cruising. One admirable feature of the elegant, notchback styling is a very capacious if high-silled boot complete with a superb toolkit that includes six spanners and a set of pliers plus spare bulbs and fuses.

That's typical of a car whose neat, thoughtful little details extend to a small torch kept charged by a socket in the glovebox. Things like that, small points in themselves, underline the impression that the 628CSi has been designed by people who really *think* about what a driver will or may need during anything from a single journey to several years of

The high-efficiency Bavarian 2.8 litre fuel injected engine bay — more power per litre than the bigger 635.

ownership. The first-aid kit with its DIY medical hints is another example of BMW's attitude.

Adrenalin started percolating through the system when I stepped out of the house that morning and started the six-cylinder engine. Beautifully smooth and responsive, a powerful punch wrapped up in a silken glove, it invariably started at the first attempt and was immediately willing to pull strongly and sweetly without a hint of early-morning indigestion.

It actually develops slightly more power per litre than the 635CSi to deliver 184 bhp at 5800 rpm and 174 lb/ft of torque 1600 revs lower down the scale. I had no opportunity to confirm or deny the 127 mph maximum claimed by BMW for the automatic, but clocked 10.5 seconds for the zero-to-60 mph sprint which was a fraction quicker than the manufacturer's figure. That's nearer adequate than spine-tingling, so there's just a hint of sheep in wolf's clothing about the car if you care to take part in the Traffic Lights Grand Prix.

I was less than delighted when BMW called to say that the car would be an automatic. It threatened to be slightly out of step with a magazine whose character is nothing if not sporting. But BMW's four-speed is in the same class

Big Black Bimmer

as that beautifully efficient engine and perfectly matched to its characteristics. The extra ratio plugs the gap and provides the additional step needed for really vivid response at open-road speeds. I can still hear passengers gasping in delight and astonishment as we trickled out of a town at 30 mph and were catapulted to twice that speed as the transmission whipped down a couple of steps as if by magic. It makes overtaking so easy you rarely need to even think about shifting manually to extract as much poke as possible from the engine.

The brakes are also superb with disc on all four wheels. The optional ABS anti-locking system may not be cheap, but you have to ponder the point that your life is almost certainly worth £939. I seemed doomed to try it on nothing more demanding than dry roads. It could not be faulted in such conditions, no matter how hard I pressed the pedal. But the system really proved its worth on the journey to Essex, when rain falling after a long, hot spell made the surface potentially lethal. Opting for ABS is like taking out an insurance policy – the wisdom is not fully appreciated until the Fickle Finger of Fate tries to poke you in the eye.

Independent rear suspension whose behaviour could become anti-social *in extremis* used to be one of BMW's few weak spots, but modifications introduced last year reduced the angle of the semi-trailing arms and solved the problem. About time too, as the less starry-eyed scribes commented. There's just a trace of understeer when you corner fast, but the line is really as near neutral as makes no difference, and coming off the power midway through results in a minor, easy-to-correct twitch rather than a full-blooded *Oh God our Help in Ages Past* loss of stability.

The 205/70VR14 XWX Michelins, mounted on 6J alloy wheels, endowed the BMW with plenty of grip, although I did take extra care during that sudden storm. When surfaces are slippery the automatic is almost too eager and responsive. It can drop an extra gear as you start accelerating from a roundabout or slow, tight corner and instantly tap torque to slide the tail and demand a quick flick of the admirably precise, well-weighted power steering.

Power and agility cut non-motorway roads down to size on the run to Essex and back, but the 628CSi really excelled as an effortless executive express on the M1 and M6. You just set the cruise control and sit back, holding the selected speed up hill and down dale without giving the right leg any longterm punishment. Lower down the scale, it can also be used to keep the Old Bill happy in built-up areas.

Sticking to a constant speed, something that's virtually impossible without a machine to do the work for you, must be good news on the frugality front. That area in general is one where BMW have concentrated a huge amount of research and development in recent years, making up in advanced electronic technology for what may be lacking in terms of aerodynamic efficiency. Visible evidence of their concern is supplied by a fuel consumption indicator whose needle recalls the finger of a fair-minded teacher, wagging in admonition when you floor the throttle but dropping to acknowledge restraint. Under the bonnet, ignition and fuel-injection systems are programmed to give optimum timing and fuel supply in relation to such factors as speed, engine temperature, air temperature, barometric pressure and, for all I know, the state of the tide at Bognor Regis.

Reduced weight is another factor. We hear an awful lot about slippery shapes nowadays, but excess poundage is there all the time. The current 628CSi is 156 lbs lighter than its predecessor – and that's the equivalent of leaving the wife at home.

I have already mentioned a figure of 18.4 mpg for a mixture of about-town driving and fast forays into Wales. The journey to Essex, followed by a battle with East End rush-hour traffic and a brisk cruise homewards, produced no less than 27.7 mpg while another check gave 26.5 over 100 motorway and main-road miles. I had warned the bank manager to expect an overall figure not much better than 20 and was delighted and impressed to average 24.5 mpg. Match that level of economy to a 14.5-gallon tank, plus a 1.8-gallon reserve, and you have a range in the

BMW 628CSi

400-mile class if the pace is kept down to within sight of John Bull's speed limits.

The BMW's interior, like its styling, is smart, functional and eschews any of the pseudo-traditional luxury touches favoured by some up-market manufacturers. Like the rest of the car, the cabin rates top marks for build quality and finish. But I wonder if some of the bigger Bavarians feel obliged to opt for transport with a little more space. I am on just six feet tall, short in leg and long in the body, and had only

Micro-chip computer — carries a 6-page list of instructions.

Timeless styling has made it a classic shape.

just enough headroom even when the driver's seat was adjusted to its lowest setting. The lid is effectively lowered by the sunroof's frame, of course, so if the *wurst* came to the *wurst* you could always forget about fresh air and put that £754 towards the £1328 air-conditioning system. It comes complete with green-tinted insulating glass.

In contrast, back-seat headroom is fair enough for a coupé and a couple of adults can travel in reasonable comfort if the folks up front are prepared to ease forward an inch or two. But I always rate driver comfort as a lot more important than that of any passenger, so it really boils down to a two-plus-one for full sized people on a long journey. What about the average family? My friend's 635CSi copes with a petite wife, two mid-teenagers and a six-year-old when they are not being ferried around by Range Rover!

I have always had my doubts about on-board computers harbouring dark visions of concentrating on pushing the buttons when I should be avoiding an unwanted short-chassis conversion courtesy of the bus that's just pulled out into the third lane. The BMW's is a beauty — so clever and so multi-functional that it was accompanied by nearly six pages of instructions. You can even set it up to make a discreet noise and flash a light when the car is half a mile from a motorway exit or some other landmark. Or so they tell me. All my attempts to get it working properly failed, as did those of my daughter who is bright enough to have eleven 'O' and five 'A' level GCE passes to her credit.

Perhaps that's typical of a Shropshire lad with straw in his hair and cow muck on his boots. Computers? We're only just getting to grips with the abacus.

For £18,710, you get in return a very classy two-door coupé with superb lines, and an instant status symbol.

BMW 635 CSi Automatic

£24,995

Max: 135 mph   0-60 mph: 7.5sec

19.7 mpg overall

FOR:
- Excellent performance
- Smooth and responsive transmission
- Potentially good economy
- High level of refinement

AGAINST:
- Traction below average
- Tricky wet-road behaviour

THIS IS the automatic version of the BMW 635 Coupé that you can buy. In basis it is the same as the specially re-engined 24-valve M635 CSi which is the subject of this week's main Autotest, but with the normal suspension and wheels, and without the deeper air dam in front.

It was something of a standing joke between *Autocar* and the Bavarian car makers' British importers that there was never any difficulty providing manual gearbox BMWs, whilst automatics were always hard to find. It is doubtful whether BMW would admit it publicly, but the ZF automatic transmissions used prior to the current four-speed one did not flatter their cars. The ZF 4HP-22 has changed all that. Even without the driver-selected three-programme added control tested in the 735i recently (*Autocar* 3 March), the "new" ZF box (it was first shown by ZF in 1982) is a standard-setter on this side of the Atlantic, the qualification being important because of the equally impressive GM four-speed, which also has an overdrive top, and the Japanese Aisin Warner type 71 used in Sweden by Volvo.

As a reminder, it is conventionally laid out, being an epicyclic with hydrodynamic torque converter (giving a maximum gearing down of 2-to-1). Its four speeds are made up of three normal ones — a 2.48 first, 1.48 second and a straight-through third — plus a numerically overdrive (0.73 to 1) fourth. There is some over-ride control over the lower gears in the usual way, in that one can prevent the car changing up above first, second or third according to the selector position used, but the box will kick down into lower gears if in 2 or 3, at up to 33 mph from 2 to 1, or 73 mph from 3 to 2 — where the programme selector arrangement allows you to lock the box in 1 or 2 or 3, for delicate slippery road driving, or a long mountain pass descent.

Fourth is entirely automatic. It is only engaged on part throttle, the box changing down to the direct third immediately one increases the accelerator pressure from that needed to maintain cruising speed. When combined with BMW's 3.07 final drive and Michelin 220/55VR-390 TRX tyres, the car is geared at a theoretical maximum in third of 23.5 mph per 1,000 rpm, which from the 135 mph mean maximum speed turns out to be typical of older BMWs — appreciably undergeared, since at 135 it is revving at 5,750 rpm, 550 rpm past the engine's power peak.

As fourth is only used for cruise loads, one cannot measure any meaningful form of acceleration or top speed in top — hence the absence of top-gear figures in our performance table. It is quite high geared, at 32.2 mph per 1,000 rpm, which is higher than the fifth usually found in manual 635s today — hence BMW's claims, understandably based on the steady speed (and therefore light part-throttle cruise in fourth) 56 and 75 mph consumption figures, that for the first time an automatic is more economical than a five-speed manual. To improve matters still further, the box automatically "short-circuits" the torque converter on part throttle at over 55 mph. This eliminates torque converter slip — the main reason for the energy, and therefore fuel, wasted by all conventional automatics.

The transmission works delightfully well, making smooth changes at part throttle and responding promptly to the driver's right foot. It is helped in its smoothness by the integration of its transmission electronic control system with that of the engine, which allows the power delivery to be reduced during any gearchange regardless of the accelerator pedal position, by momentarily retarding the ignition. This gives impressive and distinctively idiosyncratic results, especially on the full-throttle change up from first to second, which is made with a slurred note from the engine as it drops speed suitably. This first-to-second change is made at 51 mph, and second to third at 84.

Add a responsive automatic to a marvellously zestful, mostly very smooth classic six in a not too heavy coupé (3,225lb, or 330lb less than a Jaguar XJ-S 3.6), and you have a wonderfully exhilarating yet refined car. As the figures show, the 635 CSi is quick, very little slower than the manual version with the same power tested earlier (*Autocar* 7 August 1982); the five-speed figures (in brackets) are only slightly better, 0 to 30 mph taking 3.0sec in the automatic (2.7 manual), 0-60 9.7 (9.6), 0-100 20.5 (19.4) and 0-120 37.2 (34.2). There is a 4 mph drop in top speed (135 against 139 mph).

The automatic's claimed advantage in economy over

Below: Automatic 635 nose slightly lifted under its excellent acceleration, Leitz-Datron Correvit speed measuring head on rear flank

Auto TEST UPDATE

BMW 635CSi Automatic

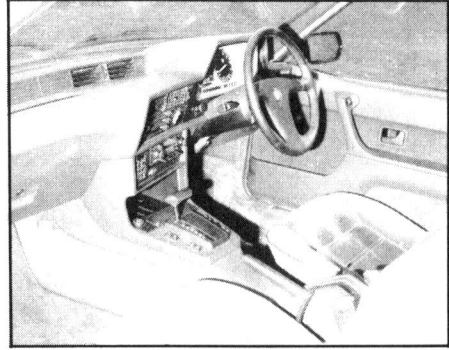

the manual car only shows up if one drives gently, which the 635 does not encourage. It is in any case a highly artificial claim, since it is only so because — inexplicably — BMW choose to make the five-speed manual lower-geared. It is surprising therefore that our overall consumption of 19.7 mpg, whilst respectable for an automatic, is nearly 10 per cent worse than for the manual (21.8). The best interval consumption we saw was 22.7 mpg, after correcting for a three per cent over-reading mileometer. Gentle owners should manage up to around 26 mpg without too much difficulty.

The rest of the car is much as before. Traction without the optional limited slip differential is not good, an enthusiastic departure from a side turning into a fast main road being too easily accompanied by a spectacular cloud of blue smoke from the inside back wheel. The car steers pleasingly, with surprisingly high-effort power assistance allowing tolerable feel. Straight stability is reasonable. The combination of near-neutral steering response and the coupé's slightly lower centre of gravity, the considerably reduced semi-trailing arm pivot angle and the power-reversal-softening effect of the torque converter make the familiar BMW lift-off tail slide in a corner a thing of the past. It is still however not a car one can easily corner fast and tidily, and in the wet it demands all the old care.

Ride is mostly good, although not so impressive when one-up, when the joggly low speed behaviour over many indifferently surfaced British roads is very noticeable, especially by taller drivers (6ft and above) for whom headroom under the sunshine roof frame is very limited. There is also a noticeable crosswise rocking motion, typical of cars with stiff-ish anti-roll bars.

The Bosch anti-lock-aided brakes are a definite blessing when the unexpected happens, and the handbrake, unusually amongst some cars of this class, holds the car easily on a 1-in-3 slope.

Like most machines of this considerable price, the 635 is extremely well equipped, with many features that make life with the car easier. The central locking system can be worked from outside via either the driver's or passenger's doors, and not just the former's. There is an unusually good sunshine roof which does not suffer badly from the organ-pipe resonant flutter customary on so many other opening tops.

Generally, the automatic 635 CSi is very pleasing indeed. It ought to be, given its very high price — you can for once buy a larger-engined Mercedes, the 500SL, for slightly less. Nevertheless, it is around the top of the automatic class in performance, and its present all-round excellence demonstrates how well it pays to wait for a model to mature in development, even when it comes from a manufacturer of good repute. □

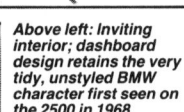

Above left: Inviting interior; dashboard design retains the very tidy, unstyled BMW character first seen on the 2500 in 1968

Above: Standard 635 engine, with impressive injection inlet manifolding. Compare with M635 underbonnet on page 50

Left: Standard air dam is not as deep as M635

PERFORMANCE

MAXIMUM SPEEDS

Gear	mph	kph	rpm
3rd (Mean)	135	217	5,750
(Best)	138	222	5,850
2nd	96	154	6,300
1st	58	93	6,300

ACCELERATION

From Rest

True mph	Time (sec)	Speedo mph
30	3.2	33
40	4.2	45
50	5.6	53
60	7.5	64
70	9.7	74
80	12.3	84
90	15.5	96
100	20.5	107
110	26.9	117
120	37.2	127
130		138

Standing ¼-mile: 15.7sec, 91 mph
Standing km: 28.7sec, 112 mph

In Each Gear

mph	3rd	2nd	1st
0-20	—	—	1.8
10-30	—	—	2.1
20-40	—	—	2.4
30-50	—	—	2.7
40-60	—	4.1	—
50-70	—	4.3	—
60-80	—	4.7	—
70-90	—	5.7	—
80-100	9.2	—	—
90-110	11.4	—	—
100-120	17.6	—	—

FUEL CONSUMPTION

Overall mpg:
19.7 (14.3 litres/100km) 4.33 mpl
Autocar formula: Hard 17.7 mpg
Driving Average 21.7 mpg
and conditions Gentle 25.6 mpg
Grade of fuel: Premium, 4-star (97 RM)
Fuel tank: 15.4 Imp galls (70 litres)
Mileage recorder reads: 3.0 per cent long

OIL CONSUMPTION

(SAE 10W/30) negligible

WEIGHT

Kerb, 28.8cwt/3,225lb/1,463kg
(Distribution F/R, 56.2/43.8)
Test, 32.2cwt/3,610lb/1,637kg
Max payload 794lb/360kg

TEST CONDITIONS

Wind: 8-12 mph
Temperature: 4deg C (39deg F)
Barometer: 29.65in Hg (1,005 mbar)
Humidity: 84 per cent
Surface: dry asphalt and concrete
Test distance: 865 miles

Figures taken at 8,600 miles by our own staff. All Autocar test results are subject to world copyright and may not be reproduced in whole or part without the Editor's written permission.

SPECIFICATION

ENGINE
Longways front, rear-wheel drive. Head/block alloy/cast iron. 6 cylinders in line, bored block, 7 main bearings. Water cooled, viscous fan.
Bore 92.0mm (3.62in), stroke 86.0mm (3.38in.), capacity 3,430 c.c. (210.7 cu. in.).
Valve gear ohc, chain camshaft drive. Compression ratio 10.0 to 1.
Breakerless ignition, Bosch L-Jetronic injection.
Max power 218 bhp (PS-DIN) (160 kW ISO) at 5,200 rpm. Max torque 228 lb.ft. at 4,000 rpm.

TRANSMISSION
Four-speed ZF 4HP-22 automatic. Lock-up torque converter, max converter ratio 2.0.

Gear	Ratio	mph/1,000 rpm
Top	0.73	32.18
3rd	1.0 - 2.0	23.49 - 11.75
2nd	1.478 - 2.956	15.89 - 7.95
1st	2.478 - 4.956	9.48 - 4.74

Final drive: Hypoid bevel, ratio 3.07

SUSPENSION
Front, independent, MacPherson strut, lower wishbone, coil spring, telescopic dampers, anti-roll bar.
Rear, independent, semi-trailing arm, anti-roll bar.

STEERING
ZF ball and nut, hydraulic power assistance. Steering wheel diameter 15.0in., 3.4 turns lock-to-lock.

BRAKES
Dual circuits, split front/rear, ABS anti-lock. Front 11.1in. (281.9mm) dia discs. Rear 11.2in. (284.5mm) dia discs.
Vacuum servo. Handbrake, centre lever acting on rear drums within discs.

WHEELS
Cast aluminium alloy, 6.5in. rims.
Various tyres (Michelin TRX on test car), size 220/55VR-390, pressures F33 R33 psi (normal driving).

DIMENSIONS
Wheelbase 103.4in. (2,626mm); track front 56.0in. (1,441mm), rear 58.5in. (1,486mm). Overall length 187.2in. (4,755mm), width 67.9in. (1,725mm), height 53.7in. (1,365mm). Turning circle 34ft. 3in. (10.4m). Boot capacity 18.7 cu.ft.

WHAT IT COSTS

PRICES

Basic	£20,062.88
Special Car Tax	£1,671.90
VAT	£3,260.22
Total (in GB)	**£25,015.00**
Licence	£90.00
Delivery charge (London)	£160.00
Number plates	£15.00
Total on the Road (exc insurance)	**£25,280.00**
EXTRAS (inc VAT)	
Blaupunkt Toronto SQR stereo radio/cassette player*	£299.50
Air conditioning and tinted glass	£1,374.00
Automatic Speed Control	£259.00
Green Tinted glass	£96.00
Electric front seat adjustment	£560.00
*Fitted to test car	
TOTAL AS TESTED ON THE ROAD	**£25,579.00**

WARRANTY

12 months unlimited mileage.

Auto TEST
BMW M635 CSi

Max: 156 mph　　0-60 mph: 6.1sec
17.0 mpg overall

FOR:
- Thrilling performance
- Excellent steering
- Good stability
- High comfort

AGAINST:
- Care needed in wet
- Thirsty – for a BMW

YOU HAVE ambled through town and village with perfect manners, the car comparatively unobtrusive, the engine typically smooth and entirely flexible in the 23.9 mph per 1,000 rpm fifth, even at 20 mph.

Now it sits, idling quietly, short aluminium alloy gearlever in first, the only uncouthness being the effort needed from your left leg (42lb) to keep the clutch pedal shoved to the floor. You don't floor the accelerator, yet: more than 3,000 rpm, and the wheelspin is dramatically lurid, snakey and wasteful of time. You ease it off at just 3,000, then back off slightly to help those meaty TRXs grip, flattening the pedal finally at around the 4,500 rpm torque peak as the car rockets up the road, the previously docile engine howling in a deliciously exaggerated accent on the usual BMW-six noise, all the way to the 6,900 rpm rev-limiter-backed red line.

Bang the lever into second just short of 40 mph with a squawk from the back tyres, revs still high at 4,100 rpm, the less easy next change into third only 3½ sec away, passing 60 only 6.1sec from the start. Third hurls you on to a shade before 100, at 15.6sec, incredible for such a big, weighty (30cwt) car; fourth through 110 before 20sec has gone, and 130 before 30sec. Snatch fifth, and 140 passes at 41.8sec as the car roars its way on to the 156 mph maximum, the clear road blurring under the broad nose.

Yet when conditions change again, no matter how quickly after flat-out, there it is again — that same coolly relaxed, refined tractability, without temperament. The power and torque peak speeds suggest a peaky delivery, with little low speed pull, and the noticeable step of valve overlap effects as

the power comes in. It *is* an exhilarating engine, with obviously more go from 4,000 rpm onwards, but it performs pretty well from as low as 2,500, and the progression from the good to the spectacular is comparatively subtle.

Although BMW plan to make the M635 available (in left-hand drive only) here eventually, it goes on sale only in Germany now, at 89,500 DM (roughly £23,000). This test was conducted in Bavaria last November, so we had some truly cold starts at below freezing. The engine started as perfectly then as all modern Bosch-injected cars do, but unlike the rest, it did need a little momentary prodding encouragement from the accelerator to keep it running whilst frosty windows were being cleared. This was its only hint of any kind of temperament, and it proved to be no more than a moment of

Continued

TECHNICAL FOCUS

THE 93.4 x 84mm 3,453 c.c. power unit dates back a long time — effectively to 1972, when BMW first showed what was eventually, in 1978, their mid-engined Group 4/5 racing cum road car GT coupé, the M1. The road car used a considerably re-worked 635 six — a stouter crank, longer connecting rods, different pistons, and dry sump lubrication to suit a classic double overhead camshaft inclined four-valves-per cylinder head, responsible for 277 bhp (PS-DIN) at 6,500 rpm and 239 lb.ft. at 5,000 rpm, instead of the normal front-engined 2+2 coupé's 218 bhp at 5,200 and 229 lb.ft. at 4,000.

After the death of the M1, the 635 engine was given a 1.4mm larger bore and 2mm longer stroke (preserving power figures but allegedly improving efficiency), but the M engine, resurrected for the CSi (and its European Touring Car championship racing), preserves the old dimensions whilst giving better output over a usefully wider spread — 286 bhp at the same 6,500 rpm, and similarly more torque, 250 lb.ft., at a lower speed, 4,500 rpm.

The exhaust system is a genuine two-pipe one from manifold ends to tail. The radator is bigger. The type 280/5 gearbox is strengthened, as is the single plate clutch, and has special close ratios allied to a much lower final drive, with partial (25 per cent) limited slip differential. Rear springs are stiffer, as are both anti-roll bars and the gas-filled dampers, and the car is set 10mm (0.4in.) lower. Ventilated front discs are 300 mm dia and 30mm thick instead of 282 and 25mm respectively, and they have four-piston calipers. Wheels are wider, with 220/55VR390 Michelin TRXs as standard equipment (with 240/45s optional, as fitted to the test car).

BMW's M1 coupé, from which the M635's engine came

BMW M635 CSi

hesitation. Otherwise there are no flat spots, nothing but entirely controllable urge, bags of it.

Overall, in fifth, it turns out to be perfectly geared, the maximum speed corresponding to 6,550 rpm, 50 rpm above the power peak. As explained in the noise section, when one is using the performance by cruising above 100, on the odd occasion the comparatively high engine note induces one to try to change up from fifth. The gearchange is mostly good, except that when hurried, it can baulk, refusing second, or when that doesn't happen, showing slight synchromesh weakness. To be fair however, this particular car, at the time only the second M635 built, had been tested by at least three other magazines, so its life had been untypically hard, as the departure of the leather-covered knob from the gearlever further suggested; we mended it successfully with "rapid" Araldite.

Economy

Most likely M635 customers will not be preoccupied with good fuel consumption figures, and if they drive the car as we naturally did during the performance measuring part of the test, they will not be surprised to return figures like 13.9 mpg, our worst. On the other hand, it wasn't difficult to achieve 22 mpg in more normal if still typically

From rear, most obvious recognition point is twin-pipe exhaust, and, an extra fitted here, the exotic looking BBS three-piece composite wheels with 240/45 section tyres

fast driving. The tank holds the standard 635's 15.4 gallons, which is a little on the low side for this sort of thirst if one is using the performance. We noted some post-over-run blue smoke from the exhaust when following the BMW, and oil consumption turned out to be unusually high for today, at 700 miles per litre (400 miles per pint).

Refinement

As a sporting — very sporting — luxury coupé, the M635 seems to our taste to have the right compromise in engine noise, pleasingly muted at low to medium speed, and only becoming anything like loud as one goes faster — but the sound is that particularly lovely one to the red-blooded driver, of a lusty straight six. It is always smooth. Stroke the M along gently, and you really would not suspect that there is anything rouse-able under the bonnet. Foot it flat, and there arises the usual delightful BMW yowl, but much

stronger than before, and accompanied by a remarkable hard mechanical edge to it, suggesting a crowd of little hammers all at work; you hear it too on the over-run. At around 130 mph, the cruise is spoilt by some heterodyning.

The car disappoints in its suppression of wind noise, which on the test car was particularly bad from around the windscreen pillar door seals.

In interesting comparison with the opening roof of the automatic 635 CSi also tested in this issue, which was equipped with a conspicuously large, fixed deflector, the unshielded opening of the M635 sunshine roof, whilst understandably wind-noisy, also does not suffer from the usual very low frequency organ-pipe flutter. Road noise is present, but not as much so as the car's stiffer suspension and tyres might make one expect.

Road behaviour

As usual with the standard 635, the assisted steering is surprisingly heavy by power standards, but not too much so; it fits the character of the special M635 steering wheel, with its thick rim. It is fairly well geared, at 3.2 turns for a minimum turning circle of 34ft 10in. dia (to the left — the test car's lock stops were very asymmetric, with a 37ft 5in. dia right-hand circle), and has all of that typically eager BMW turn-in, without feeling unstable in most circumstances. It is accurate, and mostly very straight-stable whatever the speed, except for what seemed to be a surprising sensitivity to any change of camber in the road, which deflected the car noticeably, in a suggestion of the way of cars of three decades ago.

The combination of the reduced semi-trail angle of the rear suspension plus the slightly lower centre of gravity of the coupé — plus in the case of the M version, the reduced wheel movement of stiffer suspension — gives the most powerful 635 even less of a tendency towards unwanted tail-out slides due to wheel camber change than the standard car, which does seem to be better than before. One can still induce such slides by deliberately clumsy driving, and the M is no car for a lead-footed novice, especially in the wet, when it has obviously to be treated with great care. But its excellently progressive throttle linkage makes it easy to drive carefully, so there is little excuse for stupidity.

Traction with a powerful car is as much a matter of keeping the amount of tyre tread contact reasonably large and constant, which is why cars with low camber change rear suspension are always less prone to wheelspin for the same rate of acceleration, power-to-weight and tyre load than cars with higher camber variation. The BMW cannot be expected to pull all of its 170 bhp per ton laden through even those back tyres without spin, but one cannot help suspecting that it could be better with, say, a de Dion rear end.

The stiffer springing is very obvious at lower speeds, with a distinctly joggly ride that contrasts noticeably with the comfort of the standard car. As there is only just enough headroom for our 6ft testers, the firmness of the ride at all speeds makes such people hit their heads too easily. But given the sort of car it is, the ride deterioration is not unacceptable.

Brakes are fitted with Bosch ABS anti-lock which once again is a damage and embarrassment-saver when the unexpected happens too close. There appears to be a small loss in ultimate braking ability, as usual, so that we could not better 0.95g, which is not as high as one might expect from this tyre equipment. ABS may well help however, especially for any driver who has not got used to the car when an emergency occurs, since the response to pedal effort is rather abrupt, with just 10lb producing a 0.15g slowing, jumping to 0.4g at 20lb, and 0.7 at 30; the curve does however flatten slightly above this, and anyway the anti-lock system is there as the ultimate safeguard beyond. Fade resistance is just good enough, in spite of the pads beginning to catch light at the end of our performance-related fade test.

Sight for sore eyes; one underbonnet, overfull of twin ohc 24-valve 3½-litre. Access to routine items is better than it looks. Engine is inclined both sideways and fore-and-aft

DATABASE

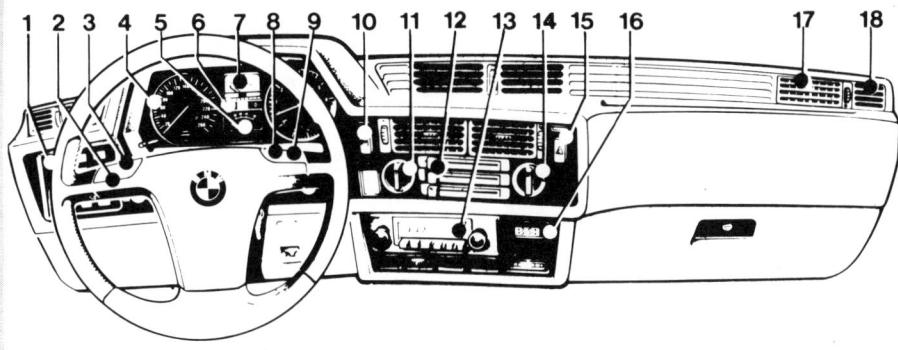

1. Check control panel, **2.** Signalling stalk, **3.** Horn push, **4.** Speedometer, **5.** Fuel gauge, **6.** Service indicator lights, **7.** Temperature gauge, **8.** Revcounter, **9.** Wash/wipe stalk, **10.** Rear window heater, **11.** Temperature selector, **12.** Distribution controls, **13.** Radio, **14.** Fan rheostat, **15.** Hazard switch, **16.** Trip computer, **17.** Face level vent, **18.** Side window demister

SPECIFICATION

ENGINE
Longways front, rear-wheel drive. Head/block al alloy/cast iron. 6 cylinders in line, bored block, 7 main bearings. Water cooled, viscous fan.
Bore 93.4mm (3.677in), stroke 84.0mm (3.307in), capacity 3,453 c.c. (210.72 cu. in.).
Valve gear 2 ohc, 4 valves per cylinder, chain camshaft drive. Compression ratio 10.5 to 1.
Bosch breakerless programmed ignition, Bosch Motronic electronic injection.
Max power 286 bhp (PS-DIN) (210 kW ISO) at 6,500 rpm. Max torque 251 lb. ft. at 4,500 rpm.

TRANSMISSION
Five-speed manual Fichtel and Sachs diaphragm spring clutch 9.45in. dia.

Gear	Ratio	mph/1,000 rpm
Top	0.81	23.87
4th	1.00	19.33
3rd	1.35	14.32
2nd	2.08	9.30
1st	3.51	5.51

Final drive hypoid bevel, 25 per cent limited slip, ratio 3.73.

SUSPENSION
Front, independent, MacPherson strut, lower wishbone, coil spring, telescopic dampers, anti-roll bar.
Rear, independent, semi-trailing arm, coil spring, telescopic dampers, anti-roll bar.

STEERING
Recirculating ball, hydraulic power assistance. Steering wheel diameter 15in. 3.2 turns lock to lock.

BRAKES
Dual circuits, diagonally split. Front 11.8in. (300mm) dia discs. Rear 11.18in. (284mm) dia discs. Vacuum servo. Handbrake, centre lever acting on drums within rear discs.

WHEELS
BBS aluminium alloy three-piece (extra), 210mm rims. Michelin TRX radial tubeless tyres (optional 240/45VR-415 TRX on test car), size 220-55VR-390, pressures F29 R32 psi (normal driving).

EQUIPMENT
Battery 12V, 90Ah. Alternator 80A. Headlamps 220/220W. Reversing lamp standard. 30 electric fuses. 2-speed, plus intermittent screen wipers. Electric screen washer. Water valve interior heater; air conditioning extra. Cloth seats, velour headlining. Carpet floor covering. Screw pillar jack; jacking points 2 each side. Laminated windscreen.

DIMENSIONS

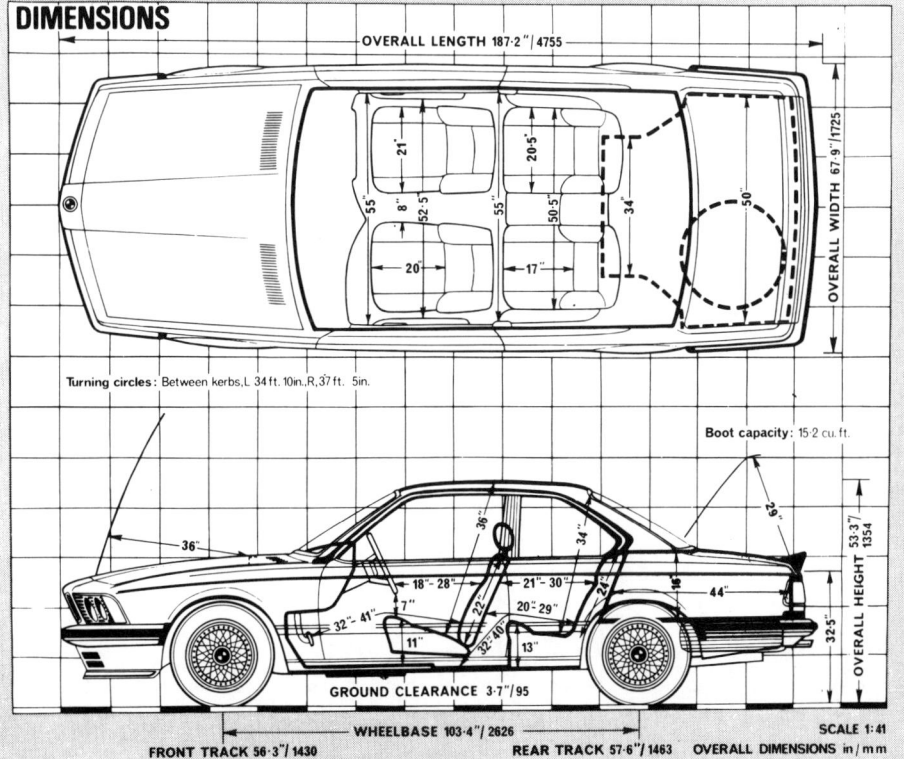

Turning circles: Between kerbs, L 34 ft. 10in., R 37 ft. 5in.
Boot capacity: 15.2 cu. ft.
SCALE 1:41
FRONT TRACK 56.3"/1430 WHEELBASE 103.4"/2626 REAR TRACK 57.6"/1463 OVERALL DIMENSIONS in/mm

PERFORMANCE

MAXIMUM SPEEDS

Gear	mph	kph	rpm
Top (Mean)	156	251	6,550
(Best)	158	254	6,600
4th	133	214	6,900
3rd	99	159	6,900
2nd	64	103	6,900
1st	38	61	6,900

ACCELERATION

FROM REST

True mph	Time (sec)	Speedo mph
30	2.4	37
40	3.7	49
50	4.8	59
60	6.1	68
70	8.6	78
80	10.5	88
90	12.7	98
100	15.6	109
110	19.1	118
120	23.4	129
130	29.7	139
140	41.8	150
150	–	161

Standing ¼-mile: 14.7 sec, 97 mph
Standing km: 26.5 sec, 126 mph

IN EACH GEAR

mph	Top	4th	3rd	2nd
10-30	–	8.0	5.4	3.3
20-40	9.6	7.1	4.8	2.9
30-50	9.6	6.6	4.5	2.7
40-60	8.9	6.4	4.2	2.6
50-70	8.8	6.4	3.8	–
60-80	9.7	6.2	3.8	–
70-90	10.1	6.0	4.3	–
80-100	10.1	6.1	–	–
90-110	10.1	6.7	–	–
100-120	11.4	7.8	–	–
110-130	14.1	10.5	–	–
120-140	20.0	–	–	–

FUEL CONSUMPTION

Overall mpg: 17.0 (16.6 litres/100km) 3.74 mpl

Autocar constant speed fuel consumption measuring equipment incompatible with fuel injection.

Autocar formula: Hard 15.3 mpg
Driving Average 18.7 mpg
and conditions Gentle 22.1 mpg

Grade of fuel: Premium, 4-star (97 RM)
Fuel tank: 15.4 Imp galls (70 litres)
Mileage recorder reads: 0.5 per cent short

OIL CONSUMPTION

(SAE 10W/30) 700 miles/litre

BRAKING

Fade (from 97 mph in neutral)
Pedal load for 0.5g stops in lb

	start/end		start/end
1	20/20	6	34-44
2	20-28	7	34-42
3	20-30	8	40-32
4	24-36	9	40-36
5	32-40	10	40-38

Response (from 30 mph in neutral)

Load	g	Distance
10lb	0.15	201ft
20lb	0.40	75ft
30lb	0.70	43ft
40lb	0.90	33ft
50lb	0.95	31.7ft
Handbrake	0.20	151ft

Max. gradient: 1 in 5

CLUTCH Pedal 42lb; Travel 6.1in

WEIGHT

Kerb, 29.7cwt/3,322lb/1,507kg (Distribution F/R, 56.4/43.6)
Test, 33.4cwt/3,742lb/1,697kg
Max payload 772lb/350kg

TEST CONDITIONS
Wind: 3-8 mph
Temperature: 11deg C (51deg F)
Barometer: 29.82in Hg (1,011 mbar)
Humidity: 97 per cent
Surface: dry asphalt and concrete
Test distance: 705 miles

Figures taken at 11,800 miles by our own staff on the Continent. All Autocar test results are subject to world copyright and may not be reproduced in whole or part without the Editor's written permission.

BMW M635 CSi

Behind the wheel

Anyone familiar with the normal 635 will feel at home straight away, since the interior differences are small. Most obvious is the thicker-rimmed steering wheel. Steering column rake adjustment on the test car didn't work, so for taller drivers the wheel was rather low, and blocked the view of the top of the speedometer and rev counter. The test car was fitted with superb Recaro seats, with plenty of firm side support, the ability to tilt the whole seat as required — although not raise it — plus the unusual feature of thigh support support adjustable for length. Pedals are perfectly arranged for heel and toe, which the delightful response and voice of the engine makes a particular pleasure. Driving position is typical German — high and commanding.

Rearward movement of the driver's seat is adequate for up to 6ft types, but as mentioned already headroom is a little tight, at any rate when the extra equipment Recaro seats are fitted.

By the standards of high-performance coupés, the view out isn't bad, although the thickness of the windscreen pillars is something to beware of, and the car lacks the wonderful all-round vision of the very slender-pillared old CSi of the last generation.

The heater is a sophisticated water-valve-controlled type, which means that the response to any movement of the temperature control is not rapid, as it is

Above: Seats complement the car, fitting and holding well

Right: Rear accommodation (right) is "tailored occasional"

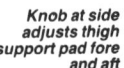

Knob at side adjusts thigh support pad fore and aft

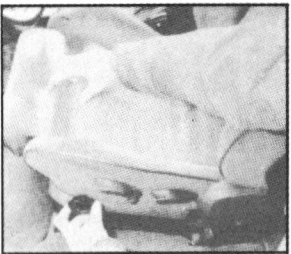

with an air-blender, but thanks to thermostatic control, one can achieve the desired temperature. It is good to find that this German car does not have warm air bled into its face level ventilation when the heater temperature is adjusted towards warm.

Living with the M635 CSi

The interior is very well planned, for two at any rate; space in the back is strictly for the occasional occupant who doesn't mind sitting hunched up for a short while. There is the usual very ingenious BMW service indicator system, whose lamps tell the driver when service is due, based not just on time or mileage, but on how he or she has used the car.

Electronic entertainment equipment fitted to the test car included a magnificent (and expensive) Blaupunkt Bavaria radio-cassette player, and BMW's equally extravagantly priced trip computer (£516 in this country, or about the same as a desk-top personal computer with some of its peripheral equipment).

Oddment accommodation is good. Electric windows are fitted, and they wind up or down in 3½ sec, which is reasonably rapid.

Open the bonnet, and there is a treat. The engine is the rare sort that owners will delight in showing off — it looks very handsome, almost as good as the way it propels the car. Detail design is tidy and neat, so that finding things is not difficult. Particular points to note are the multi-pipe exhaust, and the long (11in.) inlet tracts. The dipstick is easy to find. Anyone unfamiliar with BMWs must note the way in which the bonnet is shut, not by slamming, but by leaving it resting in the shut position, then locking it from the cockpit by moving the lock lever under the dash.

WHAT IT COSTS

PRICES

Not on sale in GB — see text.
Price in West Germany 89,500DM.

SERVICES & PARTS

Service and parts information and prices not available.

PRODUCED BY
Bayerische Motoren Werke AG
8000 Munich 40
Petuelring 130
West Germany

SPECIFICATIONS FROM AUTOCAR BUYERS' GUIDE

	£ Price	Insurance Group Company/Lloyds	Doors	Seats	Engine Capacity (c.c.)	Driven Wheels	Engine Position and Type	Bore and Stroke (mm)	BHP at RPM	Torque (lb. ft.) at RPM	Valve Position	Carburettors or Fuel Injection	Gearbox Speeds 4	5	A	Tyre size (in.) and Type
BMW M635 CSi	—	-/-	2	4	3,453	R	F6IL	93.4x84.0	286/6,500	251/4,500	2 Ohc	i	—	●		220/55VR390
635 CSi	24,995	9/OA	2	4	3,430	R	F6IL	86.0x92.0	218/5,200	228/4,000	Ohc	i	—	●	●	205/70VR-14
Aston Martin V8	42,498	9/OA	2	5	5,340	R	FV8	100.0x85.0	—	—	2 Ohc	4	—	●	●	225/70VR-15
V8 Vantage	47,499	9/OA	2	5	5,340	R	FV8	100.0x85.0	—	—	2 Ohc	4	—	●	●	275/55VR-15
Audi Quattro	20,402	9/OA	2	4	2,144	4	F5IL	79.5x86.4	200/5,500	210/3,500	Ohc	iT	—	●		215/50VR-15
Ferrari 308 GTB Qv	26,898	9/OA	2	2	2,927	R	MV8Tr	81.0x71.0	236/7,000	191/5,000	2 Ohc	i	—	●		220/55VR-390
Mondial Qv	30,710	9/OA	2	2+2	2,927	R	MV8Tr	81.0x71.0	240/7,000	192/5,000	2 Ohc	i	—	●		240/55VR-390
BB 512i	48,749	9/OA	2	2	4,942	R	M12H0	82.0x78.0	340/6,000	333/4,200	2 Ohc	i	—	●		240/55VR415
Jaguar XJ-S HE	21,752	9/OA	2	4	5,343	R	FV12	90.0x70.0	299/5,500	318/3,000	Ohc	i	—		●	215/70VR-15
Lamborghini Jalpa	28,450	8/OA	2	2	3,485	R	MV8	86.0x75.0	255/7,500	235/3,250	2 Ohc	4	—	●		205/55-225/50VR
Countach LP500S	56,600	8/OA	2	2	4,754	R	MV12	85.5x69.0	375/7,500	302/4,500	2 Ohc	6	—	●		205/50-345/35VR
Lotus Turbo Esprit	19,980	9/OA	2	2	2,174	R	M4IL	95.3x76.2	210/6,500	200/4,000	2 Ohc	iT	—	●		195/60-235/60VR
Mercedes-Benz 500SL	26,090	9/OA	2	2	4,973	R	FV8	96.5x85.0	231/4,750	298/3,200	Ohc	i	—	—	●	205VR-14
Porsche 911 Turbo	33,878	9/OA	2	2+2	3,299	R	R6HO*	97.0x74.4	300/5,500	318/4,000	Ohc	iT	●	—		205/55-225/50VR
928S Series 2	30,679	9/OA	2	2+2	4,664	R	FV8	97.0x78.9	310/5,900	295/4,500	Ohc	i	—	●	●	225/50VR-16

ABBREVIATIONS ● = Standard or optional at no cost. ○ = Optional at extra cost. **Engine:** F = Front, M = Mid, R = Rear, Ro = Rotary, IL = In Line, Tr = Transverse, V = Vee, HO = Horizontally opposed, * = Air cooled. **Valve Position:** O = Overhead camshafts per cylinder bank.

COMPARISONS

THE RANGE of cars available to the man who is interested in this sort is limited. If one restricts oneself to the over 150 mph rarities, then the variety is wide in one sense — price and type — and narrow in another — the small number of cars that have such a top speed capability. In our overall list, we have broadly speaking concentrated on machines which are more of the BMW ilk, which means the bigger, more comfortable, luxurious Grand Tourers which have the performance, the size and the space. For this reason, Porsche are represented only by the 928S, and we have omitted the 911 Turbo, practical and immensely fast and efficient as it is (162 mph; 5.1sec to 60; 16.4 mpg overall). Lotus are borderline cases, their 2+2 cars being not quite quick enough, so that the Turbo Esprit — which is highly competitive in acceleration if not absolute top speed — is their only representative, although it does not compare in space. Ferrari need to be considered. Their 3-litre cars have not come our way for proper Autotest since the adoption of the four-valve combustion chambers, nor has the injection Boxer; the 400i is not strictly a coupé.

Concerning the cars that are listed as possible rivals, the 928S is the most obvious, from the BMW's native country at any rate. Its bigger engine and greater horsepower for similar weight just give it the advantage in speed, and it is a fascinating car. It has ultimate handling, at a very high limit, that is a little tricky, since if it does break away at the rear, its high polar moment of inertia can make it rather a handful to catch. Its power steering is even less light than the BMW's.

The Lotus is of course a very different device, remarkably rapid, and blessed with immensely good manners and a very good standard of ride for its type. Its obvious limitation, especially in this company, is its space and convenience. Audi's Quattro offers wonderful stability, and of course traction that is barely rivalled only by the rear-engined Porsches. It is not in quite the same performance league as the others, and its ultimate handling demands respect. The big Aston Martin is a larger-engined machine from a similar mould to the BMW; only tested in automatic form (although there is a manual which should match the M635 more or less), the "ordinary" Aston is beautifully mannered, with no handling quirks in spite of its size and weight — and there is always the Vantage if arguments are really to be settled. Jaguar's XJ-S in original 5.3-litre V12 form only comes, most regrettably, in automatic form; it is still an extraordinary blend of power, handling, ride, refinement and a competitive price.

Verdict

None is perfect, of course — but there are degrees of imperfection acceptability. The Lotus disqualifies itself only because of its size, which is not comparable with the BMW. We are very fond of the Quattro, whose stability, ability to put down all of its remarkable engine performance on to the road, and practical body make it a very amiable car. It is however just a turbocharged, relatively small engine, whose five-cylinder layout makes it less than refined. The big Aston has many of the BMW's attributes; it is a similar type of car, if bigger, and its manners are excellent. But efficiency is not one of its virtues.

Although the automatic transmission of the Jaguar may seem to put it out of contention, the price, performance and behaviour of the V12 XJ-S insist that it must be considered against the ultimate front-engined BMW. It is not quite as accelerative, although in between 70 and 110 it matches the manual BMW very closely, and its fuel economy is also very similar. Its handling up to its limit is more reliable than the BMW's, but the German car probably has marginally better grip in optimum conditions, thanks to its wider tyres. The BMW also wins in the feel of its steering, which does not take experience to appreciate as the Jaguar's does. For the performance-minded driver, it will certainly be the M635 CSi that scores highest of its type, with that wonderful power unit giving it such a glorious character, without the ultimate refinement of the XJ-S. We would be split between the two if it came to a decision, and would ideally prefer to take both, to suit differing moods.

TESTERS' SHORT LIST

BMW M635 CSi *see text* — OUR CHOICE
156 mph; 0-60 mph 6.1sec; 17.0 mpg

Jaguar XJ-S HE (A) £21,752 — OUR CHOICE
153 mph; 0-60 mph 6.5sec; 16.0 mpg

Aston Martin V8 £42,498
146 mph; 0-60 mph 6.6sec; 13.3 mpg

Lotus Turbo Esprit £19,980
148 mph; 0-60 mph 6.1sec; 18.0 mpg

Audi Quattro £20,402
Model not tested

Porsche 928 S Series 2 £30,679
158 mph; 0-60 mph 6.2sec; 16.6 mpg

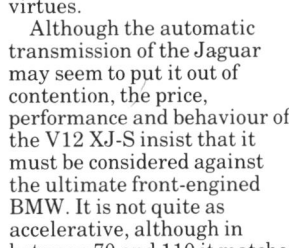

AUTOTEST PERFORMANCE DATA

Brakes	Power Steering	Fuel Tank (galls)	Wheelbase in.	Overall Length in.	Width in.	Height in.	Boot Luggage Capacity cu.ft.	Maker's Unladen kerb Weight lb	Maximum Payload (luggage + passengers) lb	Main service ,000 miles		Legroom front/rear (fully back) in.	Maximum speed mph	0-60 mph sec	Overall fuel consumption mpg	When tested
DS†	●	15.4	103.4	187.2	67.9	53.3	15.2	3,307	772	V	**BMW M635 CSi**	41/32	156	6.1	17.0	4/84
DP†	●	15.4	103.5	187.2	67.9	53.7	18.7	3,153	926	V	635 CSi	41/32	135	7.5	19.7	4/84 (Auto)
DS		23.0	102.8	183.7	72.0	52.2	8.6	3,864+	1,008	5	**Aston Martin V8**	43/31	146	6.6	13.3	7/82 (Auto)
DP†		23.0	102.8	183.7	72.0	52.2	8.6	3,892	1,008	5	V8 Vantage	43/36	170	5.4	13.5	4/77
DS		19.7	99.4	173.4	67.8	52.9	13.8	2,866	1,014	10	**Audi Quattro**	—	—	—	—	—
DS	—	16.3	92.1	166.5	67.7	44.1	8.6	2,835	—	6	**Ferrari 308 GTB Qv**	—	—	—	—	—
DS		19.1	104.3	180.3	70.5	49.2	10.6	3,152	717	6	Mondial Qv	—	—	—	—	—
DS		26.4	98.4	173.2	72.0	44.1	4.9	3,307	386	6	BB 512i	—	—	—	—	—
DS	●	20.0	102.0	186.8	70.6	50.0	15.0	3,900	772	15	**Jaguar XJ-S HE**	42/30	153	6.5	16.0	4/82
DS		17.5	97.0	168.0	72.0	45.9	8.0	2,900	—	6	Lamborghini Jalpa	—	—	—	—	—
DS		26.4	96.5	163.0	78.7	42.1	8.5	3,263	—	6	Countach LP500S	42/-	164	5.6	14.6	10/82
DS		18.7	96.0	165.0	73.0	44.0	5.5	2,690	466	5	**Lotus Turbo Esprit**	40.5/-	148	6.1	18.0	5/81
DS*	●	18.7	96.7	172.8	71.5	50.8	9.2	3,395	926	12	**Mercedes-Benz 500SL**	—	—	—	—	—
DS		17.6	89.4	168.9	69.9	51.6	7.0	2,886	838	12	**Porsche 911 Turbo**	43/28	162	5.1	16.4	4/83
DS†	●	18.9	98.4	175.7	72.3	50.5	7.0	3,197	926	12	**928S Series 2**	—	158	6.2	16.6	Autocar figures (5-spd)

Carburettors: Figure = Number of carburettors, i = Fuel injection, S = Supercharged, T = Turbocharged. **Gearbox:** * = Overdrive standard, A = Automatic transmission.
Brakes: † = Anti-lock standard, * = Anti-lock optional at extra cost.

Brief Test

BMW's 6-Series coupé has been steadily developed during its eight-year life. We discover whether the 628 CSi has the dynamic qualities to match its elegant looks

BRITAIN REMAINS an oddly receptive market for both expensive and high-performance cars, and last year some 800 fortunate people took delivery of £19,000 plus BMW 6-Series coupés. The styling of this elegant two-plus-two has changed little since its launch in 1976, but the Bavarian company have continually updated the engine, suspension and equipment specification.

The 3.3-litre 633 CSi has come and gone, while in 1980 the 635 model received a stablemate in the form of a 2.8-litre version powered by the straight "six" from the 528i saloon. Although the 628 CSi has proved popular in Britain — accounting for one-third of 6-Series sales last year — it was really intended for the Continent, where it fits neatly below 2.8 or 3.0-litre tax breaks.

In late 1982, both models received considerable modifications under the skin, though the 218 bhp 635 CSi benefited from the lion's share of engine improvements. The £25,000 top-of-the-range model gained a fully integrated engine management system, while the 628 simply went to a later generation of the Bosch L-Jetronic fuel injection; the ignition is only transistorised. Power output remained unchanged at 184 bhp at 5,800 rpm, but the torque dropped fractionally from 177 to 174 lb ft at 4,200. The sweet revving in-line "six" retained its 2,788 cc capacity 86/80 mm bore/stroke, chain driven single overhead camshaft, and aluminium cylinder head mounted on a cast iron block.

In early 1983 BMW offered the 628 with a four-speed automatic for the first time, but it was left to the 635 to boast an advanced switchable automatic with economy and performance modes. The 628 we test here is the normal manual five-speed transmission with a direct fourth gear and 0.813 overdrive fifth. The close ratio five-speed gearbox is only available on the 635.

We liked the secure high speed handling of the car when we tested it previously, but that has not stopped BMW taking a fresh look at the suspension. The familiar MacPherson struts at the front now have a double-jointed lower wishbone to reduce steering kickback and prevent the steering pulling to one side in tricky conditions. The steering remains a power-assisted ZF worm and nut system. At the rear, the trail angle of the semi-trailing arms has been reduced to 13 degrees to minimise the camber changes which can cause unpredictable oversteer and a small link between the rear cross member and trailing arm effectively stiffens the suspension in roll and also helps limit camber change.

BMW 628 CSi

There is an anti-roll bar at the front. The standard-fitment tyres are Michelin XWX 205/70 VR14, with the chunky Michelin TRXs reserved for the 635 and the options list for the 628 (£601).

Inside the car those familiar with the earlier model will notice the driver is now treated to a service interval indicator, fuel consumption indicator, comprehensive dashboard computer, and active check for faults such as headlamp failure.

Chief rival for the BMW is undoubtedly the recently launched Jaguar XJS 360 which is marginally more expensive at £20,756 but offers more bhp than the £25,000 635 CSi. Also, it has air conditioning as standard — something which adds £1,374 to the £19,275 basic price of the 628. The cost is raised further with options like the worthy ABS anti-lock braking (£972), on-board computer (£516) and sun-roof (£626) which pushed the price of our test car to £22,763. The problem is that most rivals — with the exception of the 911 Porsche — are cheaper and accelerate faster. The under-rated Opel Monza has the edge on performance when lined up against the BMW, yet is handsomely cheaper at £13,801.

On the Millbrook banked circuit the 628 CSi found its 129.6 mph top speed (fourth gear) easily in hot conditions. This speed is identical to that achieved by the Opel, but surpasses only the Lotus Excel (126.6 mph) as the Audi Quattro and Jaguar will manage in excess of 135 mph while, of course, the Porsche is in another class altogether. In fifth, the BMW achieved 125.8 pulling 4,700 rpm.

If the 628's top speed can be judged up to the class average then its acceleration cannot. It proved mildly disappointing. Its 0-60 mph time of 8.5 sec would be beaten by over half a second by a Sierra 2.8 XR4i, while of its chosen rivals only the Opel fails to break the 8.0 sec mark (8.4 sec). Turbo power takes the Audi to 60 mph in 6.5 sec, while even the big Jaguar manages a creditable 7.2 sec time. The BMW trails even further behind when accelerating in the gears. Its 30-50 mph time in fourth gear of 10.7 sec is a full two seconds slower than the Opel, the Jaguar manages an excellent 6.9 sec for the increment. The long fifth gear on the car is fine for high speed cruising, but penalises the performance badly. The BMW completes the 50-70 mph increment in 15.7 sec, rather slower than the Opel (13.2 sec) and Jaguar (11.5 sec) while the Quattro and Porsche almost halve the BMW's time.

On the road these disappointing figures are largely borne out. The tall gear is not particularly suited to the super-smooth engine, which delivers most of its punch above 4,000 rpm. It is no hardship to rev the BMW engine to this speed, as it never becomes rough or shows signs of losing its superb refinement. But, unless the gearbox is used energetically the performance is disappointing. Jump out of the 628 into the Jaguar and the German car's disadvantage seems marked, lacking the low-speed torque of the Coventry 24-valver.

For those customers not particularly interested in tyre-burning performance the 628 could well be an ideal form of transport, as it offers excellent fuel economy. We managed an overall consumption of 25.0 mpg — excellent, considering that its rivals have trouble bettering 20 mpg. The touring consumptions are much closer together, though.

Bearing in mind the performance and torque the gearbox has to handle its light feel and easy action is very good and contributes much to the enjoyment of driving the BMW. Its slight rubberiness is a small price to pay for its slick action.

The relatively dated tyres have no ill effects on the car's handling, which is invariably neutral, offering the driver excellent feel and precision in fast sweeping bends. In slow corners there is certainly enough power to send the tail out even in the dry, and on wet roads the traction away from junctions is inferior to most of its rivals. For the enthusiast, in particular, the steering deserves much praise, offering firm weighting at all speeds with the level of feel and information from the road surface few power assisted set-ups achieve.

The brakes are ventilated discs at the front, solid at the rear, with the addition on our test car of ABS. They are powerful and fade free, with better pedal feel and progression than we have come to expect of either ABS or the normal BMW servoed system which often suffers from sticking-on of the servo action.

Between the comfort of the Jaguar's ride and the contrasting firmness of the latest Quattro, the BMW is an excellent compromise. It offers a good balance between firm control and good damping for secure high-speed driving, with enough comfort to make the 628 a pleasant long distance touring car. The ride found favour with all our testers.

As expected, the BMW struggles to match the impressive quiet experienced in the Jaguar. Not that it could be classed as noisy, but the pleasant engine note remains in the background while wind noise builds up past 80 mph due, mainly, to the sealing around the frameless doors.

There is no shortage of legroom for drivers and the fore-and-aft adjustment on the steering wheel makes it easier to find a comfortable driving position, yet it is difficult to make it perfect because the sunroof badly restricts headroom. This leaves the driver wishing for a slightly higher seat, to get a better view over the large steering wheel. The seat height adjuster is not terribly sophisticated, as it only adjusts the angle of the squab not its actual height. The seats are more comfortable than is perhaps expected of the traditionally overly-firm German seating and the excellent support they offer is rather more subtle but hardly less effective than Recaros.

The BMW's instrumentation is among the clearest on offer in any car, by virtue of the generous size of both the speedo and rev counter. These are supplemented by only two gauges — fuel and water temperature — while we judged the fuel economy display unnecessary in this class of car.

The ventilation is good, aided by a useful number of facia vents. We found the simple rotating knob for heating temperature selection good too. Not so good is the fact that, unlike the Jaguar, the air conditioning is not integrated into the normal set-up and is brought into action by a separate switch. Neither can you simply select an ambient temperature for the cabin and let the air conditioning get on with the job.

If you are buying a coupé, then accommodation is unlikely to be your prime concern. Front legroom is on a par with the Quattro and only slightly down on the Jaguar, but the BMW and Audi score a big advantage on rear accommodation. The Opel's packaging is worse.

To its credit, the BMW also has the biggest boot by some margin (12.5 cu ft) with the Audi having the smallest at 7.6 cu ft, measured with our Samsonite luggage.

The BMW complements the interior space with tasteful decor. The absence of polished walnut and chrome couldn't provide a bigger contrast with the XJS, yet its appeal comes from the considerable attention to function — matched to subdued trim colours.

As we have already mentioned, our test car had many additional equipment items — most of them standard on the 137 mph BMW 635. On the 628, the alloy wheels are standard, but a glance down the options list reminds us that items like the headlamp wash wipe *should* be a normal fitting on a £19,000 car.

What, of course, the buyer does get for his money is a fine-looking vehicle, superbly built and finished. From behind the wheel, the BMW offers a highly appealing blend of refinement and sporting feel to provide a high-class long-distance cruiser. But the flaw is the performance which, though derived from a lovely smooth engine, is a disappointment when similarly-priced rivals are taken into consideration. What it does have to offer is excellent economy — and there lies the choice for the purchaser with £19,000 to spend on a desirable car like the 628 CSi.

MOTOR ROAD TEST BMW 628 CSi

PERFORMANCE

WEATHER CONDITIONS
- Wind: 7-22 mph
- Temperature: 77 deg F/25 deg C
- Barometer: 28.5 in Hg/966 mbar
- Surface: Dry tarmacadam

MAXIMUM SPEEDS

	mph	kph
Banked Circuit (4th gear)	129.6	208.5
Best ¼ mile	130.5	210.0
Max in 5th gear	125.8	202.4
Terminal speeds:		
at ¼ mile	84.1	135.3
at kilometre	107.9	173.6
Speeds in gears at 6,500 rpm:		
1st	36.3	58.4
2nd	63.0	101.4
3rd	99.2	159.6

ACCELERATION FROM REST

mph	sec	kph	sec
0-30	2.8	0-40	2.3
0-40	4.5	0-60	4.3
0-50	6.4	0-80	6.4
0-60	8.5	0-100	9.4
0-70	11.6	0-120	13.2
0-80	15.0	0-140	17.7
0-90	19.1	0-160	24.8
0-100	25.3		
0-110	31.9		
Stand'g ¼	16.7	Stand'g km	30.5

ACCELERATION IN TOP

mph	sec	kph	sec
20-40	14.5	40-60	8.9
30-50	15.3	60-80	9.4
40-60	15.5	80-100	9.5
50-70	15.7	100-120	10.4
60-80	17.2		

ACCELERATION IN 4TH

mph	sec	kph	sec
20-40	10.6	40-60	6.7
30-50	10.7	60-80	6.6
40-60	10.4	80-100	6.2
50-70	10.2	100-120	6.7
60-80	10.6	120-140	6.9
70-90	10.6	140-160	8.1

FUEL CONSUMPTION
- Overall: 25.0 mpg / 11.3 litres/100 km
- Govt tests: 19.9 mpg (urban) / 41.5 mpg (56 mph) / 32.1 mpg (75 mph)
- Fuel grade: 97 octane / 4 star rating
- Tank capacity: 16.3 galls / 74.1 litres
- Max range*: 460 miles / 740 km
- Test distance: 760 miles / 1,223 km

*At estimated 28.4 mpg touring consumption

NOISE

	dBA	Motor rating*
30 mph	64	10
50 mph	64	10
70 mph	71	17
Maximum†	78	28

*A rating where 1 = 30 dBA and 100 = 96 dBA, and where double the number means double the loudness
†Peak noise level under full-throttle acceleration in 2nd

SPEEDOMETER (mph)

Speedo	30	40	50	60	70	80	90
True mph	28	37	46	55	67	74	86

Distance recorder: 1.7 per cent fast

WEIGHT

	cwt	kg
Unladen weight*	28.3	1,438
Weight as tested	32.9	1,671

*with fuel for approx 50 miles

Performance tests carried out by *Motor's* staff at the Motor Industry Research Association proving ground, Lindley.

Test Data: World Copyright reserved. No reproduction in whole or part without written permission.

GENERAL SPECIFICATION

ENGINE
- Cylinders: 6 in-line
- Capacity: 2,788 cc
- Bore/stroke: 86.0/80.0 mm
- Cooling: Water
- Block: Cast iron
- Head: Aluminium alloy
- Valves: Single OHC
- Cam drive: Chain
- Compression: 9.3:1
- Fuel system: Bosch L-Jetronic fuel injection
- Ignition: Breakerless electronic
- Bearings: 7 main
- Max power: 184 bhp (DIN) (135 KW) at 5,800 rpm
- Max torque: 174 lb ft (DIN) (240 Nm) at 4,200 rpm

TRANSMISSION
- Type: 5-speed manual
- Clutch dia: 9.4 in
- Actuation: Hydraulic
- Internal ratios and mph/1,000 rpm
- Top: 0.813/26.2
- 4th: 1.0/21.3
- 3rd: 1.398/15.3
- 2nd: 2.202/9.7
- 1st: 3.822/5.6
- Rev: 3.705
- Final drive: 3.45:1

BODY/CHASSIS
- Rust warranty: 6 years against perforation of the bodywork, checked during annual inspection

SUSPENSION
- Front: Independent by MacPherson struts; coil springs; double-jointed lower wishbones, anti-roll bar.
- Rear: Independent by semi-trailing arms; co-axial coil springs and telescopic dampers; anti-roll bar.

STEERING
- Type: ZF ball and nut
- Assistance: Yes

BRAKES
- Front: Discs, 284 mm dia
- Rear: Discs, 284 mm dia
- Park: On rear, internal drum
- Servo: Yes
- Circuit: Diagonally split
- Rear valve: Twin
- Adjustment: Automatic

WHEELS/TYRES
- Type: Aluminium alloy, 6J rims
- Tyres: Michelin XWX 205/70 VR14
- Pressures: 32/32 psi F/R (normal) 33/36 psi F/R (full load/high speed)

ELECTRICAL
- Battery: 12V, 66 Ah
- Earth: Negative
- Generator: Alternator, 80 Amp
- Fuses: 30
- Headlights
  - type: Halogen
  - dip: 110 W total
  - main: 110 W total

COMPARISONS

	Capacity cc	Price £	Max mph	0-60 sec	30-50 sec	Overall mpg	Touring mpg	Length in.	Width in.	Weight cwt	Boot cu ft
BMW 628 CSi	2,788	19,275	129.6	8.5	10.7	25.0	28.4	187	68	28.3	12.5
Audi Quattro	2,144	20,402	138*	6.5	8.2	19.9	24.7	171	68	24.8	7.6
Jaguar XJS 360	3,590	20,756	136.8	7.2	6.9	18.9	24.2	188	71	32.5	10.9
Lotus Excel	2,174	14,990	126.6	7.9	7.0	19.6	26.4	172	72	22.2	8.9
Opel Monza	2,969	13,801	129.6	8.2	8.4	20.8	26.4	187	69	28.5	10.7
Porsche 911 SC	3,164	24,532	151.1	5.3	5.6	21.2	26.7	169	64	22.9	9.8

*estimate

Make: BMW **Model:** 628 CSi.
Maker: Bayerische Motoren Werke AG, 8000 Munich 40, Petülring 130, W. Germany.
UK Concessionaires: BMW (GB) Ltd, Ellesfield Avenue, Bracknell, Berks RG12 4TA.
Price: £15,471.58 basic plus £1,289.29 Car Tax and £2,514.13 VAT equals £19,275.00. Options fitted to test car: ABS anti brake-locking system £972.00, manual sliding roof £626.00, air conditioning £1,374.00 and on-board computer £516.00. Total as tested: £22,763.00.

ROAD TEST

It's like a Rolex watch, this special coupe from BMW: solid, immaculately crafted, and terribly exclusive. But there's one difference. The M635 CSi has no difficulty justifying its steep price.

Climb into the snug leather seats and spend a few miles in this machine, and you'll agree: It's simply one of the finest cars in the world. It does virtually everything well, and leaves its driver with that satisfied glow one feels in the presence of rare competence.

Sad to say, darn few of us will ever have the chance to enjoy a few miles in a BMW M635, much less park one of these for-Europe-only jewels in our own garage. But through the magic of the magazine page—and the cooperation of Mike Sheehan at European Auto Restoration in Costa Mesa, California (714/642-0054)—we can offer you a private showing of this Teutonic masterpiece.

BMW uses the M badge (for Motorsport) to designate its high-power, high-sport, high-lust special editions, recalling the phenomenal M1 supercar that bowed in 1979. Though that mid-engine exotic, created to be an all-conquering racer, fell through cracks in the rulebook and withered on the vine, the M635 has inherited its legacy. Plus a good measure of its performance. Because the roadgoing coupe took a lot more than its first initial from the 163-mph M1. The fuel-injected twincam 24-valve wundermotor that used to snarl behind the tensed shoulders of M1 drivers now lives and breathes—oh does it!—under the hood of the M635 CSi.

A tour through the M-coupe properly begins with that straight-six powerplant. Originally a race engine for the old 3.5 CSL Group 2 cars, a detuned version landed in the M1 when BMW's on again/off again 4-liter V-12 project was "off." The 3453cc cast iron block had grown out of the 2500/2800/3000 six of the late '60s and '70s. Atop it sat a gleaming aluminum cylinder head with four valves set in each pentroof chamber at a 38° included angle. Two camshafts, chain driven, operated the 24 valves. The design entailed 240 separate valvetrain parts. With Kugelfischer fuel injection, dry-sump lubrication, and Marelli electronic ignition, the M1 engine produced 277 hp (DIN) in its homologation road tune and 470 prepared for Group 4 competition.

At about this same time (1978), down some Munich hallway, the engineers in charge of the svelte 633 coupe found themselves in a top-

speed race with the fellows over at Mercedes. And the 5-liter 450SLC from Stuttgart was leading the pack. Thus the 3.5-liter engine (with a street-going single-cam 2-valve head again) landed in the 6-series coupe, its 218 hp good for 140 mph in the spoilered and air-dammed 635 CSi.

A 1983 change in racing regulations brought the M635 one step closer to reality. New Group A touring car rules limited modifications, chopping into Porsche's horsepower advantage and giving the BMW 6 coupe its first racetrack possibilities. Private and semi-factory 635s did well all year, and at the 24-hour enduro at Spa-Francorchamps, Belgium, Hans Heyer held off the V-12 Jaguars to clinch the 1983 championship for BMW.

In its excitement over this win, the factory decided to get serious and upgrade the homologated street specifications. Back came the old killer cylinder head from the M1, this time with new intake and exhaust plumbing, modern Bosch Motronic black boxes, and 10.5:1 compression (up from the M1's 9.0:1). Street-legal horsepower was now 286 (DIN) at 6500 rpm, with a lusty 251 lb-ft of torque at 4250 rpm.

And with that, BMW—and all the enthusiast world—had itself an M635 CSi.

Now, we don't test a lot of gray-market cars, since there can be so many inconsistencies, imponderables, and unpredictables where *ex post factory* federalization is going on. And when we do drive such cars, we generally want them after the conversion process, so we can see how they perform in their final U.S.-legal configuration.

But give us a break just once, okay? Mr. Editor Swan had burned some kilometers at autobahn speeds in an M635 (Jan. '84), and came back from the continent bewitched, bothered, and bewildered by the German-market cruise missile. So when Mike Sheehan asked if we wanted to wait out the modifications and dyno runs to have this car in legal form, or drive it now before the EPA assaulted its performance virginity, well, what could we say?

And what can we say now, after subjecting its every aspect to our cold, professional scrutiny? After wringing it out in American driving conditions? After whipping it unmercifully to expose its tiniest weakness? What do we say?

"Praise be to Woden, this sucker's *strong*!"

Yes, everyone who drove the car was fairly smitten by it, and the buttery, electrical, turbine-esque power delivery of that long 24-valver was

There are certain imponderables in *ex post factory* federalization

BMW M635 CSi

In this sleek coupe body
beats the twincam, 4-valve, 286-hp
heart of an M1

by Kevin Smith
PHOTOGRAPHY BY WILLIAM CLAXTON AND BRIAN WAINGROW

ROAD TEST

the impression that stuck. The engine will run cleanly and without audible complaint from a mere 1500 rpm—in top gear! Work the bottom half of the tach's operating range, up to 3500 rpm, and it's smooth and quiet; the essence of cultured motoring. Whack open the throttles as the needle sweeps past 4000 rpm and some beast opens its eyes and goes to work with stunning enthusiasm. All the way up to an honest 155 mph.

There are automobiles that will accelerate from 0-60 faster than the M635. Even some U.S.-market sports cars can do it: the Corvette, Lotus Esprit, Porsche 911. But that doesn't matter. This isn't a drag racer. It isn't even a sports car. BMW's M635 CSi is a 4-place touring coupe with luxurious comfort, loads of luggage space, and a little 211-cu-in. 6-cylinder motor. And a power rush like *Discovery* leaving Cape Canaveral.

Naturally, one brilliant stroke doth not a masterpiece make. While the M1-derived engine dominates the experience of driving this car, the M635 offers plenty of other delights. Like, for example, a ride/handling profile that seemingly defies the laws of physics. This middleweight (3300 lb is more than many sedans but less than a 928 or SEC) rides taut and flat, assuring you it will change direction without drama or hesitation anytime you will it. Yet it also swallows road disturbances uncannily well, apparently grading off the tops of bumps and filling in the hollows as it rolls. You are left with just a satisfyingly intimate sense of road texture, one that lets you "read" the tire contact patches.

And rather substantial patches those were in the case of our test car. Michelin's largest TRX radials, 240/45VR415s, stuck this coupe to the road. The alloy BBS wheels carrying them measured 415 x 210 mm—about 8¼ in. wide, 16⅓ in. in diameter. Factory literature specifies slightly smaller 220/45VR390 tires on 390 x 165 wheels; the 240s were a retrofitted upgrade. We've not been great TRX fans in the past, but these wide-section, short-sidewall examples, with this chassis, work beautifully.

Underpinnings follow standard 6-series practice, with only minor

The M badge recalls the phenomenal M1

tweaks and tuning changes to suit the higher power, speed, and traction of the M-car. BMW's "double pivot" geometry (borrowed from the 7-series sedan) with MacPherson struts and lower wishbones, coil springs, and an anti-roll bar takes care of front suspension, with semi-trailing arms, coils, and another anti-roll bar out back. Brakes are discs all around of course, with BMW's excellent antilock mechanism that always delivers the best braking possible—with full stability and steering capability—for the available tire traction. Steering is a mildly heavy but positive recirculating ball system, with a variable ratio and power assist.

Inside our M635 was a layout nearly identical to U.S.-market 633s, in basic black on black. This design is clean and utterly without visual confusion, unless you count the German labels on the trip computer's buttons: GESCHW., BERBR., REICHW., und zo on. Some small switches are different from what BMW sends here, but it's essentially the same classy environment.

Our car had the sport seats and steering wheel that we first found—and enjoyed immensely—in a 325e we drove last summer. Pronounced side bolsters hug an occupant's body and dictate a total of one position in the seat, but the seats' shape, padding density, and adjustability (reach and rake, rock up and forward, pivot at the back, extend front cushion) add up to fine comfort for short stays or long. And, of course, in fast cornering, lateral location in the seats is not even a matter to notice because it's so complete and comfortable.

We had a factory-imported (white market?) 633 CSi to drive back to back with the M635 for comparison. While specific differences between the two cars were vast, it's a credit to BMW and the 6-series design that the American-market car matched up as well as it did. This is a gem of an automobile in either configuration. Naturally, the detoxed 3210cc engine comes up well short on maximum power alongside the 24-valve 3.5, but in the casual operating range up to about 3000 rpm, the more mildly tuned 2-valve engine seems to have at least as much instant torque. It may also be a touch smoother and quieter.

In freeway ride quality, the Amerispec car comes out on top, though at

the expense of some chassis response. It feels floaty compared to the M-coupe, giving a less immediate sense of bumps and ripples, and fractionally slower reaction to steering input. It's still a fine-handling automobile, understand, but without the M635's fine edge. That car's wider wheels, short tire sidewalls, and stiffer spring rates and shock valving produce a tauter snap in its moves, at a small (and, we think, entirely acceptable) cost in road isolation.

Selfishly, we wish another cost—the one to purchase this car—were smaller than it is, but all considered, its pricetag is not unreasonable. Any BMW 6-series coupe embodies quality, sophistication, and determined development, and all those cost money. The M635 CSi offers tremendous performance and exclusivity besides, for not a lot more. Mike Sheehan talks about $50,000 (actually $47,000-53,000 depending on equipment) for federalized M635s, which compares amazingly well with the $40,705 BMW charged for an '84 American-market 633. True, gray-market cars offer no warranty, and parts supplies can be tricky. But we're talking fantasy here so those realities don't have to bother us. And at least this particular fantasy is German-engineered.

For an automobile with striking good looks, rewarding comfort, stunning power, and remarkable flexibility, $50,000 isn't really so bad, is it? Makes more sense than giving your Rolex man $3000 just to know how late you are. **MT**

DATA
BMW M635 CSi

POWERTRAIN
Vehicle configuration......Front engine, rear drive
Engine configuration......L-6, DOHC, 4 valves per cylinder
Displacement...................3453 cc (211.0 cu in.)
Max. power (SAE net).....273 hp @ 6500 rpm
Max. torque (SAE net)....239 lb-ft @ 4250 rpm
Transmission.....................5-sp. man.
Final drive ratio 3.02:1

CHASSIS
Suspension, f/r...............Independent/independent
Brakes, f/r.......................Disc/disc
Steering..........................Recirculating ball
Wheels............................415 x 210mm TR-type alloy
Tires 240/45VR415

DIMENSIONS
Wheelbase ....2624 mm (103.3 in.)
Overall length................4755 mm (187.0 in.)
Curb weight...................1500 kg (3307 lb)
Fuel capacity..................70.0 L (18.5 gal)

PERFORMANCE
0-60 mph ....7.16 sec
Standing quarter mile .....15.58 sec/92.2 mph
Braking, 60-0 ....124 ft
Skidpad..........................0.83 g

ROAD TEST

BMW 635CSi

Use it as a fast car is used in Europe: as a small plane.

• In many ways, the 635CSi is a superlative car. It produces prodigious speed, thanks to its new 3.4-liter engine. It exhibits much better handling than BMW's early 6-series coupes. It is put together with rare care. Its ergonomics are generally excellent. And now it has Bosch ABS anti-lock braking equipment for help during emergency stops. All told, the 635CSi is a very accomplished piece of work.

On the other hand, it's not perfect.

BMW's glassy coupe has been around for eight years. The 635 remains a very clean design, but flush glass and headlights, integral drip channels, and aerodynamic wheels do not yet grace its charms. The speed it makes is due to muscle, not aerodynamic slickness. This year, the 635CSi does appear more determined, thanks to a body-colored air dam that houses two fog lights.

The rolling stock has also been upgraded to handle the larger engine. Fat Michelin TRX tires, size 220/55VR-390, are mounted on 6.5-by-15.4-inch alloy wheels (a move from 70-series tires on 6.5-by-14-inch wheels). We found that the new combination is worth 0.79 g on the skidpad, a clear improvement in adhesion.

The engineers have also tuned the driveline for a more aggressive attack on the competition. Unfortunately, the 635's Getrag 260 five-speed gearbox is plagued with notchiness, and second gear balks when cold. BMW long ago dumped smoother-shifting, weaker-synchroed ZF

gearboxes in favor of the notchier Getrags. This transmission asks for a heavy hand on the lever, and even the 635's clutch demands extra attention for smooth shifts. (For minimal disruption, a ZF four-speed automatic is optional, for $795.) Lighter shifting is available with Detroit's five-speed, high-output V-8s, which deliver much more torque than the big BMW.

Still, we're talking 3.4-liter, high-revving, seven-league boots here. BMW has punched out the 3.2-liter's bore from 89 to 92mm, and the torque has increased from 195 to 214 pounds-feet. This single-overhead-cam six is old, but it remains atop the pack in basic design. Its aluminum hemi head is fed by Bosch Motronic fuel injection. Crossflow intake and exhaust ports handle the breathing, and a new, adaptive Lambda-sensor control sets the fuel-air mixture more precisely through a wider range of operation. To allow maximum spark advance at all times, BMW has not yet resorted to a knock sensor, and the 3.4-liter motor's compression ratio remains a relatively low 8.0:1. The added displacement has bumped up the horsepower only from 181 to 182. Tuned instead for midrange response, the big six's power peaks at 5400 rpm, 600 rpm earlier than in the 3.2-liter. A 3.45:1 final-drive replaces the 633's taller 3.25 gear, boosting the throttle response throughout the rpm range, but a rev limiter clips the power smartly at 6200 rpm.

Additional power or no, stepping hard on the throttle will hang the 635's tail out with ease. BMWs were once known for hanging out their tails for altogether different reasons. Lift-throttle oversteer whipped the cars sideways if you suddenly backed out of the throttle when cornering hard. Their semi-trailing-arm independent rear suspensions provided supple adhesion over poor roads, but criticisms leveled against their ultimate instability finally motivated BMW into a partial redesign: the semi-trailing arms were angled less so they

BMW 635CSi

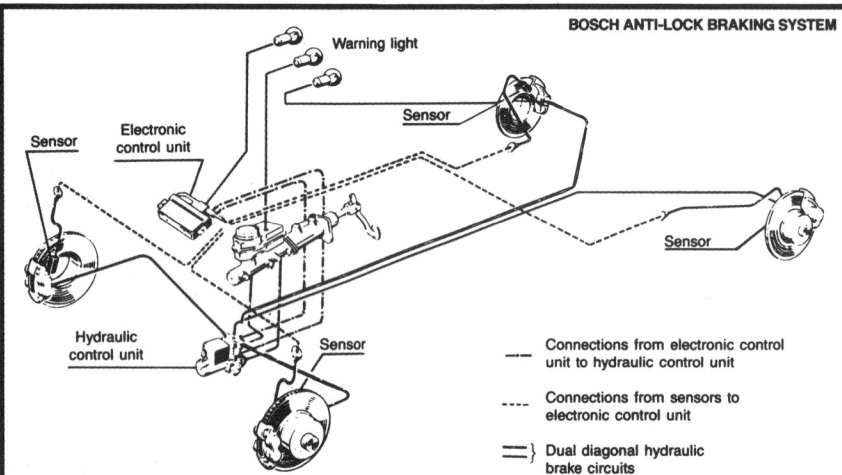

BOSCH ANTI-LOCK BRAKING SYSTEM

would steer the rear wheels less when the throttle was closed, and compound outboard pivots, called Trac Links, lowered the rear roll-center height and reduced jacking effects during cornering. Even earlier, BMW had built a MacPherson-strut front suspension with double pivots that improved steering feel and response. When it put both these fixes into its cars, they became at once nimble and stable. The 6-series got this good stuff during the 1983 model year, and it has felt terrific ever since.

Best of all, at long last (*praise be!*), ABS braking arrives on the American market. It is standard equipment on all 3.4-liter 5-, 6-, and 7-series BMWs (as well as assorted Mercedes-Benzes), albeit twenty years after Bosch started development and seven years after Euro-market BMWs got it. Remarkably, ABS detects the imminent locking of any of the four 11.2-inch discs and immediately reduces the hydraulic pressure to that brake, releasing it to turn, whereupon the pressure immediately pumps up again for maximum braking. The cycle continues until the pedal effort is relaxed or the car stops. The result is maximum braking with no loss of steering or stability. The driver has only to stand on the brake pedal and steer.

ABS will not save you if the situation is too far gone, but BMW says it will stop the car ten to fifteen percent shorter on dry pavement and 25 to 40 percent shorter in rain, snow, and ice, a whale of a payoff from a system that weighs less than 40 pounds. ABS employs individual wheel-speed sensors, an electronic control unit, and a hydraulic control unit that's integral with an electrically driven hydraulic power assist. A quick-footed driver *might* pump the brakes four times per second. ABS pumps each front brake separately and both rear brakes together up to fifteen times per second. What you feel is a pulsing through the pedal. The 635 exhibits a softer pulsing than ABS-equipped Benzes we've tried, perhaps because of differences in software or hardware, or both. In every case, however, the stopping ability is nothing less than sensational. ABS monitors itself, provides a status light, and automatically cuts off if something goes awry. And ABS is only there when you need it: the pedal normally feels like any other in even the hardest driving. Our standard dry-pavement stop from 70 mph needed only 189 feet in the new 635CSi, although a number of non-ABS cars do as well or better, at least in a straight, dry line.

As with any high-performance car, proper support in the driving position would be a great help during the hard acceleration, braking, and cornering that the 635CSi encourages. If you're broad across the beam and opposed to exercise at the wheel, you may love these seats. But if you're skinny everywhere but in the wallet, you may not love them. They are wide, flat, and slick with leather. No fabric is available in the 635, nor are any of the lumbar and bolster adjustments that are stuffed into nickel-and-dime Japanese and American GT cars. Roominess up front is so-so, and the two rear cuplets are ridiculously tight, but, hey, this is a GT coupe by BMW's definition; the company offers roomier sedans, including the lighter and even faster 535i, for sport delivered in boxes. Thanks to a total of twenty (!) console buttons with complex graphics, dedicated occupants can power the 635CSi's seats into almost any position, but BMW should swallow its pride and copy the simple Mercedes 3-D controls, designed in the shape of a seat.

The 635 has power everything, plenty of storage, and a carpeted, tool-kitted trunk. The electronic stereo seems plenty competent but sounds choked by its smallish speakers. The air-conditioning system is not a set-and-forget climate control, but it's very adaptable, thanks to good flow, fine controls, and four vent levels. The otherwise sensational black-leathered steering wheel is handicapped by small horn but-

tons but enlivened by a dab of BMW Motorsport colors. In BMW analog tradition, the 160-mph speedo, the 7000-rpm tach, the fuel and temp gauges, and the instantaneous-mpg scale are crisply legible. The wraparound dashboard is both exotic and practical. The cruise control is flawless. A nine-function dash computer handles everything from day tripping to tripping up would-be thieves. And everything is put together as if lives depended on the coupe's solidity.

A good thing it is, for the 635CSi is a quick car, if not blinding in acceleration. Its ripe torque, short final-drive, and early rev limiter conspire to produce both wheel hop off the line and a time-consuming shift to third gear at only 59 mph. The result is a 0-to-60 time of 8.2 seconds, a half-second slower than the last 633CSi we tested. The quarter-mile performance of 16.0 seconds at 85 mph matches the 633 in elapsed time but is 1 mph slower in trap speed. Top speed, however, is up from 124 to 132 mph, and real-world driving demonstrates considerably more poke, even though this particular car is 95 pounds heavier than the 633. The engine feels great, but many cars, including the new American VW GTI, the Honda CRX, the Porsche 928, and the European-spec 635CSi, have compression ratios of 10.0:1. With a measly 8.0:1 in this 635, and much less than one horsepower per cubic inch of displacement, we can't help feeling that something is missing. Lately, it seems as if BMW's efforts to block the gray-market importation of higher-performance European models have exceeded its interest in building state-of-the-art machinery for the American market. A mandate should be sent to the engineering department: send more torque *and* more horsepower to the U.S.

In BMW's defense, the 635CSi is so well balanced that it is capable of easily returning hair-raising average speeds without ever standing an occupant's hair on end. For a car of its size and weight, the 635 is quite the gymnast—and, for our money, the best-handling BMW since the M1. It is poised (if occasionally porpoisy and undershocked) when called to heavy duty, and it's stable over endless stretches of deep throttle. Feed it hard and fast with the strong inputs it calls for, and it rushes back for more.

But don't face off this $42,000 BMW with the $25,000 Corvette (which appeals to a different crowd but is also *much* faster), and be careful around IROC-Zs, Trans Ams, Mustang GTs, Merkurs, and sixteen-valve Saabs. And for God's sake, don't get mowed down by the 32-valve, monster-motor Porsche 928 (which does appeal to this crowd, costs less than $10,000 more, and is 100 bhp stronger). Just use the 635CSi as any fast car is used in Europe: as a light plane. Quick and to the point.

—*Larry Griffin*

BMW 635CSi

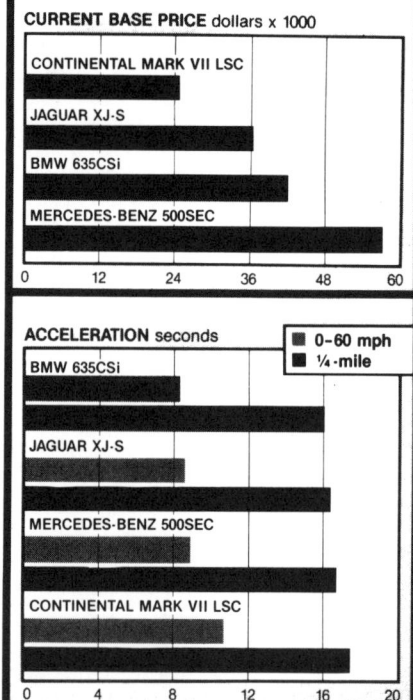

Vehicle type: front-engine, rear-wheel-drive, 4-passenger, 2-door sedan

Price as tested: $41,705

Options on test car: base BMW 635CSi, $41,315; limited-slip differential, $390.

Sound system: BMW AM/FM-stereo radio/cassette, 4 speakers, 13 watts per channel

ENGINE
- Type .... 6-in-line, iron block and aluminum head
- Bore x stroke .... 3.62 x 3.39 in, 92.0 x 86.0mm
- Displacement .... 209 cu in, 3430cc
- Compression ratio 8.0:1
- Engine-control system .... Bosch Motronic
- Emissions controls ..... 3-way catalytic converter, feedback fuel-air-ratio control
- Valve gear .... chain-driven single overhead cam
- Power (SAE net) .... 182 bhp @ 5400 rpm
- Torque (SAE net) .... 214 lb-ft @ 4000 rpm
- Redline .... 6300 rpm

DRIVETRAIN
Transmission 5-speed
Final-drive ratio .... 3.45:1, limited slip

Gear	Ratio	Mph/1000 rpm	Max. test speed
I	3.82	5.5	34 mph (6200 rpm)
II	2.20	9.5	59 mph (6200 rpm)
III	1.40	14.9	92 mph (6200 rpm)
IV	1.00	20.9	125 mph (6000 rpm)
V	0.81	25.8	132 mph (5100 rpm)

DIMENSIONS AND CAPACITIES
- Wheelbase .... 103.5 in
- Track, F/R .... 56.3/57.5 in
- Length .... 193.8 in
- Width .... 67.9 in
- Height .... 53.7 in
- Ground clearance .... 3.7 in
- Curb weight .... 3375 lb
- Weight distribution, F/R 55.0/45.0%
- Fuel capacity .... 16.6 gal
- Oil capacity .... 6.1 qt

CHASSIS/BODY
Type .... unit construction with 1 rubber-isolated crossmember
Body material .... welded steel stampings

INTERIOR
- SAE volume, front seat .... 50 cu ft
- rear seat .... 34 cu ft
- trunk space .... 12 cu ft
- Front seats .... bucket with 8-way power assist
- Recliner type .... infinitely adjustable
- General comfort .... poor fair **good** excellent
- Fore-and-aft support .... poor fair good **excellent**
- Lateral support .... poor **fair** good excellent

SUSPENSION
F: .... ind, MacPherson strut, coil springs, anti-sway bar
R: .... ind, semi-trailing arm, coil springs, anti-sway bar

STEERING
- Type .... recirculating ball, power-assisted
- Turns lock-to-lock 3.3
- Turning circle curb-to-curb .... 33.1 ft

BRAKES
- F: .... 11.2 x 1.2-in vented disc
- R: .... 11.2 x 0.4-in disc
- Power assist .... hydraulic with anti-lock control

WHEELS AND TIRES
- Wheel size .... 165 x 390mm, 6.5 x 15.4 in
- Wheel type .... forged aluminum
- Tire make and size .... Michelin TRX, 220/55VR-390
- Test inflation pressures, F/R .... 32/32 psi

CAR AND DRIVER TEST RESULTS

ACCELERATION Seconds
- Zero to 30 mph 2.5
- 40 mph 4.1
- 50 mph 5.7
- 60 mph 8.2
- 70 mph 10.7
- 80 mph 14.0
- 90 mph 18.1
- 100 mph 24.0
- 110 mph 33.0
- Top-gear passing time, 30–50 mph 10.1
- 50–70 mph 10.6
- Standing ¼-mile .... 16.0 sec @ 85 mph
- Top speed .... 132 mph

BRAKING
- 70–0 mph @ impending lockup .... 189 ft
- Modulation .... not applicable
- Fade .... none moderate heavy
- Front-rear balance .... not applicable

HANDLING
- Roadholding, 300-ft-dia skidpad .... 0.79 g
- Understeer .... minimal **moderate** excessive

COAST-DOWN MEASUREMENTS
- Road horsepower @ 50 mph .... 16.5 hp
- Friction and tire losses @ 50 mph .... 7.0 hp
- Aerodynamic drag @ 50 mph .... 9.5 hp

FUEL ECONOMY
- EPA city driving .... 16 mpg
- EPA highway driving .... 22 mpg
- C/D observed fuel economy .... 17 mpg

INTERIOR SOUND LEVEL
- Idle .... 49 dBA
- Full-throttle acceleration .... 77 dBA
- 70-mph cruising .... 72 dBA
- 70-mph coasting .... 71 dBA

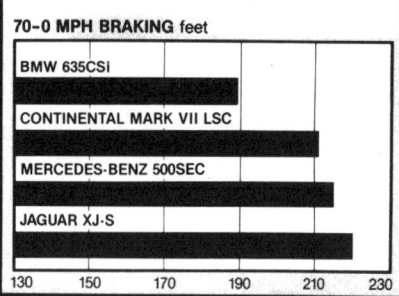

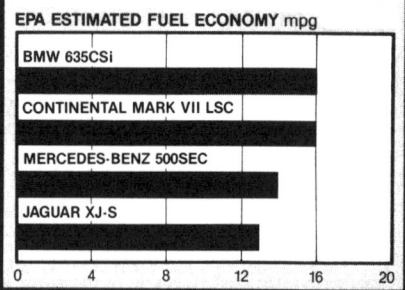

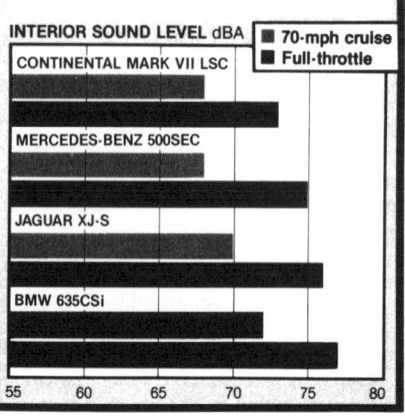

EVEN at prices beyond £30,000 BMW in Britain have customers queuing up for the latest version of the nine year old big BMW coupe, the extraordinary 24-valve M635 CSi. Complete with 286 hunky Bavarian horsepower, Performance Car was the first British magazine to fully fifth wheel test the "M for Missile" 6-Series and published its results in February 1984: a timed kilometre in 150 mph, 0-60 mph in 6.5 seconds and the standing start quarter mile devastated in 15.1s.

This is *the* coupe in which you change from fourth to fifth in the heavy duty (precision selection) Getrag gearbox at 131 mph, looking for three Shredded Wheat and a fast lane to munch for brekkie …

Essentially all current BMWs comprise tried and proven components in a variety of cocktails stirred at frequent intervals and the M635 is no exception. The basic floorpan and wheel-base is that of the 5-Series (but an M535 has the 218 bhp, 12-valve six), carrying virtually unchanged 6-Series two door body and an engine descended from that of the fabled mid-motor M1 of 1979.

To suit today's conditions BMW Motorsport engineers – the group who squeeze up to 850 bhp regularly from the 1½ litre GP engine – specified Bosch ML-Jetronic injection and a 10.5:1 cr in place of the M1's non-Motronic injection and 9:1 cr.

They also had to install the engine at the usual BMW slant, rather than upright, and they dispensed with the M1's dry sump lubrication in favour of a beefed up production system, generally using many more mass production parts than the original. However, the 3453cc capacity was kept from M1, whereas today's 635/735s actually use a smaller bore 3.4 litre.

The results were extra power and fuel conscious torque bonus. Instead of the M1's 277 bhp there were 286 M635 horses and 250 lb ft torque was a more manageable street combination than the original's 244 lb ft at 5000 rpm. We returned 19 mpg overall in our first encounter, which is not bad for a 3300 lb missile of 150 mph capability.

Motorsport have also attended thoroughly to the chassis of the M-Coupe and managed to provide a better ride than a standard RHD 635 taken along for comparison!

Key suspension components include Bilstein gas damping, increased front and rear anti-roll bar diameters (the rear with a progressive action), 15 per cent stiffer springs, about half an inch from the ride height and the presence of an additional helper spring within each front strut "to counter body roll."

More obvious changes include the TRX dimension alloy wheels which can optionally (in some markets) carry those uniquely sized Michelins of up to 240/45 VR.

Brakes are better too with 300mm × 30mm vented discs served by four piston calipers to the old M1 spec instead of the standard 280 × 25mm units.

In action the results are sensational. Not just in the kind of figures recorded – galloping to 100 mph takes little longer than many small cars take to reach 60 mph (actually some 17.9s is Macho Coupe's 0-100 trip) – but in the sensations reaching the parts other BMW Coupes cannot reach.

For the M635 tolerates towns instead of gliding through in the refined manner of less highly strung 6-Series. Aside from driving it in glorious sunshine, we also used it to commute around Munich and mutter up to the BMW plant at Dingolfing, over an hour's wet driving on crowded two lane roads.

In such use it was obvious that the car required a lot more effort and involvement from the driver than many British BMW drivers will expect. Many UK BMW owners, even of the overtly sporting 6-Series, opt for automatics. The mechanical satisfaction that comes from shifting a race pattern 5-speed "across the gate" from first to dogleg into second, plus a heavyish competition clutch, would come as an unpleasant surprise to such a BMW owner.

If you like masses of front engine rear drive power, delivered with a thrilling blend of gruff low rpm dependability and a shriller 6900 rpm, this is probably the BMW you have been waiting for.

No longer need you suffer the insolence of passing Porsches, for the M635 has the muscles to at least stay in touch, whilst living within a shell that sports only slightly flared wheel-arches as major clues to the reason for those discreet M prefix badges. In fact BMW had the fleetness of the latest 928S and the 3.2 litre 911 when they put this package together. The UK price of M635 is £1297 less than the latest UK 928 and £5650 over that of the UK Porsche 911.

Nothing will ever obscure the memory of some of the most exciting miles any of us have enjoyed secure within M635's Recaro embrace.

Likes: Fabulous engine of inspired clout and eye-popping performance. Predictable oversteer handling under power.

Dislikes: Tiring in everyday use and quite noisy. At first the fuss is fun, but it is at odds with suave 6-Series image.

Price: £32,194

BMW M635 CSi

BMW 635 CSi Jaguar XJ-CS 3.6 Cabriolet

NOT SO long ago BMW versus Jaguar would have been a classic confrontation for journalists alone. The buying public had made up its own mind. While British businessmen stood on hard shoulders and cursed company considerations that had forced a patriotic purchase, BMW was accelerating hard towards the top of world sales charts.

Not any longer. We've all read – endlessly it seems – about the resurgence of the Coventry Cats and if the story seems to have something of the quick-drying PR gloss, the sales figures, especially in the USA, climb like the unconquered face of Everest. BMW isn't sweating yet but it's breathing hard as a spate of new or improved models illustrates.

So is the confrontation now a classic? It's certainly intriguing: two coupés, both of them long in years but what a different history they have had. The 6-series BMW, nine years old now, has been steadily smoothed and refined to something that, despite its age, shows off the current state of BMW's sophisticated technology.

The Jaguar XJS, on the other hand, was allowed to drift through almost a decade untouched by any significant modification until, under the new regime, it became quite suddenly the oyster in which Jaguar hopes to develop its pearl – a new 3.6-litre, six-cylinder engine to power next year's new saloon range into the '90s, and at the same time, its ugly top was also sliced off and a tidy cabriolet roof replaced it.

So a state-of-the-art BMW 635 CSi takes on a Jaguar XJS 3.6, hardly even the state of seventies art save for its AJ 3.6 engine, which may yet prove a more significant portent for the future than all BMW's technical wizardry on its 635.

Two cars chasing the same buyers could scarcely be less alike: the Jaguar is the elegant outcome of classical engineering theory allowed its head in manufacture; the BMW much more the refinement of simple and sometimes compromised mechanical principles.

Motive force of the 635 is its iron-blocked and alloy-headed six-cylinder engine, now measuring 3,430cc and slightly over-square at 92 × 86mm. It has a simple, two-valves-per-cylinder head (the M635CSi gets a four-valve head) with the single overhead camshaft being chain driven. Compression is 10.0:1.

Contrast that simple lump of a power unit with the relative complexity of the AJ6 Jaguar unit: it's all alloy to begin with, 3,590cc and almost exactly square, 91 × 92mm in dimensions, and uses a four-valve-per-cylinder cross flow head, with two overhead camshafts – the unit being inclined at 15 degrees to give sufficient under-bonnet clearance.

Power and torque figures of the pair are very similar, though: the BMW has a peak 218bhp at 5,200rpm, the XJS 225bhp at 5,300rpm and the BMW's torque best is 224lb ft at 4,000rpm compared with the Jaguar's 240lb ft at 4,000rpm. Look to the engine electronics, however, and it is the BMW which is by far the more advanced. Bosch L-Jetronic fuel injection is linked to the Motronic engine management system for complete digital electronic control of fuel and ignition. Tied into that system are fuel consumption and service interval monitoring, too. The Jaguar has Lucas digital electronic injection but not, as yet, any engine management electronics, relying instead on a simple breakerless electronic ignition system.

BMW's advanced electronics show off in the transmission system as well – the 635 comes with manual or auto-boxes at the buyer's choice and ours had the latest ZF four-speed programmable automatic that features alternative "sport" or "economy" settings and torque converter lock-up. Fourth in this 'box is a long 32.2mph/1,000rpm overdrive aimed at maximising cruising economy. The 3.6 XJS has a manual Getrag 'box only – a five-speeder that is also topped with a long-legged overdrive giving 28.9mph/1,000rpm.

The 635's suspension might reasonably be described as a triumph of experience over hope. A seemingly simple layout of independent MacPherson struts with lower wishbones and anti-roll bar at the front coupled to independent semi-trailing arms and anti-roll bar at the rear holds more perils than promises. But the set-up has been substantially revised in recent times, especially at the rear, where alterations to the pivot angle of the semi-trailing arms and camber compensating links counter the old tendency to liftoff oversteer.

Alongside this, the Jaguar offers a suspension system of greater complexity but fewer compromises: the front end uses double wishbones, coil springs and an anti-roll bar, while the independent rear has lower wishbones, uses the driveshafts as upper links, twin concentric coil spring/damper units each side and an anti-roll bar. It's the same system that is used to great effect throughout the range – but will such costly sophistication find a place on the new model or will it be nearer the Euro-norm of BMW?

Steering and braking also highlight important differences: the Jaguar uses rack-and-pinion, the BMW ball-and-nut – both with power assistance. Jaguar brakes are all-disc, the fronts being ventilated; the BMW uses solid discs all round but has the considerable advantage of ABS anti-locking as standard equipment. Wheel and tyre choices are closer – the Jaguar opts for 215/70 VR 15 Pirelli P5s on six inch alloys while the BMW comes on 220/55 VR 390 Michelin TRX and 6½in alloys.

Different solutions under the skin, but there is little to choose between the two cars in overall size; both are approaching 16 feet long, with the Jaguar just half an inch longer and the two are but an inch apart in wheelbase – the BMW an inch longer at 103in.

However, despite its alloy engine the XJS still scales in a flabby 400lb heavier than the BMW – a relative lightweight by luxury class standards at 28.8cwt. The 635 weighs much more heavily on the bank balance, though – especially after price rises forced by recent exchange rate movement. It now sells for a whopping £26,195, which makes it second only to the M635 in the BMW price league, and gives the XJS 3.6 at £19,248 for the Cabriolet model in this test, the look of something approaching a bargain. But is it?

BMW 635CSi

LIKE MOST BMW's, the 635 understates its case. There's a simple elegance about its lines but little of the swank or glamour that would make a Porsche or an XJS stand out even in the most affluent of golf club car parks. Not such a bad idea, perhaps; the well-cut cuff rather than the chunky gold bracelet approach. The 635 driver finds he can hustle discreetly along without upsetting the sensibilities of the less affluent wheels around him or attracting untoward attention when the desire comes to cruise at more than 2,200rpm. Diplomatic discretion can be a useful virtue, for this BMW is not short of the necessary tiger to tackle the Coventry cat. Like a well-dressed bodyguard, the cut of the suit hides but doesn't cramp the muscle underneath.

The engine is silkily smooth throughout its range, but more than that, it is crisp and eager to rev. There is a liveliness and enthusiasm about its response that belies its size and gives the big coupé something of the urgency of a smaller machine.

Surprising to some, possibly, may be the comment that the automatic transmission contributes as much – perhaps even more – than the engine to the liveliness of the 635. Indeed, over the sort of fast but serpentine roads which can bring out the best in both car and driver, it is probably fair to say that the auto-box makes the BMW quicker and easier to handle than manual gears. It's that good.

Normally, switched into economy, it will change up readily into overdrive top and need a forceful, kicking down shove on the pedal to change back down. Try sport when the going gets twistier, and the 'box forgets about top and kicks down more smartly as well. The result, on an unfamiliar road with blind bends couldn't be better: see the exit of a corner, onto the throttle and an instant, blur-smooth down-change has the 635 accelerating its impressive hardest.

Of course a sensitive 'box like this one can change gear at the wrong moment if the throttle gets an injudicious squirt – BMW sorts that one by using obvious pedal pressure 'steps', or you can play 3-2-1 and use that switch to lock the 'box in any chosen one of the lower ratios.

So, while 0-60mph acceleration in 7.5sec and to 100mph in 20.5secs are the stuff of any fast lane, more exhilarating – and more relevant day-to-day – are times like 60-80mph in 4.7sec or 80-100 in only 9.2sec.

The overdrive fourth speed is very high, even by big-car standards and the BMW certainly won't reach or hold its top speed in that ratio. Instead, its 135mph maximum sends it just over the power peak in third. But the overdrive does economy some good, especially on long, fast cruising runs: on one measured leg of our photographic run to the Pennines the dash computer read more than 24mpg – the XJS just bettered 19. Overall, mpg was a more than tolerable 21.8mpg.

Enough of such talk, though: this is a car to drive and relish. Settle in behind the wheel and it is obvious that driver and driving have the highest priorities. A complex dashboard dissolves into well ordered, ergonomic simplicity; the view from the seat is clear and commanding but, best of all, a hand on the wheel finds that pin-sharp, eager and accurate response which characterises every BMW.

Power assisted it might be, but the steering still has plenty of weight and no lack of feel: the 635 can be turned in to a corner with surgical precision. With taut damping and relatively modest cornering roll, it thrives on sinuous cross-country roads that give full rein to the eager engine and the crisp handling. On the truly tight bits or in the wet, some of the old vices rear their heads: brutal throttle work can send the tail sliding – especially on a damp road – and unless you pay extra for a limited slip diff, traction can sometimes suffer too.

Brake action is progressive and impressive – it needs a dire emergency indeed on a dry road to provoke the ABS anti-locking into action.

Even at this stratospheric price level there are compromises, though. The ride can be knobbly, especially around towns or on poorly surfaced country roads, but it smoothes out well at speed. Disappointingly, wind noise is then building up and there is quite a roar when travelling fast.

Despite the fine facia, the interior is something of a compromise too. At least anyone in the back seat would think so for it is strictly for the kids or the uncomplaining – head and legroom are not in abundance.

Headroom isn't overly generous in the front either, where the electric sunroof causes the loss of some height inside. But there is plenty of hip room on wide, firm seats, while the driver finds seat height and wheel rake adjustments as well as immaculate pedal positioning there's no reason to be uncomfortable. The check panel, computer (lately simplified for us dunderheads) and an array of effective heating and ventilation controls swell the dash, around a central core of precisely styled instruments. Yet, despite the price tag, there are still desirable extras to be added: air conditioning, electric seat adjustment, sound system and cruise control would all help edge the bill up nearer £30,000.

JAGUAR XJS 3.6

THE XJS is undeniably an experience. That long, low body almost always rates a second glance from outsiders, and there's a sense of occasion for those inside, too, as they gaze out over a gleaming aircraft-carrier deck of a bonnet from the low seating position beloved of British sports car makers. An experience, certainly; ostentatious – probably – but pretty? Hardly.

Then look again; aside from the chance of a suntan, there is a very good reason for choosing the Cabriolet version of the 3.6 – as we have here. Those heavy, 'flying buttress' rear flanks are gone, leaving a rather simple and elegant line below. There's a price to be paid, though – even if this soft-top model is a surprising £1,500 cheaper than the Coupé. It's yet more of an unmitigated extravagance, for the back seats, tiny enough as they were in a near sixteen feet long car, have disappeared completely. Two-seater convertibles don't come any bigger.

Unfortunately, the driving

With its roof and 'buttresses' chopped off the XJS (bottom) has emerged as a surprisingly elegant design. Both cars, in fact, look suitably refined and well-bred despite being notably long-in-the-tooth in automotive terms.

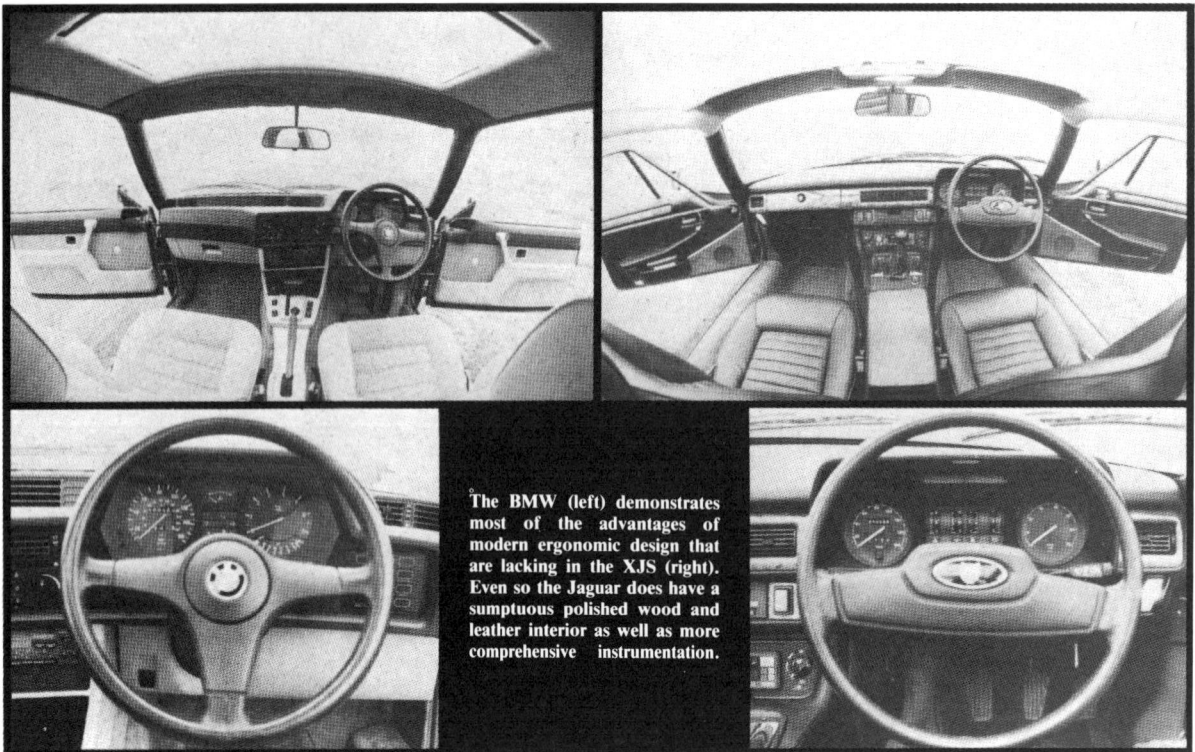

The BMW (left) demonstrates most of the advantages of modern ergonomic design that are lacking in the XJS (right). Even so the Jaguar does have a sumptuous polished wood and leather interior as well as more comprehensive instrumentation.

experience doesn't quite live up to the glamour. A comparison with the fabulous V12 engine is instinctive, though obviously unfair, but you don't need to have sampled that delight to appreciate something is lacking in this new six cylinder unit.

Simple comparison with the BMW six tells that. A rev counter red-lined at a conservative 5,800rpm is another clue though; in practise, the AJ6 is warning off its driver well before that point. It never sounds or feels as sweet as the BMW's engine; there's a raspy edge and slight coarseness to its revving that comes as something of a surprise. But right at the top end things get worse; a clearly discernible convulsion of coarseness seems to shudder through engine and driveline during the last 750 revs before the red line. If that sound melodramatic, let us qualify it by saying that the 3.6 is an engine which seems to be only average when one expects any Jaguar engine to be superb.

Refinement aside, the 3.6 performs as well as can be expected — which means it is good but not outstanding. Designed for what will be (one hopes) a significantly lighter car, it hasn't the output to make the XJS feel a flyer. It is particularly hamstrung by that high fifth ratio, which makes it feel very inflexible when trying to gain speed on a motorway, for instance — 70-90mph would take 14.1sec without down-shifting to fourth.

So, while it has a very close edge on the 635 in outright acceleration — 0.9sec faster to 100mph — the XJS frequently feels slower on the road. Out of fifth, though, the engine reveals itself as strong and flexible, pulling strongly from below half revs.

Maximum speed of 137mph is, again, very close to the BMW's best, but there is an understandable din from the cabriolet roof, well made as it all is, which won't encourage prolonged motoring near those speeds.

The five-speed Getrag 'box has a chunky precision: it feels stiff when cold or unless the long clutch is fully depressed, and there is some irritating jerkiness in the driveline at times. Despite its high top, fuel economy of the heavyweight XJS is disappointing — around 25 per cent worse than the 635 on our back-to-back run and just 18.5mpg overall.

While the German car errs on the side of sportiness in its suspension settings, the Jaguar chooses comfort as a first priority. As a result, the handling just isn't as sharp. Its steering is the strongest deterrent to forceful motoring: the power assistance robs the system of any feel or progression, so though the XJS responds sharply to steering inputs, the driver cannot rely on any feeding-back of information from the front wheels — it's not an easy car to drive quickly.

A shade soft, in the cause of comfort, it also leans more and squirms around more on a poor road where this lack of body control becomes the final limiter on cornering. Roads the BMW thrives on just throw the XJS out of its usually composed manner. Over rippling country roads, it has a nervous, pattering manner under the surface of what is a generally composed and tidy ride — as if it's skimming over the tarmac rather than hugging it, BMW-fashion. Frustrating. For if you can start to come to terms with the lack of road feel provided by the steering and damping, there is an impressive suspension system underneath which gives better roadholding and handling than you usually dare try for.

But the XJS is still much happier in its role as an elegant, two-seater tourer. It takes motorways at the glide and, if the 3.6 hasn't the utter refinement of the V12, it still drives quietly and smoothly at speed.

The shining chromework, the elm burr veneers and the leather of the interior are as attractive in their way as BMW's high-tech cockpit. True, the XJS interior has its charms, but it has its flaws too. The seats are the worst flaw: they are thin and flat, offering little support against cornering forces and not a lot of comfort on a long journey.

The facia, too, is quite a muddle, with poorly styled minor instruments, cheap looking and fiddly-to-use switches. A ring-full of different keys is another annoyance. But, it's an old machine and one hopes some lessons will have been learned on graphics and switchgear for the XJ40.

The boot is big, though spare wheel, battery and stowage bag for the roof panels take up valuable space. There is more room where the rear seats used to be and since they were only fit for small children, added luggage space is probably a more realistic use. The roof goes down — and up again — swiftly and is a snug fit. As well as a handsome price-saving over the German car, the XJS throws in air conditioning as standard and a stereo system, so you can still enjoy yourself when it's raining outside and the roof has to stay up.

VERDICT

THE BIGGEST rival the XJS faces is not the 635 CSi but the XJS HE. It would be hard to persuade anyone whose budget could stretch as far as the BMW into settling for the 3.6 model when they could have the V12 and still be making a saving on the German car. Compared, as it always inevitably will be, with its glorious bigger brother, the 3.6 is an anti-climax and it is hard to find compensating virtues. Despite performance-blunting high gear ratios, fuel economy is still poor — certainly not good enough to tempt many away from the thirsty V12. And the engine, despite its complex, twin-cam 24-valve configuration, seems neither startlingly quick nor outstandingly smooth – the BMW's simpler unit is markedly superior.

True, the Cabriolet body has at last made the Jaguar look elegant – even if it has made it an even more extravagant consumer of road space by reducing it to only two seats. And in that guise, as a stylish, sometimes open-topped, inter-city express the XJS has a certain limited appeal.

Not enough, though, to match the all-round excitements offered by the BMW. Expensive it might be and a trifle bland in outward appearance, but there is nothing bland about its driver appeal. It is a poised, sure-footed, extremely competent machine which, above all, is enjoyable and engaging to drive fast. It hasn't the Savile Row elegance of the Jaguar; instead, it is an effective, carefully detailed car. Perhaps in the end that makes it a slightly soulless one. It is overwhelmingly a BMW – a 635 only by degree of price and performance. The XJS, on the other hand, could be nothing else – it is an individual, eccentric in its own right. Shine its chrome, polish its walnut and love its character – warts and all.

A tempting proposition that would probably sway us if the V12 was under consideration but doesn't offset the deficiencies of the 3.6. And there stands a big problem for Jaguar in the future.

GENERAL DATA

	BMW 635 CSi	Jaguar XJS 3.6 Cabrio
Price	£26,195	£19,248

ENGINE

Cylinders	In-line 6	In-line 6
Capacity, cc	3,430	3,590
Bore/stroke, mm	92/86	91/92
Valves	Sohc	Dohc, 4 valves/cyl
Compression ratio	10.0:1	9.6:1
Fuel system	Bosch Motronic injection	Lucas electronic injection
Max power, bhp/rpm	218/5,200	225/5,300
Max torque, lb ft/rpm	224/4,000	240/4,000

TRANSMISSION

Type	Rwd, 4-speed auto, switchable	Rwd, 5-speed manual
Internal ratios and mph/1,000rpm		
Fifth	—	0.67/28.9
Fourth	0.73/32.2	1.00/22.0
Third	1.00/23.5	1.39/15.8
Second	1.48/15.9	2.06/10.7
First	2.48/9.48	3.57/6.2

SUSPENSION, STEERING, BRAKES

Front	MacPherson strut; lower wishbone; coil springs; anti-roll bar	Double wishbones; coil springs anti-roll bar
Rear	Semi-trailing arms; coil springs; anti-roll bar	Lower wishbones; driveshaft upper link; coil springs; anti-roll bar
Steering	Assisted ball and nut	Assisted rack and pinion
Brakes, front/rear	Disc/disc with ABS	Ventilated disc/disc

WHEELS, TYRES

Wheels	Alloy, 6½J × 390	Alloy, 6JK × 15
Tyres	Michelin TRX 220/55 VR 390	Pirelli P5 215/70 VR 15

DIMENSIONS

Length, in	187.2	187.6
Width, in	67.9	70.6
Height, in	53.7	49.7
Wheelbase, in	103.4	102
Front/rear track, in	56.2/58.5	58.0/58.5
Fuel tank, gall	15.5	20.0
Kerb weight, cwt	28.1	32.7

PERFORMANCE

Maximum speed, mph	135	137
Speeds in gears	(at 6,300rpm)	(at 5,800rpm)
First	58	36
Second	96	62
Third	135 (5,750rpm)	92
Fourth	—	128
Acceleration through gears, sec		
0-30mph	3.2	2.6
0-60mph	7.5	7.2
0-100mph	20.5	19.6
Acceleration in fourth (kickdown for BMW), sec		
30-50mph	2.7	6.9
50-70mph	4.3	7.4
70-90mph	5.7	7.1

FUEL CONSUMPTION

Overall test average, mpg	21.8	18.5
Government test figures, mpg		
Urban cycle	19.3	15.7
Steady 56mph	41.5	36.1
Steady 75mph	32.5	29.4
Maker/importer	BMW GB Ltd Bracknell, Berks	Jaguar Cars Ltd Allesley, Coventry

Performance tests carried out at Millbrook Test Track, Bedfordshire.

Feature: The 1985 Touring Car Champion

Inside JPS-Team BMW

We take a ride in the new Australian Champion tourer to see what makes it a winner

by Michael Stahl

photography by Kem Mears

JIM RICHARDS is one man who doesn't mind working weekends. His part-time job takes place behind the wheel of one of Australia's quickest — if not most powerful — Group A Touring Cars, and his efforts throughout the 1985 season would suggest that the 37-year-old Kiwi is currently the best at what he does.

Very rarely does anybody get to share Jim's "office" with him. The JPS-BMW 635CSi isn't exactly the place for the kids to spend their school holidays. It has no rear or front passenger seats, no soundproofing, no comforts for anyone save the man on the job. His comfort lies in the 150kW, 3.5 litre six-cylinder engine that propelled him to the title of 1985 Australian Touring Car Champion.

Two days before Jim sealed his title at Sydney's Amaroo Park circuit, Modern MOTOR was scheduled to take a ride with the champ-in-waiting before practice for the ninth ATCC round. The frustration of correspondent Peter McKay in July 1984 — it rained the day he drove the works Nissan EXA Turbo race car — really hit home when Richards' flight from Melbourne was delayed by an industrial dispute.

Even when unforeseen circumstances such as these arise, Richards has the fortunate position of being able to step into his "office" and know that the boss has everything in order. Frank Gardner, JPS-BMW Team Manager and holder of a truckload of Australian and international trophies, looks after the testing and development of the team cars for drivers Richards, Neville Crichton and Tony Longhurst.

"I couldn't change anything in the

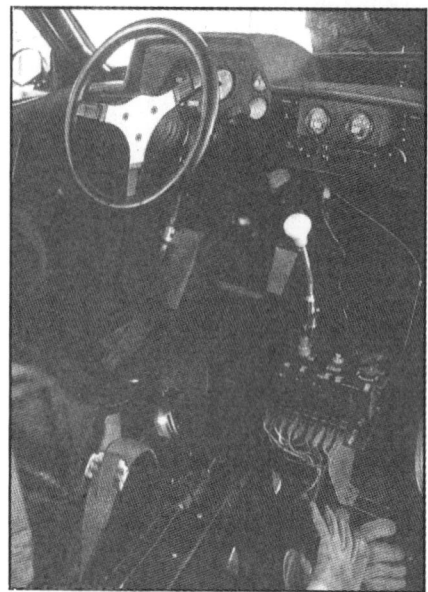

way Frank sets up the car", Richards says. He's previously mentioned the similarity of his own driving style to Gardner's, and the similarity in their speed at this circuit is closer than you might think.

With racers impatiently crowding pit lane for Friday's unofficial practice session, Gardner was strapped in, a passenger seat quickly bolted in, then I bolted in. Gardner sits facing an impressive array of instruments, made less daunting by the ergonomic efficiency of the BMW layout and Frank's familiarity with their function. A large tachometer — rotated 45 degrees so its 6800rpm redline stands vertically — is surrounded by gauges for fuel level, water temperature, oil temperature and oil pressure.

Three one-cent-sized lights in green, red and orange provide instant warnings for water, alternator and oil. To the side of the main cluster are two further gauges for the gearbox and differential.

"We'll have to do a few laps to warm up the diff and gearbox first. I like to make sure the thing's hanging together

Battle station in the lhd BMW; Richards settling down to work. Injected 3.5 litre six is a sweet unit.

right", Gardner advises before approaching the starter button on the transmission tunnel. Flanking it are the fuel isolating switch and the front-to-rear brake bias adjuster, the three backed up by a long row of circuit breakers.

Aside from the controls already mentioned, there's bugger-all else in the big black Bimmer. Its door and ceiling trimmings are intact, but its floors and firewall are completely bare. Its black colouring lends to the impression of sitting in a burnt-out wreck.

Try sitting in a burnt-out wreck sometime and get eight of your mates in steel-capped boots to kick the doors in. This exercise will provide some understanding of Sensory Overload Number 1, which happened when Gardner hit the starter button.

The Group A car's two-valve engine should be no louder than the old Group C's four-valver, but the absence of seats and other "luxury trim" items makes

for an echo-box on wheels. At idle it's a raspy crackle, rising to a coarse growl as we moved off on the first of our three warm-up laps.

"Mmflgumph ecky phtang" said Gardner on the second run up Bitupave Hill; that's at least what it sounded like through a full-face helmet and the roar of the engine. As he also had earplugs to contend with, I bothered not to reply, noticing instead the shift pattern of the five-speed ZF racing gearbox which places first gear to the left and back, and the remaining gears in a conventional H-pattern.

We surged ahead on the first of our fast laps, the BMW not providing a shove in the back but a smooth and obviously substantial power delivery. It's an exaggerated example of most road-going BMW sixes, but deceptively powerful through its subtlety. It'll rev beyond its 6800rpm redline, but power apparently drops away after 6600.

Gardner's feet fox-trotted from the clutch to the left footrest, the brake to the angled accelerator pedal which allows fluent heel-and-toeing. On throttle lift-off there's a backfire that runs up your spine and kicks your head in.

Hands were constantly jiggling the wheel, partly due to road bumps transmitted through the steering but mostly through continual corrections to get the smoothest possible line for the Pirelli P7 D3 265/40 VR16 slicks. Gardner's characteristic shuffling of his hands on the steering wheel through the corners is still evident.

This BMW sits even flatter on the track than the Group C version did; a car in which I also had a ride with Gardner at Amaroo a couple of years ago. There's minimal delay in response through the low-profile slicks, stiff suspension and lightweight moulded racing seat to the bones in your butt. Actually, it's only when one starts hopping ripple strips that the bump-to-bum interaction becomes really noticeable, but having tried a couple of single-seater racing cars it came as no surprise.

Through Amaroo's downhill Dunlop Loop the car assumed a drifting stance, Gardner maintaining this slight attitude through steering and throttle adjustments. One is hardly aware of it, though a couple of times we got just a tad out of shape on the exit where shade had left the surface damp. We float through right-handed Mazda House with a slight lift-off, before the charge down to the left-hander at Honda.

Hard on the brakes, quick downshifting and Gardner smooths the JPS car around and flattens it for the run into Stop Corner, the slowest turn on the Amaroo Circuit. Again, the lateness of braking possible with slicks will surprise the uninitiated. Another surprise when you've been bracing yourself against the uninsulated firewall is finding your feet beginning to fry.

The camber of the track falls away on the exit from Stop Corner, so power is gently fed at first for the short sprint into Wunderlich Corner. Again the car is balanced through the turn, drifting out close to the Armco on the exit where one of the JPS-BMW crew has been adding to the atmosphere by hanging out a pit board to count our laps.

We climb through the gears, past the Control Tower, flat through the kink up Bitupave Hill where the track invisibly falls away to the left over the long crest at the top. After that, a downshift and dab on the brakes sets it up once again for the Loop. It's funny how 160 kW can shorten a circuit.

A racesuited Richards awaited us in the pits, while other anxious competitors took to the track for their own adrenalin bath. I must have looked much like Jim did when he got the

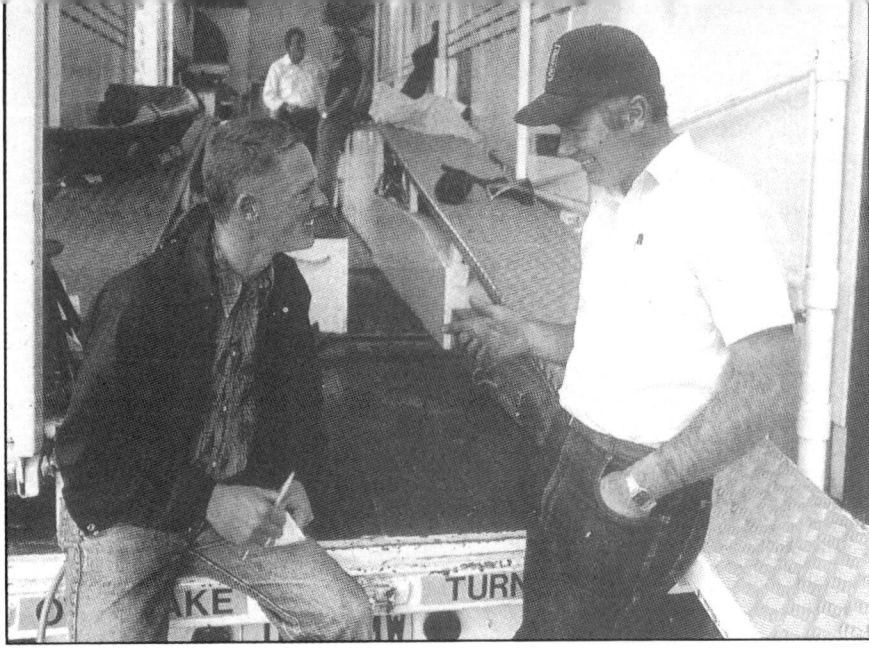

Richards explains the finer points of rallying a Morris Marina.

phone call from Frank Gardner back in 1982 — and all I'd done was ride in a racing JPS-BMW. Jim, by that time, had already paid his dues in NZ rallying and racing, persevering through the financially lean years — if rich in success — with his Sidchrome Mustang sports sedan.

Jim had packed up his family and Mustang in 1975 and left his home in Weymouth, New Zealand, bound for Melbourne and the Australian branch of his Sidchrome sponsorship. Despite victories in several major Sports Sedan races over the next two years — including two on the trot in the well-contested Tasmanian 10,000 at Baskerville — the Sidchrome Mustang's competitiveness was on borrowed time.

Perhaps Frank Gardner, at that time campaigning the formidable Chev Corvair sports sedan, made a mental note of the Kiwi newcomer back then. Gardner and others certainly took note when he appeared with the Big M Falcon in 1978, finishing second in the Sports Sedan Championship to Allan Grice, who had assumed the seat in the Corvair.

On a visit to Bathurst back in 1974, Jim and his compatriot the late Rod Coppins had worked their Torana SLR5000 into the lead from ninth place. The race that year was appallingly wet, as were the first few events with the Mustang in '75. Jim quickly earned a reputation as a rainmaster, though as things stand, the man doesn't appear to be too tardy in the dry either.

His efforts garnered him a share in the Holden Dealer Team effort at Bathurst in 1978. It was to be the first of a hat-trick of victories at The Mountain with Peter Brock.

Victory under his own steam has so far eluded Richards in the Great Race. The former Group C regulations did few favours for the JPS-BMW team, though the 3.5 litre Bavarian baby always put in a fair showing against much larger opponents. "It was frustrating, really", Richards reflects. "The team put in just as much work in preparing

the car then as they do now, I drove it just as hard, but all we managed were fifths and sixths. Exactly the same amount of effort went into those placings, and it was pretty hard on the whole team".

The coming of Group A and The Cars That Frank Built made all the difference. "It's a beaut thing to drive", says Jim of the Group A car. "The harder you drive it, the faster it goes. With the Group C car, you'd push it and the tyres would go off and you'd end up going slower. The Group A car is about 90 kW down on the old one but its lap times are much the same".

Jim's had to make few concessions to adapt — but hey, here's a guy who has stepped from a Morris Marina rally car into a Mustang V8 with no perceptible problems. "Because it doesn't have the torque of the Group C car, the new one is a little different to get off the start. If you go for the big clutch dump it'll either bog down or the clutch will start to slip, as it's only been homologated with a single-plate clutch. I use between 6000 and 6500rpm and let the clutch in gently. In Europe there's no problem with the single-plate unit because they have rolling starts."

Although Jim hasn't raced there, he is sure Gardner's cars are the equal of any of Europe's top Group A BMWs. "I don't feel as though I need to do Le Mans or the European Touring Car

"You should have thought of that before you put your Nomex on".

Gentleman Jim.

Championship. I have a family and a business in Melbourne and I'm happy to stay there. I don't need to go to Europe to prove something to myself. It would be nice to win Bathurst".

A good clue to Richards' chances was the penultimate round of the ATCC that weekend at Amaroo. He needed only 12th place or better to hold back Dick Johnson and take the title. "Unless something goes wrong, I guess we should be looking pretty good for it", Richards had characteristically understated.

Jim started from the second row while Johnson took off to establish an early lead. One wondered at Jim's tactics — surely he'd hang back, keeping it off the Armco and cruising into a safe placing. No, Jim Richards is a racer's racer and he went for it, taking ground from Johnson's tyre-hungry Mustang with every lap and looking annoyingly tidy in the process. He dived past the green machine in Mazda House and romped away for a win, and the Championship.

In third place, behind Johnson was Richards' teammate Tony Longhurst in the little BMW Motorsport 323i. The former water ski champ — another mas-

ter in the wet! — is the hot tip to partner Richards at Bathurst this year. The young Queenslander's drive at Amaroo defied description.

It was yet another well-deserved success for Richards, Gardner and the JPS-BMW team, which operates out of a headquarters in Terrey Hills, a northern suburb of Sydney. Gardner's job is as much that of an accountant as a Team Manager — he looks after the books for the team's 50/50 activities in racing and building BMW M-Sport road burners.

The money spent on the race team is very closely monitored. "A lot of racers tee up a big sponsorship deal, then turn up at the first race with a bunch of new leather jackets and the same cruddy car", Gardner says. "The art in running a team like ours lies in cutting costs without cutting quality".

The quality image is enhanced by the team's black-and-gold, $85,000 Mercedes-Benz transporter. "Yeah, it makes life easier, particularly for the interstate events, but we don't use it when we're just coming out for a testing session. We've got a little Mazda traytop and trailer for that. Likewise, you don't put new tyres on just for a test session or for unofficial practice, or you're throwing your money away".

On race days, the transporter carries two sets of wheels and tyres for each of the cars, along with one spare 635CSi engine and a couple of gearboxes and diffs. Each car will use about 45 litres of fuel during practice, and the same figure again for most races.

Gardner has a full-time staff of six at Terrey Hills, most of whom go to every race along with "a couple of weekend warriors". He estimates the cost of competing in each interstate round at $10,000, and that's with his eagle eye looking over accommodation, tyres, entrant's fees, fuel and other expenses.

Obviously, not everyone operates on the scale of this top-flight tourer team, but both Gardner and Richards have seen the other side of the fence — Jim in NZ and Australia, and Gardner in Europe. "I remember being so poor over there I couldn't afford to buy a blanket, so I moved to Sicily where it was warmer", jokes Gardner.

"I don't know how young blokes can afford to start off these days", Richards comments. "It seemed a lot cheaper to get into motor sport when I was starting off in the late '60s, but then again, I suppose it's all relative. I used to run to collect my starting fee as soon as the interstate meetings were over, so I could afford to get home again. These days, the younger blokes are lucky if they get a hand with fuel expenses".

"I suppose if you really love the sport, you just keep on doing it. Nobody told me 10 years ago I'd end up in a plum drive with JPS-Team BMW". □

POWER-FIX 6

BY JOHN McCORMICK

HAVE NO doubt that the BMW M635 is fit to join the supercar league. Its 240 km/h plus performance puts BMW in exclusive company in Australia — Aston Martin, Ferrari, Lamborghini, Jaguar and Porsche sell cars of similar ability. In terms of price, too, the M635 will align with its supercar rivals.

Although the M635 uses the Six Series bodyshell its engine is a development of the 24 valve twin cam six that powered the famed M1. That mid-engined exotic flowered briefly at the end of the '70s and then was dropped in mid-'81 after a limited production run. However, the fabulous 207 kW engine was kept alive and appeared, modernised and with more power, in the M635 as originally launched in Germany two years ago.

Unlike the M1 unit, which used mechanical fuel-injection, the 3.5-litre six is managed by a Motronic system. As a result power is up by 7 kW over the M1 to 214 kW at 6500 rpm and torque is increased marginally to 340 Nm at 4500 rpm. Suspension changes over the standard 635 include gas dampers, tauter springs and wider alloy wheels shod with 220/50 VR390 Michelin TRX tyres. The overall gearing of the long-legged 635 transmission is lowered considerably to suit the car's sporting nature, a 25 per cent limited slip differential is fitted and the disc brakes are enlarged.

Visually BMW's Motorsport division has done little to distinguish the M635 from the standard car. The most obvious giveaway are the superb-looking BBS alloy wheels. The only other changes are a deeper front spoiler, bootlid spoiler, colour-coded door mirrors and discreet M badges.

Despite having 1588 kg of body weight to shift and poor aerodynamics (Cd is 0.39) to contend with, the M635 will scorch from standstill to 97 km/h in 6.2 seconds, just before you reach maximum revs in second gear. Third gear will run to 160 km/h at the 6500 rpm power peak (a fuel cut-out operates at 6800 rpm), and fourth is good for 210. In fifth it's not too hard to see an indicated 240 km/h but pushing on to the claimed 254 maximum would need several kilometres of clear road.

Although the M635 feels most impressive for its urge above 160 km/h, it is also very strong in its mid-range acceleration. For instance, using third gear 65 to 160 km/h takes a mere 10 sec. But the M635 doesn't have to behave like a racer all the time. Despite the lowered final drive ratio fifth gear is still a relatively long 40 km/h per 1000 rpm, and it's possible to cruise at 120 with just 3000 rpm on the tacho.

The M635, in fact, can be quite

refined transport but BMW's portrayal of the car as a luxury coupe is stretching the truth. Unlike the smooth and quiet engine in the standard 635 the 24 valve version always makes its presence felt. Floor the throttle and a rumble at idle turns into a deep, satisfying growl and finally a full blooded roar as the engine nears peak revs. The pleasing thing is that while the engine noise is always audible it never becomes harsh.

With such a massive tyre footprint the M635, as you might expect, has impressive roadholding. The handling is improved by the firmer suspension settings but it's still at the mercy of the semi-trailing arm rear axle design. The car understeers noticeably to start with but if you push further the handling will pass through a neutral phase into familiar BMW oversteer. The huge reserves of power make it possible to kick out the tail more or less at will although in the dry you need to be moving at considerable speed or rounding a hairpin bend to do it. Body roll is well contained but the BMW doesn't corner as flatly as a 928 or Mondial, nor does it react so kindly to throttle changes.

The ZF power steering, however, is superb. It's surprisingly heavy but extremely communicative and responsive, and stability at very high speed is beyond reproach.

Sophisticated though the M635 is in some respects, its ride quality is not the finest. At high speeds it's not too bad but around town or over badly surfaced country roads the suspension reacts harshly, with poor bump-thump absorption. In wind noise and tyre noise, too, the M635 is lacking for what is presented as a luxury tourer.

Elsewhere the car benefits by being a BMW, with slick gearbox, good driving position, comprehensive electronic monitoring systems and roomy-feeling cabin. In fact, limited head and legroom make the M635 a two-plus-two at best, though visibility out of the car is excellent compared with its supercar rivals. Fuel consumption is good with an overall return of 7.1 km/l (20 mpg), but prolonged high speed cruising — when economy drops to not much more than 5 km/l or 15 mpg — shows up the 70 litre tank to be meanly sized. Who wants to stop unnecessarily for petrol, after all, when before you is a continent coverer of the M635's potential?

QUANDARY A

COMPARISON R&T ROAD TEST

FURNACE CREEK:

BMW 635CSi, JAGUAR XJ·S, MERCEDES-BENZ 560SEC

No surprises here. These are three of the best cars in the world. They know it. We know it. You know it. The quandary (assuming you can own just one) is deciding which best is best for you.

BY T C BROWNE

LEAFING THROUGH CRUTCHFIELD'S wonderful stereo equipment catalog is nothing like listening to *Also sprach Zarathustra* through a pair of Klipsch horns positioned in the corners of your nearest empty bowling alley; and reading about three of the best cars in the world is nothing like driving them through Death Valley in the false dawn at 140 mph.

More about that later. First, we should explain that the opaque title was inspired by a staffer who created one similar just for her own convenience in identifying the paperwork package that grows like tree fungus around each *Road & Track* test. This apt designation expresses her quandary when she was obliged to choose one of these three cars for a short trip in the line of duty. Her problem was the mirror-image of the one with which our testing team was confronted as the desert games ended and the time came for the reckoning. "The Best Car" turned out to be the one you were driving. Different as they are, each of these standard-bearers is the ne plus ultra of the factory from whose loins it sprang. One is quite long in the tooth; another boyish and impetuous; the third—and newest of them all—is anything but youthful. It must be regarded as the most mature of the group.

The pricing in this category of motorcar is not exactly popular. One is tempted to conclude that anyone able to own one of these cars could probably afford all three, thus avoiding the trauma of having to choose among them. Sober reflection, however, argues that this facile conclusion misses the point. Most people able to write a check for one of these treasures tend to be more circumspect about such expenditures than those who acquire it on the never-never. And there is reason to believe that at least one of our trio is within easy reach of your average Yuppie household now that two incomes and zero children have become so common.

Another curious development, and one becoming more conspicuous, is the tendency of different marques to dominate the upmarket in individual communities. It's no secret that one Mercedes or another has been *de rigueur* in Beverly Hills for just ages; but not everyone is familiar with the penetration of the Jaguar XJ6 into the green hills of Palos Verdes. The enterprising BMW, meanwhile, has staked a claim to certain sections of the demographic rather than the geographic market. The Bavarians' skill in wedging a tempting array of models between Mercedes and Porsche—while claiming virtues of both marques in their advertising—has rewarded them with hegemony over a multitude of mini markets. The speed with which this has been accomplished should have created a good deal more uproar than it has.

In 1960, the citadel of the blue-and-white prop spinner was facing its own *Götterdämmerung*. The excellent and archaic motorcycle cost as much as a VW automobile, and the handwriting on the wall was becoming so clear that even the most optimistic Münchener could read it.

The rescue party hired a marketing wizard. He launched the modest, stalwart 1500 economy 4-door, and it caught the public's fancy. So did the 1600; and the 1800; and the 2000, in all its appealing guises. Then came the 1602/2002. Here was the ancestor of what we would one day call Eurosport.

Since then, the proliferation of BMW models has been breathtaking. And the eagerness of Americans to buy whatever the BMW managers wish to sell is in the nature of miraculous.

It is puzzling to contemplate that, for 1986, nothing from BMW will come into this country with fewer than six cylinders. Of course, nothing will come in with more than six cylinders. Yet, there is a total of 10 models to be offered. While giant Daimler-Benz is offering the same number of choices, the range of their engines destined for America runs from fours, through eights, inclusive, omitting only sevens.

The early marketing of Mercedes in North America was the work of the maladroited. From the puppet-master, Max Hoffman (who introduced BMW and had a shot at Jaguar, as well), through the awful Curtis-Wright and Studebaker nightmares, the 3-point-

GENERAL DATA

	BMW 635CSi	Jaguar XJ-S	Mercedes-Benz 560SEC
Base price	$43,055	$37,800	$62,110 (east & Gulf) $62,530 (west)
Price as tested[1]	$43,705	$40,400	$64,030 (west)
Layout	front engine/rwd	front engine/rwd	front engine/rwd
Engine type	sohc inline-6	sohc V-12	sohc V-8
Bore x stroke, mm	92.0 x 86.0	90.0 x 70.0	96.5 x 94.8
Displacement, cc	3430	5343	5547
Bhp @ rpm, SAE net	182 @ 5400	262 @ 5000	238 @ 5200
Torque @ rpm, lb-ft	214 @ 4000	290 @ 3000	287 @ 3500
Fuel injection	Bosch Motronic	Bosch-Lucas L-Jetronic	Bosch KE-Jetronic
Transmission	5-sp manual[2]	3-sp automatic	4-sp automatic
Gear ratios: 5th	0.81:1		
4th	1.00:1		1.00:1
3rd	1.40:1	1.00:1	1.44:1
2nd	2.20:1	1.48:1	2.41:1
1st	3.83:1	2.48:1	3.68:1
Torque-converter stall ratio		2.4:1	2.2:1
Final drive ratio	3.25:1	2.88:1	2.47:1
Steering type	recirc ball, power assist	rack & pinion, power assist	recirc ball, power assist
Brake system, f/r	11.2-in.vented discs/11.2-in. discs; hydraulic assist, ABS	11.1-in.vented discs/10.3-in. discs; vacuum assist	11.3-in. vented discs/11.0-in. discs; vacuum assist, ABS
Wheels	forged alloy, 390 X 165TR	cast alloy, 15 x 6J	cast alloy, 15 x 7J
Tires	Michelin TRX, 220/55VR-390	Pirelli P5, 215/70VR-15	Michelin XWX, 205/65VR-15
Suspension, f/r	MacPherson struts, double lower links, tube shocks, anti-roll bar/ modified semi-trailing arms, coil springs, anti-roll bar	unequal-length A-arms, coil springs, tube shocks, anti-roll bar/ lower A-arms, fixed-length halfshafts, trailing arms, dual coil springs & tube shocks, anti-roll bar	unequal-length A-arms, coil springs, tube shocks, anti-roll bar/ semi-trailing arms, coil springs, tube shocks, anti-roll bar, torque compensator

[1] Price as tested includes: For the 635CSi, Gas Guzzler tax ($650); for the XJ-S, electric sunroof ($1300), Gas Guzzler tax ($1300); for the 560SEC, Gas Guzzler tax ($1500)
[2] 4-sp automatic also available

Top to bottom: BMW, Jaguar and Mercedes.

ed star has survived in the USA, nourished from deep roots in the *Vaterland*. While the automobile is hardly a sideline with the people who are expending much effort in 1986 to convince you they invented it, Daimler-Benz AG, unlike Bayerische Motoren Werke AG and Jaguar plc, has wide-ranging interests elsewhere (the other two points on the star).

Now, Mercedes marketing in North America is in the hands of magicians. There has been much speculation—particularly among drivers who will never own the car—as to what the magic is. How do they get all that money for one automobile? This is one of the toughest issues for researchers to get their hooks into. The last guy you want to ask why he owns a Mercedes is a Mercedes owner. But here's an educated guess: Getting all that money for one car *is* the magic. While the occasional car freak may own a Mercedes, it is in the ranks of the untutored that the great market penetration has taken place. In Beverly Hills, turbulent crucible of all status-seeking nonsense, one-upping your peers requires two essentials: 1) The symbol has to be extravagant; 2) the peers have to know it. If

BMW 635CSi.

Jaguar XJ-S.

Mercedes-Benz 560SEC.

you just had your hair done at Sassoon, not everyone will be able to tell. But, by now *everybody* knows that the 3-pointed star means wanton excess, darling. This cachet could not have been perpetuated had the Mercedes marketing mavens failed to raise prices, regardless of inflation rates, money-market relationships or improvement in the product. Thus, one of the great motivating myths of capitalism has been thrown into reverse by the wizards of Untertürkheim (more likely New Jersey). Rather than rising to the top of the heap by delivering a better product at a steadily declining price, M-B of N-J has managed the unimaginable by following this dictum: As long as they keep buying your stuff, raise the price. And, so long as America produces affluent consumers at a rate greater than Stuttgart's old-world Turkish craftsmen can crank out cars, the music needn't stop.

And what of the most venerable of our test subjects? But for a geniune love of Jaguar cars in the USA, the good humor of the workers on the shop floor and the inspired leadership of John Egan, there might not even be a Jaguar for us to test or for you to buy. And that would be a real pity. For here is that rare, happy marriage of traditional elegance with modern performance. The moment you settle into the driver's seat, you know it's a Jaguar. Oh, the smell of it! Okay, so you banged your head on the roof as you entered. Classic GT conformation doesn't come free. And the view out the rear may be a tad limited. Those flying buttresses were erected back when Pininfarina thought they were cute. Observe the lovely polished burl bolsters and don't ask.

The XJ-S is an improbable Jaguar, anyway. It certainly doesn't fit the sports car tradition first established by the prewar SS100 and carried on so nobly by the revolutionary XK120 and the spectacular E-Type. It doesn't belong with the Mark VII and Mark X saloons that finally found coherence in the XJ6, either. Never mind. You want a V-12? This is where you get it, and who knows for how long? During one of those cosmic spasms through which the once mighty British Motor Industry insisted on putting itself over the past decade, there was a plan to phase out the V-12 in the early Eighties. And during the oil crisis, a badly addled segment of Jaguar management had the notion to replace it in the XJ-S with the venerable 6-cylinder engine, for goodness' sake! For sure, we would have a 2-car comparison test if that had come to pass.

Mechanical features of the XJ-S are a unique amalgam of the future and the past. While the cunning Michael May head is on the cutting edge with an impressive 11.5:1 compression ratio that boosts output to 262 bhp at 5000 rpm on unleaded regular, the archaic air injector pump is still used to control HC and CO emissions. And while just about everything else in the class has a 4-speed automatic transmission—either standard or optional—the XJ-S is making do with the GM 400 3-speed, period. No manual box is available (although the new, fuel-frugal XJ-S 3.6 coupe and cabriolet, coming soon to these shores, have the Getrag 5-speed). Same story with the brakes. While the 635CSi and 560SEC have anti-lock braking standard, the XJ-S stops okay. Always has.

So, what we have in the Jaguar, then, is a luxury 2+2, with the second 2 for kiddies, that moves with graceful, silent alacrity.

But what are the cars really like?

DEATH VALLEY may be a great place to test cars, particularly high-zoot cars, but it's not a *real* place. "Moonscape" is a term often used to describe such desolation, and most of Death

Valley is indeed desolate. Having acknowledged all that, we should point out that the moon lacks first-class paved highways where one may place all sorts of outrageous demands on a modern automobile in eerie solitude, leading to a kind of desert euphoria in which the driver gradually achieves immortality and the car becomes a chariot of the gods.

This is no way to accumulate hard-nosed comparisons on living with these high-tech marvels in the real world, but it is great fun. And while it wasn't actually planned this way, the validity of the test increased greatly as it became obvious that getting to know the real differences in these cars did require just such a setting.

Here might be a place to note a few interesting similarities in these discrete carriages before the prominent differences are examined. Single overhead-cam engines are common to all, as is a front-engine/rear-drive layout. All are fuel injected, yet each uses a different Bosch system. All wheels are alloy, independently suspended and braking is with discs on all wheels. The major instruments are remarkably similar in size, layout, color and function, although the Jaguar keeps its unique and not universally appealing vertical instrumentation for minor monitors.

While you sit on leather in each car, the Jaguar also surrounds you with real hide. It might be worth mentioning that while the Mercedes has the exclusive claim to heated front seats and an electrically adjustable steering column (with a memory of *your* favorite position), the other two combatants have the onboard trip computers. So yours is the choice of jerking yourself back and forth and in and out in the Mercedes, or frustrating the average mileage readings by fancy right-foot work in the BMW or Jaguar.

The test cars are near perfect expressions of what each represents, and in Death Valley they seemed omnipotent. Like approaching the stunning stranger in the Astor Bar, it wasn't so much whether the cars were good enough for the drivers, but whether the drivers would measure up to the cars.

Entering the BMW seems like stepping into the next century. Everything is high-tech, ergonomic, non-reflecting, purposeful. And ready to be turned loose.

The approach to the M-B 560SEC recalls a bank vault of the Thirties, displayed right out front where the customers could just feel the security.

Insinuating yourself into the Jaguar is like stooping to enter the fort you and your pal built in the vacant lot when you were nine. Once inside, it's just as private and cozy.

The BMW turned out to be the sports car of the three, the 5-speed manual box filling much of the gap created by its smaller engine. Slightly stiff, the 635 was taut, purposeful, waiting for your instructions, always pleased to carry them out. The flat-black inte-

STANDARD AND OPTIONAL EQUIPMENT

	BMW 635CSi	Jaguar XJ-S	Mercedes-Benz 560SEC
5-speed manual gearbox	S	NA	NA
4-speed automatic transmission	O	NA	S
3-speed automatic transmission	NA	S	NA
Cruise control	S	S	S
Power-assisted steering	S	S	S
4-wheel disc brakes	S	S	S
Anti-lock braking system	S	NA	S
Alloy wheels	S	S	S
Foglights	S	S	S
Electrically adjustable mirrors	S	S	1
Heated outside mirrors	S	NA	S
Central locking	S	2	S
Adjustable steering column	S	S	3
Eight-way electrically adjustable front seats with memory	S	NA	S
Full leather upholstery	NA	S	NA
Leather seating	S		S
Heated front seats	NA	NA	S
Supplemental restraint system	NA	NA	S
Tinted glass	S	S	S
Electric window lifts	S	S	S
Trip computer	S	S	NA
Rear-window defroster	S	S	S
Air conditioning	S	S	S
Automatic climate control	NA	S	S
Tilt/slide electric sunroof	S	O	S
Anti-theft AM/FM stereo/cassette system with electric antenna	S	S	S
Theft-deterrent system	D	D	S

S = standard; O = optional; NA = not available; D = dealer-installed; 1 = right-hand mirror only (driver's mirror manually adjustable from inside); 2 = electric door locks; 3 = electrically adjustable, included in seat memory system.

PERFORMANCE

	BMW 635CSi	Jaguar XJ-S	Mercedes-Benz 560SEC
Acceleration:			
Time to distance, sec:			
0–100 ft	3.5	3.4	3.4
0–500 ft	9.1	9.3	8.5
0–1320 ft (¼ mi)	16.3	16.5	15.5
Speed at end of ¼ mi, mph	86.0	89.5	92.0
Time to speed, sec:			
0–30 mph	2.8	3.4	2.9
0–60 mph	8.5	8.7	7.0
0–80 mph	14.1	13.9	11.5
0–100 mph	23.1	21.4	18.6
Top speed, mph	131	141	145
Test fuel economy, mpg	17.5	13.2	14.5
Brakes:			
Stopping distance from:			
60 mph	149	162	149
80 mph	263	287	267
Pedal effort for 0.5g stop, lb	18	31	15
Fade: % increase in effort, 6 stops from 60 mph @ 0.5g	30	nil	nil
Overall brake rating	very good	good	excellent
Handling:			
Lateral acceleration, g	0.80	0.75	0.77
Slalom speed, mph	58.4	60.0	56.4
Interior noise, dBA:			
Idle in neutral	53	50	56
Maximum, 1st gear	74	76	68
Constant 30 mph	62	57	56
50 mph	65	63	61
70 mph	73	69	67

rior, with a padded, adjustable steering wheel and firm, leather-covered seats, carried out the sporting theme. The impressively simple environment was constantly inviting the driver to fool around and see what he could do. What he could do was pretty much as he pleased. The ABS braking was about perfect; the BMW people are terribly proud that *their* cars' power assisted brakes have hydraulic rather than vacuum assist.

In the BMW, you feel more in control of the car; in the Mercedes, the car feels more in control of you; the Jaguar leaves all that nicely up in the air.

If one abandons the BMW to enter the Jag, the transition is more abrupt than if the move is Mercedes to Jaguar, this because of a more nearly equal comfort quotient between the latter pair. The English car continues to charm all who cross its portals with the traditional, Edwardian luxury once associated with those much-mourned saloons of yore. The wood on the doors seems a bit much, but the stereo performance is impressive, and the silent engine is so torquey that the somewhat uncertain kick-down performance of the GM 3-speed automatic doesn't seem to matter. The trip computer's responses are much quicker than those of the one in the BMW.

The Jaguar XJ-S is smooth, quiet, luxurious, soft, elegant, fast, strong, comfortable, reassuring and—best of all—cheap. Well, relatively, anyway. But it lacks the sporting character of the BMW, or that certain "We're going to take care of you, no matter what happens" feeling of the Mercedes. If the design is dated, then so is the Cellini saltcellar and your grandmother's Fabergé egg.

The Mercedes 560SEC requires terms not yet in the lexicon to describe its excellence. If it's possible to have a car that thinks for you, this may be it. That it patronizes you at the same time may also be true. What a study in contrasts!

Suppose you are caught in the infamous freeway rush hour, and there's no sneaky escape route. What you can do in the 560 is wind up all the windows (easy), light off the a/c (not so easy), surround yourself with a little Vivaldi from the stereo (not at all easy) and picture yourself in Ravenna's Basilica of San Vitali, with the chamber orchestra sawing away as you look up at the mosaic of Theodora on her way to the Holy Land (piece of cake). If you feel a little inferior from being clumsy when you try to operate things, not to worry. You and the car are getting off on the right foot.

Suddenly, all the other cars have peeled off and the *Autobahn* before you is empty. You punch the pedal, the gearbox downshifts, the only way you know the engine revs are mounting is to watch the tach, the upshift comes at 120, and you settle back to resume your trip at 140 mph, with no more noise or fuss than there was at 60, which is to say—none. The question is: When the time comes, what are they going to improve? It must be an awesome thing for an engineer to discover that he and his colleagues may have created something so near perfection that no one can find anything better to do.

If you enjoy esthetic juxtapositions, the Mercedes might associate with Richard Wagner and Peter Paul Rubens; the Jaguar would feel more comfortable with Henry Purcell and Gustav Klimt; the BMW would surely seek out the company of Igor Stravinsky and Frank Frazetta.

After all that hyperbole, a summation hardly seems necessary, but for those among you who need one: Use all three cars gently and the Mercedes takes a slight lead over the Jaguar. Drive them all *con brio* and the BMW barely edges out the Mercedes. If your image is important to you, the Mercedes is certainly senior, the Jaguar a decent second place, and the BMW may label you *au courant*. Needless to say, the Jaguar is the Best Buy.

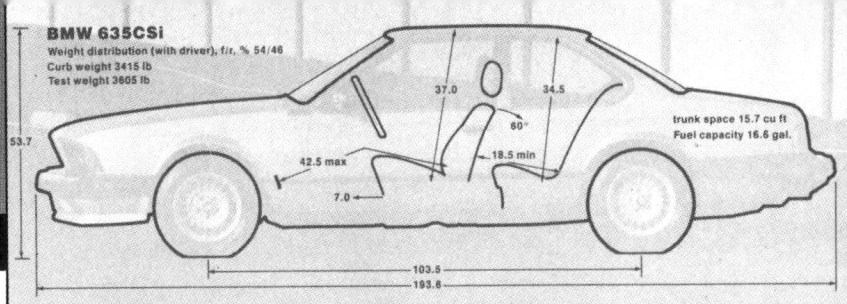
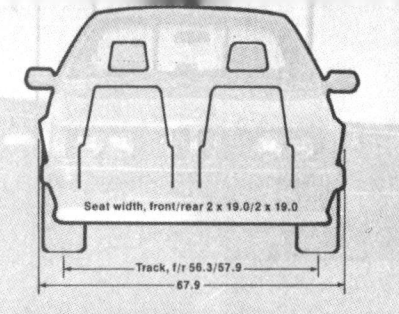
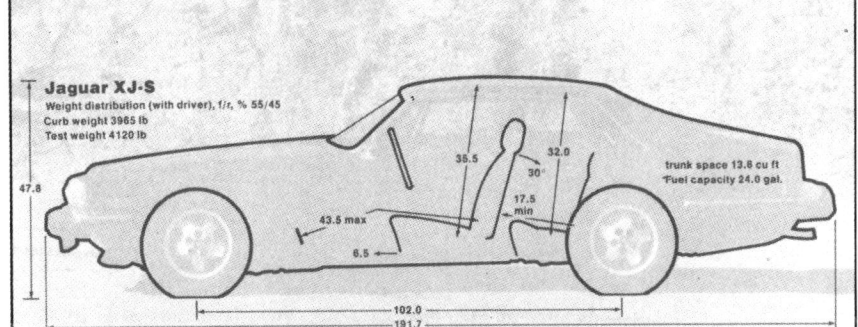
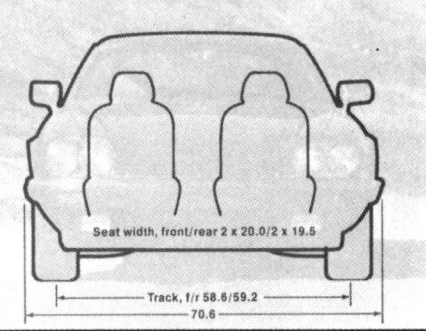
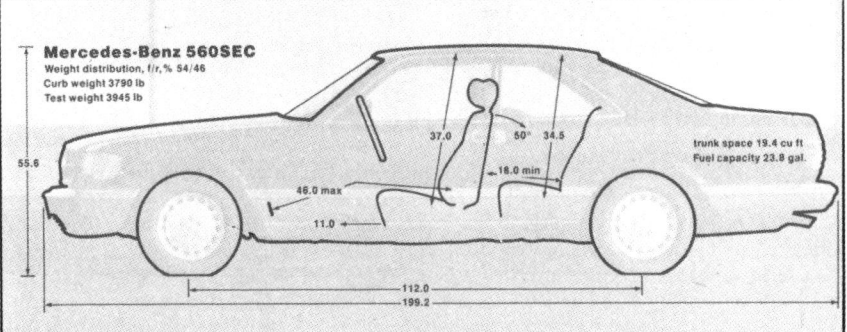
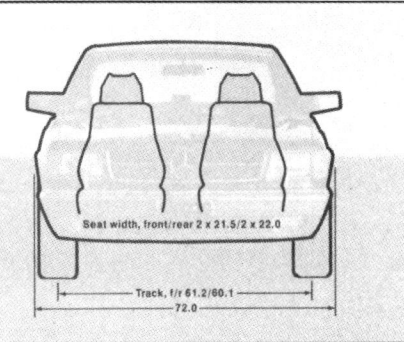

CUMULATIVE RATINGS—SUBJECTIVE EVALUATIONS

	BMW 635CSi	Jaguar XJ-S	Mercedes-Benz 560SEC	Comments
Performance:				
Engine	8.0	8.7	9.3	5.5-liter engine makes 560SEC the runaway performance winner.
Gearbox	7.7	6.3	8.7	Only the BMW has a "gearbox"; of the automatics, Mercedes' is superior.
Steering	9.0	8.3	8.0	Mercedes' steering is heavy, Jaguar's light; BMW's is best compromise.
Brakes	9.3	8.0	8.3	With ABS, BMW & Mercedes are state-of-art. Jaguar not competitive here.
Ride	8.0	8.7	8.7	BMW's ride is the most sports car-like, Mercedes' the most civilized.
Handling	9.3	7.3	8.7	The payoff: BMW is tautest, turns in best skidpad times.
Body structure	9.3	8.7	10.0	All are staunch by world standards, but Mercedes is THE world standard.
Average	8.7	8.0	8.8	
Comfort/Controls:				
Driving position	8.0	8.0	8.3	All good, but Mercedes' electric super-adjustability wins by small margin.
Controls	7.7	7.7	9.0	Mercedes' supremely logical approach wins.
Instrumentation	8.0	8.3	9.0	Jaguar's instrumentation is most complete, Mercedes' most legible.
Outward vision	9.0	7.0	8.3	BMW's outward view is best; Jaguar's roof buttresses block rearward vision.
Quietness	7.7	9.0	9.7	Mercedes has combined pavement-blistering performance with remarkable silence.
Heat/vent/air cond	8.0	8.3	8.3	Jaguar & Mercedes have easy-to-use automatic climate-control systems.
Ingress/egress	9.0	7.7	9.7	Ease of ingress & egress is in inverse proportion to their heights.
Front seats	8.3	8.0	7.7	BMW's electrically adjustable sports seats are tops.
Rear seats	6.7	5.0	7.0	Mercedes most closely approaches full 4-passenger seating.
Luggage space & loading	8.7	8.3	9.3	Mercedes has the largest trunk, Jaguar the smallest.
Average	8.1	7.7	8.6	
Design/Styling:				
Exterior styling	8.0	8.7	8.3	Despite bulbous sides & dated details, Jaguar exudes old-world character.
Exterior finish	9.0	8.7	9.0	All three are beautifully finished, the Germans a bit more so.
Interior styling	8.3	9.0	9.0	Jaguar and Mercedes are both elegant—in their British & Teutonic ways.
Interior finish	9.7	9.7	9.7	All excellent. Jaguar uses leather everywhere, not just on seats.
Average	8.8	9.0	9.0	
Overall Average	8.0	7.7	8.3	
Staff Members' Preferences:[1]				
Price-independent	4	6	9	
Price-dependent	5	9	4	

[1] Three staff members' preferences, 1st choice = 3 points through 3rd choice = 1 point. One voter awarded a tie for 2nd place in price-independent choice.

THE STRUGGLE between Mercedes-Benz and BMW for German — perhaps even world — automotive supremacy is an intricate, multi-layered affair, fought across a battleground which covers all models and intense at every level. Three series versus 190, Fives against 230/300E and Sevens eyeball-to-eyeball with the 420 SE/560 SEL, they match almost model for model.

At the peak of this rivalry are the coupes, the indulgent, two-door occasional four-seater models for the very rich who have no need for the practicality of four doors, who can aspire to something more stylish, more exclusive than a sedan, to cars some might consider the best sets of wheels in the world.

Mercedes-Benz says of the coupe, "this concept has always embodied the successful combination of perfect engineering and exceptional aesthetic appeal. The pace setters for refined, high-performance cars, because they also incorporate all the other characteristics which make a Mercedes."

BMW puts it this way, "By its very character, the 635 CSi is the type of car only BMW can build. It unmistakeably combines the merits of motor racing achievements with a technically outstanding concept of engineering. The BMW coupe is not merely derived from a similar saloon version. Instead, it is more compact, dynamic, nimble and agile than a saloon could ever be."

In Australia the latest expressions of this coupe competitiveness are the Mercedes-Benz 560 SEC and the BMW 635 CSi. There is nothing better, nor more expensive, from either car maker's range. In the case of Mercedes the 560 SEC is indeed the ultimate expression of the three pointed star. For BMW there is one higher level but the problems and cost of certifying the M635 CSi, with its 213 kW, 24-valve twin cam six, mean that it's a non-starter.

Any student of German cars will know that the two door coupes have played an important role in creating today's popular image of both makes. BMW has had similar models since the 327 of the '30s, the 503 of the '50s, the 2000 CS of the '60s, the 2.8 CS/3.0 CSi of the '70s to the current 635 CSi that was first released in smaller engined form as long ago as 1976, and which isn't due to be replaced until 1988.

The Mercedes lineage is just as long. In 1938 there was the 320 Coupe, 1955 the 300SC, 1957 the 220 Coupe, 1969 the 280 SE 3.5 and in 1971 the SLC, leading to the SEC development of the present S-class sedans. The heritage of a fast, refined, luxurious large two door coupe is unmatched by any other car maker.

BMW's 635 CSi is making a comeback to Australia. It was last listed, as the automatic-only 633 CSi, in our New Car Prices columns in March '81. The return has been made possible by the use of the unleaded 3.5 litre six from the 535i sedans in the coupe body. This engine can be mated to either of three gearboxes: an automatic, of course, and two manual five speeders, the close-ratio Getrag box from the M535i or the taller, wide-ratio BMW-built box used on the test car.

BMW Australia has been keen to sell the 635 CSi ever since Jim Richards began beating the Commodores in touring car races, and especially once he won the '85 Australian Touring Car Championship. Of course BMW is hoping the old Ford adage "win on Sunday, sell on Monday" applies to the CSi. Even in these tough times the company is convinced it can move 200 coupes during 1986, 80 percent of them automatics, 15 percent with the wide ratio five speeder and just five percent — 10 cars — with the ultra-sporty gearbox. Regardless of the transmission the price is $86,000 and there are just two options; an electric sunroof adds $2200, and superb buffalo leather upholstery sells for $2490, to take the list price of the car we drove to $90,690.

And yet there is a whopping difference in price between the two cars, because the 560 SEC carries a towering $160,162 tag, or just about 70 big ones more than the BMW. If you want to be objective about it that means you could have both a Six and a Seven series BMW for the price of one SEC, and still have enough for a first class round the world air ticket.

These are huge prices, but

DUEL OF THE TITANS

BMW 635 CSi-v-Mercedes-Benz 560 SEC

PHOTOGRAPHY WARWICK KENT

fairly consistent with the figures quoted in Germany for the same models with very similar equipment.

At the time of its release in February (and before the dramatic fall in new car sales) Mercedes predicted it would sell 100 560 SECs during '86 — as well as 200 SELs and 100 plus SLs — a figure that now looks somewhat optimistic. Still, it is the flagship and there are some (enough?) people who will settle for nothing less.

Performance

If both cars are essentially conventional in mechanical layout, their solutions to the problem of powering a high performance coupe take distinctive routes. BMW follows its traditional answer and relies upon a medium sized straight six as the motive power for the relatively smaller and lighter (1495 kgs versus the SEC's 1785) 635 CSi. Mercedes is confident that nothing can match sheer engine capacity and is happy to use a massive single overhead cam 5.6 litre V8 with vast amounts of torque to shift its massive bulk.

BMW's now well known 3430 cm^3 six has full Bosch Motronic digital electronic fuel injection and is in the same state of tune as the engine in the 535i, with 136 kW at 5400 rpm and 290 Nm at 4000 rpm on a compression ratio of 8.0:1.

In contrast the Mercedes bent eight is a long stroke version of the old alloy five-litre engine that was never officially sold in Australia. It has a full electronic engine management system, a mechanical-electronic fuel injection system and — even in unleaded form — it still produces 182 kW at 4800 rpm and a huge 400 Nm at 3500 rpm. The resulting performance (and remember the BMW has the clear advantage of a five speed manual gearbox) is remarkably close.

The heart of any BMW is the engine, and engines don't come any sweeter than the 3.5 litre six in the CSi, despite the fitting of a catalytic converter and a subsequent loss of 27 kW in its unleaded form. It remains supremely flexible, for it will pull the 40.7 km/h per 1000 rpm fifth gear from 1000 rpm; yet it remains virtually vibration free all the way up to the ignition cut-out at 6300 rpm. It will trickle through traffic in the higher gears at very low rpm, yet is responsive to the throttle although it is not until the crankshaft reaches 4000 rpm that it truly comes alive. Then it simply soars with that hard, businesslike hum that is a delightful howl to the ears of any BMW enthusiast. Off the line there is wheelspin aplenty as the car rockets forward to cross the standing 400 metres in just 15.6 seconds, 0.3 sec quicker than the slightly lighter M535i, with the close ratio gearbox, tested in the June issue. The extra gear change required in the sedan must account for the difference.

Fast though the CSi is, the Mercedes with all its cubic grunt is faster, despite its additional weight, its automatic transmission and very tall gearing. With the latest S-class, Mercedes has set the auto so that it always starts in first gear — previous 380 models started in second on anything less than full throttle — and it changes up so close to the redline that you just know that the engineers in Stuttgart have set it up to shift at the most efficient point. As befits its clearly less sporting character, the engine is much quieter at idle and in normal motoring than the BMW six.

Excessively long throttle travel disguises the SEC's potential, because at first touch of the accelerator nothing happens. It is a purely false impression for once the linkages have taken up the slack there is never any doubting the sheer power of this engine. No, it won't spin the rear wheels on a coarse bitumen surface but, automatic or not, it thrusts the big car down the strip at an astonishing rate, the spread of power seemingly linear in the way it is delivered from 1500 rpm to the redline of 6100 rpm, or 1300 rpm over the point at which maximum power is developed. There is always power on hand even if it means kicking down into

a lower gear. The speedometer says it all for the change-up points are at an indicated 70 km/h, 120 km/h and a breath-taking 205 km/h. The markings are serious; so is this car.

The 560 SEC is easily the fastest-accelerating automatic this magazine has ever tested. We've yet to run figures on the four-valve 928S Porsche, but nothing else currently sold in Australia would come remotely close. A standing 400 m time of 15.4 seconds is impressive by any standards; the fact that it can be repeated simply by flooring the throttle means that no special driving technique is required. The car's potential is seemingly endless.

The irresistible strength of this mighty engine is placed in its right perspective when you realise that its direct drive fourth is geared to give 48.1 km/h per 1000 rpm. So tall is the gearing that 110 km/h is just under 2300 rpm. At this speed the engine is barely idling along and can hardly be heard above the tyre noise that seems inherent in most German cars.

It is its dual character that makes the 560 so desirable,

for it is just as happy meandering as it is running hard. Either way the level of performance is deceptive and it is all too easy to exceed the speed limit without consciously doing so. Serious drivers will find the comfortable cruising speed is set more by road conditions and legality than any dynamic

Interior of the BMW is more obviously sporty, in the overall style of the car itself. There's real room in the rear if the front seats aren't too far back. Front seats are very snug, vision is good, and the buffalo hide truly luxurious

limit of the car. Two hundred kilometres per hour — still just 4200 rpm — is effortless and quiet enough for conversation to continue at normal levels, the SEC running arrow-like with marvellous stability and the beginnings of wind rustle from around the A-pillars. Beyond 200 km/h there is still strong acceleration up to the top speed of 225 km/h.

If this is not the best automatic gearbox in the business then we've yet to try any better for the changes slur from one ratio to another and on light throttle openings it requires a constant eye on the tachometer to confirm that a change has actually taken place. First gear can now be held by knocking that beautifully precise gear selector sideways when in B-position, an advantage in very slippery conditions or when towing a caravan. Most owners, of course, will be happy to leave the transmission alone and rely upon the engine's vast torque.

BMW's "overdrive" gearbox gives the 635i a super slick gear change, perfectly complemented by an ultra-smooth clutch. Most importantly, however, it means the car has the long legs and the effortless cruising ability that is so necessary in such a vehicle. The close-ratio box, which made the 535i sedan seem quite unnaturally undergeared, produces the same results in the coupe, suitable perhaps for racing but not for the road.

The overdrive gearbox errs in the other extreme, though we have no doubts that for most people it will be

preferred. Fifth gear in the close ratio box gives only 33.0 km/h per 1000 rpm so the engine will easily reach the redline in top — which thus artificially limits maximum speed to 208 km/h. In the overdrive gearbox, fourth gear is the same as fifth on the close ratio option while the fifth gear of 0.81 gives 40.7 km/h for a top speed of 217 km/h at just 5300 rpm, and with it a clear advantage in economy.

Those really keen drivers who wind out this lovely engine over their favourite mountain pass will immediately notice that there is a penalty to be paid for this cruising ability, for the lower indirect ratios are very widely spaced with a huge gap between the 94 km/h second gear and 149 km/h third. More than once when accelerating out of a winding uphill corner we had the engine on the ignition cut-out in second or, after a change up, struggling to climb up on the cam in third.

In outright performance, then, the Mercedes is a clear but close winner although those who enjoy the pleasures of a manual gearbox would never consider anything but the BMW.

Where the BMW beats the Benz hands down is in

fuel economy. There is just no way 5.6 litres of bent eight in a car weighing 1785 kg can hope to match a 3.5 litre six pushing 1495 kg, and so it proved during our tests. The 635 CSi's overall average of 7.5 km/l (21.2 mpg) is truly excellent given its performance, while the Mercedes' 6.0 km/l (16.9 mpg) must also be considered outstanding for a car of this specification. Despite a tank capacity of 70 litres — 20 below the 560 — touring ranges are virtually identical at around 480 kms.

Dynamics

Price aside, the biggest distinction between these two German coupes is the way they feel from behind the wheel. This is never more obvious than in the first few kms in the driver's seat, when the differences seem almost overwhelming. The BMW is altogether more compact and sporting, firmer in the way it rides with more precise steering and far more overtly an enthusiast's coupe, making all the right sounds. And it has, of course, that manual five-speed box.

Nobody could pretend the Mercedes is in any way compact. Despite being a significant 220 mm shorter in the wheelbase than the very similar 560 SEL (though the SEC shares not one exterior panel with the sedan) it is a further 220 mm longer in wheelbase than the 635 and it feels that way, wider with a vast bonnet, softer to drive and obviously more directly related to a sedan than the purpose-built BMW coupe.

These impressions diminish with familiarity, but they never entirely disappear. Maybe, we found ourselves wondering, these cars are so different in their essential character that a Porsche 928S or Jaguar XJS should have been included in this comparison. Look deeper into their specification and marketing objectives, however, and you realise that the 928S and XJS, competent rivals though they may be in so many other ways, are not to be taken seriously as practical as four seaters.

BMW's handling philosophy is well known: its engineers assume that BMW drivers know what opposite lock is and how to apply it correctly. They have needed to know.

The latest 635 CSi incorporates the front and rear suspension modifications first seen on the Five and Seven series sedans, the bottom strut of the front suspension being located fore-and-aft by an articulating link which effectively reduces offset. At the rear, changes to the semi-trailing arms mean they now run at a reduced angle (13 degrees) and have compensating links to prevent toe-out and severe changes in balance during hard cornering.

In the dry the new CSi handles superbly up to a point which few drivers will exceed. On most corners there is just a hint of understeer and the fat 220/55 Michelin TRX tyres provide plenty of grip (and whine) in dry conditions; in the wet it's not as predictable. In long sweeping bends the car's stability and balance is excellent and it is capable of generating impressive lateral g-forces. In tight conditions the understeer builds up quite strongly under power; lift off and the BMW does oversteer, but only marginally. Keep the power on and in some situations the car reaches a point where the steering becomes unpleasantly vague and it swings into easily manageable oversteer, with the inside rear wheel spinning momentarily as it loses traction. Seconds later the tail has regained its composure and the car then will simply rocket forward.

Bumps never throw it off line and the ride, while firm, is never harsh and has a remarkable pliancy that means it is comfortable and beautifully controlled, though oc-

ENGINE	BMW 635CSi	Mercedes-Benz 560SEC
Cylinders	in-line six	vee-eight
Valves	single ohc	single ohc
Induction	electronic fuel injection	mechanical/electronic fuel injection
Comp ratio (to one)	8.0	9.0
Bore/stroke (mm)	92 x 86	96.5 x 94.8
Capacity (cm³)	3430	5547
Max power (kW/rpm)	136/5400	182/4800
Max torque (Nm/rpm)	290/4000	400/3500

PRICES AND EQUIPMENT		
Base Price	$86,000	$160,162
Price as tested	$90,690	$160,162

PERFORMANCE	BMW 635CSi	Mercedes-Benz 560SEC	
Top speeds (km/h/rpm)		Drive	Held
First	54/6300	78/6000	80/6100
Second	94/6300	117/5900	121/6100
Third	149/6300	200/6000	204/6100
Fourth	208/6300	225/4700	
Fifth	217/5300		

Standing start (secs)		
0-50 km/h	2.8	3.1
0-60 km/h	3.6	3.8
0-70	4.6	4.6
0-80	5.6	5.4
0-90	6.8	6.4
0-100	8.1	7.7
0-110	9.8	9.2
0-120	11.5	10.9
0-130	13.4	12.7
0-160	20.7	19.6
400 m	15.6	15.4

In the gears (secs)	2	3	4	5	Kickdown
30-60	3.3	4.9	6.9	9.7	1.9
40-70	3.0	4.8	6.9	9.3	2.0
50-80	2.9	4.7	6.8	9.5	2.4
60-90	2.9	4.6	6.6	9.7	2.9
70-100		4.6	7.0	10.0	3.2
80-110		4.6	7.3	10.8	3.5
90-120		4.8	7.5	11.2	4.3
100-130		5.0	7.8	11.9	4.9

BODY		
Claimed Cd	NA	0.34
Exterior (mm)		
Wheelbase	2630	2850
Front track	1430	1555
Rear track	1460	1527
Length	4755	4935
Width	1725	1828
Height	1365	1407

casionally it tramlined on undulating bumps. There is always plenty of tyre whine and once or twice over particularly deep potholes the suspension was felt to bottom although most of the shock was absorbed by the suspension/bump stops.

BMW persists in using powered ZF recirculating ball steering. If it's not quite as super-sharp at the straight ahead as the very best rack and pinion set-ups few people will notice, for the BMW turns into corners confidently and crisply, with a feel and accuracy that few other steering systems provide. By power-assisted standards the steering is even heavy, but the weight of the system is perfect for this type of car and with just 3.2 turns lock to lock seems ideally geared to give the car an agile feel.

Many right-hand-drive-converted BMWs suffer from a spongy brake pedal, with long initial travel followed by a heavy, solid feel. Unfortunately, the 635 carries on the tradition. It is impossible to gently dab the brakes on entry into a corner to set the car up, the pedal simply isn't progressive enough for that. Even the inclusion of an ABS anti-lock system as standard equipment can't make up for the disappointing spongy pedal feel. Once through this almost delayed action the brakes work exceptionally well, but that is really only in a crash stop situation.

The sheer size of the Mercedes counts against it in any comparison which places priorities on agility and precision. The SEC always feels as if it needs more road than the CSi, the steering — again a power-assisted recirculating ball system — is less accurate at the straight-ahead and lacks on centre feel, a perception the huge steering wheel does its best to emphasise. First impressions are that the Mercedes can't be driven as quickly as the BMW, but with experience the driver comes to realise the enormous reserves of the 560 — although its smaller 205/65 Michelin MXV tyres probably mean it doesn't have quite the same ultimate grip as the 635. The handling, however, remains more consistent with a gradual build up in understeer, especially under power.

Keep the power on and in tight corners it is possible to induce a touch of oversteer that requires no more than rolling wrists to keep in check. The steering is always lighter than the BMW's, though the SEC doesn't change direction with the same precision, even that delicate accuracy (yes, despite the BMW's relative heaviness), although experience soon indicates the Mercedes has high limits of adhesion and, with its remarkably resilient suspension, can absorb any punishment.

Its brakes, like the

BMW's, are huge four wheel discs with ABS and are very powerful, and with far less pedal travel than the CSi they are much more pleasant to use.

Point-to-point there is very little between the two but it seems for pure driving pleasure that the BMW, for all its flaws, is to be enjoyed more on winding roads, while the SEC is unbeatable as a high speed express. Each, however, should be mightily respected in the other's territory.

Accommodation

How enticing is the prospect of driving either of these beautifully finished machines. The cockpits are superbly appointed, styled with understated elegance, though there is nothing subdued about the rich aroma given off by the soft buffalo hide upholstery in the 635. These cars are both genuine four seaters provided the rear seat occupants are under

CHASSIS	BMW	Mercedes-Benz
Drive	rear	rear
Front suspension	MacPherson struts, double-joint linkage, coil springs, anti-roll bar	Wishbones, coil springs, anti-roll bar
Rear suspension	Independent, semi-trailing arms, coil springs, anti-roll bar	Independent, semi-trailing arms, coil springs, anti-roll bar
Brakes	disc, ABS	disc, ABS
Steering	power assisted, recirculating ball	power assisted, recirculating ball
Turning circle (m)	11.2	11.55
Turns lock to lock	3.2	3.0
Kerb mass (kg)	1495	1785
Fuel capacity (l)	70	90
Tyres	Michelin TRX 220/55 VR390	Michelin MXV 205/65 VR15

TRANSMISSION		
Ratios/Km/h per 1000 rpm		
First	3.83/8.6	3.68/13.1
Second	2.20/15.0	2.41/19.9
Third	1.40/23.6	1.44/33.4
Fourth	1.00/33.0	1.00/48.1
Fifth	0.81/40.7	—
Final Drive	3.46	2.47

FUEL CONSUMPTION		
km/l (mpg)		
Best	7.6 (21.5)	6.4 (18.1)
Worst	6.9 (19.6)	5.4 (15.2)
83 km test loop	10.7 (30.1)	7.6 (21.5)
Overall	7.5 (21.2)	6.0 (16.9)
Minimum range (km)	483	486
Test distance (km)	802	711

SERVICE REPAIRS		
Warranty	12/20,000	12 months/unlimited distance
Insurance rating[1]	9	5
Major service Distance	NA	20,000
Minimum cost[2]	NA	$281
Cost of replacement parts[3]		
Front bumper	$978	$2,037
Front mudguard	$862	$474
Grille	$394	$466
Headlight glass/assembly	$250	$42
Tail-light glass/assembly	$382	$296
Windscreen	$430	$930
Full exhaust system	$1977	$5004
Set front disc pads	$75	$90
Radiator	$545	$949
Air filter	$24	$66
Door mirror	$303	$431

● Notes: [1]Insurance ratings from the GIO on a scale between 1-9 [2]Minimum service cost made up of labour and replaced routine parts [3]Replacement parts prices include tax

1.8 m high, only head room being a problem. There is far more rear seat leg-knee room in the Mercedes but if the front seat occupants are willing to compromise the BMW makes the grade with two superb individually-shaped rear bucket seats.

Drivers of either vehicle will want for very little, so complete are the equipment lists for both cars. The heritage of the respective marques is as clear in their interiors as in their styling.

The BMW's snug-fitting and marvellously comfortable bucket seats grip their occupants very securely; you most assuredly sit *in* these seats, not on them. Their electrically-operated adjustment is controlled by a total of 10 small buttons, located on the centre console, for each seat. It is an embarrassingly complicated arrangement that is in complete contrast to the simplicity of the Mercedes arrangement. This has door-mounted controls shaped like a seat which you simply push in the direction you want the seat to take. The BMW's seat has three memory positions, the Benz just two.

The BMW's lovely leather-bound, three-spoke steering wheel is adjustable for height but not reach and always seems too high. Raising the seat cushion only produces headroom problems for tall drivers. The pedals are too high off the floor and the turn indicators are still on the left of the steering column, Mercedes goes to the expense of changing them over to the right for right-hand-drive cars.

The SEC has an imitation suede upholstery; it "breathes", and it looks indistinguishable looks from the real thing. Its bucket seats offer more support and are more shaped than those of any other Benz in recent memory, but they are far bigger than those in the BMW and just as comfortable over a long trip. There is an ambient temperature gauge, and the steering wheel — why does Mercedes persist with such enormous wheels? — is leather bound, while the simple yet shapely dashboard seems high for visibility, one area in which the CSi is a clear winner. One of the flaws with the modern, rounded aerodynamic shapes is that visibility suffers and it takes a car like the BMW to bring home the point. You can see all four corners in the CSi; its C-pillars are narrow by comparison with the SEC.

Both have vast boots, with more than enough room for the combined luggage of four adults. The BMW has that fine touch of a tool kit mounted in the boot lid.

Conclusions

For all its faults, there is something very desirable about the BMW. It feels so solid and yet responsive, it is far more reassuring to the enthusiast when he first slides down behind the wheel. It stimulates and excites.

The Mercedes almost intimidates the driver with its sheer size and power and it takes time to realise that it is both more refined and quieter, and that its soft ride doesn't limit its roadability.

The BMW has too much wind noise, and that brake pedal modulation problem needs fixing; but it is mechanically smooth and efficient and looks the part. The Mercedes is a better car with few obvious faults though its character is less blatant. Both are civilised and serious about their roles and the price difference — huge though it may be to mere journalists — seems to make sense to the buyers. □

THE M6 ALTERNATIVE

RATHER THAN FLYING, GEORG KACHER USES BMW'S VERY FAST YET COMFORTABLE M635CSi TO DASH FROM MUNICH TO TURIN

THERE ARE OCCASIONS WHEN A car is a better means of transport than a plane, even on a longer journey. Especially if the car is BMW's M635CSi coupe. Take the trip from Munich to Turin. There is no direct flight, although you can wing it to Milan and then take the bus, which is tedious, slow and hot. Or you can fly via Frankfurt but that plane is always chock-a-block and arrives in Turin after all the taxi drivers have gone home. No, the most rewarding way to get to Turin is by road.

And what a road it is. It starts with a 40mile stretch of autobahn which runs from Munich to Eschenlohe near Garmisch. We are leaving early to beat the holiday traffic, but the big six-cylinder engine barely reaches its working temperature before we are slowed down by one of those trial 100km/h speed limit zones. You can safely waft along at an indicated 125km/h – about 80mph – but there are so many emission measuring points and radar traps that every lead-footed driver will eventually get his ticket. The reason for this nationwide go-slow is, of course the pollution problem. If the scientists decide by the end of the year that exhaust gas emission is radically lower in the 100km/h zones, a rigorous speed limit may be imposed in Germany as early as spring 1986.

This 25mile purr, when the needle of the rev counter barely touches the 2500rpm mark in fifth, gives us time to check out the BMW's cabin. Despite its age – the Six Series was first shown in 1976 – the car's ergonomics are still good. In terms of design and function, this dashboard still deserves full marks. The only way to tell its age is by examining the kind of plastics used, which vary considerably in texture, colour and grain. These rather cheap-looking synthetics are particularly noticeable in a cabin clad with leather where the hand-stitched hide contrasts sharply with its plastic environment.

At the end of the 100km/h autobahn, we take the secondary road to Garmisch, Mittenwald and eventually to the Austrian border. It has started to drizzle, and the flimsy wipers are struggling with the raindrops. Above 85mph, they begin to lift off, and generally they are noisy, slow and not very efficient. There is nothing wrong with the overall visibility in this coupe, though. The glass area is generous, the pillars are unusually slim, and you don't need a swan's neck to overlook all four corners of the car either. The driving position in the M635CSi is good but not faultless. The adjustable steering wheel certainly helps, and the bucket seats are also fully adjustable in reach, rake, tilt and height, and yet it is not easy to find a comfortable driving position since the tuning of all these levers and handwheels is far too coarse. There is also surprisingly little elbow and shoulder-room, while headroom is restricted by the optional sunroof.

The customs officers at Scharnitz are still half asleep, but the combination of a German numberplate and an Austrian and a British passport (photographer Richard Davies is my passenger) instantly arouses their interest. 'Where are you going? Anything to declare? Gute Fahrt!' We briefly consider giving two blonde hitch-hikers with Swedish flags attached to their rucksacks a lift, but they wouldn't feel too comfortable in the back of our coupe. With the front seats pushed back all the way, the rear passenger compartment is barely big enough to accommodate two children. The boot is not exactly huge either (in the M635CSi it is even 20percent smaller than in the other Sixes), but then it is usefully long and wide as well as relatively easy to load.

The road from Scharnitz to Innsbruck leads us down the steep and winding Zirler Berg where skull and crossbones, low-gear warning signs and a dozen escape lanes indicate that this gradient *is* serious. We stay in third gear for the first 500yards, but since the road is glistening wet, the first bends are tackled in second. The brakes warm up completely, and soon even the slightest dab on the pedal will produce stunning deceleration. For the M version of the BMW coupe, the brakes were beefed up with four-piston fixed calipers in the front and with a standard antilock device. While the front brake discs are inner-ventilated, the solid rear discs feature integrated hand brake drums. Modulation and brake balance are virtually perfect, but hard use will substantially increase the required pedal pressure. Brake fade is minimal.

The 25mile-long Brenner autobahn is probably the most expensive stretch of toll road in the world. The return ticket costs a whopping 32Marks (£8.20), but it is money well spent since the alternative is twice as long, narrow, in poor condition and dotted with blind summits, dangerous corners and greedy Austrian cops. There is more than one mobile radar trap on the Brenner motorway, but this Saturday we get through unscathed, and this time, even the formalities at the border crossing point into Italy are surprisingly painless. 'La Autostrada di Brennero a Modena' is a very memorable piece of road. Memorable firstly for the usual five-mile tailback at the Sterzing toll barrier where single-file traffic, TIR trucks in low gear and malfunctioning toll ticket machine make your hair turn grey very fast.

We are lucky and lose only 10minutes, but then we lose another 20 buying petrol coupons. After just 270miles, the 15.5gal fuel tank of our red M635CSi is almost bone-dry and

the consumption works out at a sobering 17.6mpg. On all Italian motorways, there is a speed limit of 140km/h or 88mph for cars with a capacity of 1.6litres or over, but that 40mile torture track between Sterzing and Bolzano is in itself the most perfect speed limit ever built. All major European car makers use this section regularly as a giant chassis test bed, and sure enough we notice a fleet of Opels with factory numberplates resting in a lay-by. The menu prepared by the Italian road chefs includes a wide selection of potholes, a cocktail of depressions, hollows and dips (some casual, some in rhythmical sequence), crossfalls, channels and furrows, massive transverse ridges and 17 badly-lit tunnels for desert.

The BMW is well equipped to cope with such obstacles. At the front, it has double-joint spring strut suspension with anti-dive geometry. The rear suspension, which boasts anti-squat kinematics, is made of a delta-shaped axle beam which holds the propshaft, the differential and the forks of two massive lower wishbones. With the exception of uprated springs and dampers, a lowered suspension and fatter anti-sway bars, the chassis of the M635CSi is identical with that of its lesser brethren.

Together with the deeper front spoiler and the small rubber lip on the bootlid, the sportier suspension improves directional stability at high speeds and in crosswinds. The M635CSi is a remarkably stable car which stays unerringly on course even under braking and during abrupt transitional manoeuvres. Those unnerving lurching movements along the longitudinal axis which still haunt the BMW Seven Series saloon above 100mph are totally absent in the M coupe. Unfortunately, a stiffer suspension not only spells better roadholding and quicker handling but also reduced ride comfort. In this case, the negative effect of the sportier springs and shock absorbers is particularly noticeable over transverse grooves and short and shallow undulations where the car feels bouncy and harsh.

The topography of the road can also be felt through the steering which is quick, precise and nicely balanced but not well enough damped. Naturally, the taut suspension characteristics are further emphasised by the extra-wide 240/45VR415 Michelin TRX tyres which are fitted to three-piece BBS alloy wheels. Especially on rippled surfaces, this blend of fat gumballs and taut damping evokes a slightly unstable floating movement which is interrupted by occasional jerks when the car encounters a really deep pothole.

Just south of Bolzano, the Etsch valley widens, and after about 40miles the high mountains drop back in favour of breathtakingly beautiful hilly countryside which leads down to Lake Garda and the outskirts of Verona. At a bizarre intersection bristling with rusty armco, pale green road signs, old street lamps still wearing their first primer coat and the ubiquitous Agip filling station, we turn right towards Brescia, Milan and Turin. This is a three-lane motorway in decent condition, and since traffic is light, the BMW gets a chance to demonstrate that the little M suffix truly makes all the difference.

The 3.5litre in-line six was first seen in 1978 when it powered the revolutionary BMW M1 supercar which died an untimely death three years later. Wedged into the engine compartment with barely an inch to spare in any direction, this M power unit is not only brawny but also visually attractive. The tidy plumbing, the partly polished rocker cover, the six intake trumpets and even the nuts, bolts and screws and the heat-resistant hoses all look durable, expensive and very professional. The all-alloy motor features two overhead camshafts, six throttle blades,

STILL EYE-CATCHING AFTER 10YEARS, BMW'S COUPE LOOKS TOUGH IN M635CSi GUISE. LOVELY 24VALVE SIX DELIVERS 286BHP

24valves and a Bosch Motronic control system which governs fuel feed, ignition, idle speed, warm-up and exhaust emission.

Power output is 286bhp at 6500rpm with 246lb ft torque at 4500rpm. Mated to a close-ratio five-speed gearbox, the engine will accelerate the 3318lb coupe from 0 to 60mph in 6.4sec, and from 0 to 100mph in 14.9sec. Top speed is 159mph. Despite digital engine electronics, a thermodynamically refined combustion chamber design and a high compression ratio of 10.5 to one, the average ECE fuel consumption of the M635CSi works out at 24.2mpg. Our test car returned 17.6, 16.4, 14.5 and 12.8 mpg per tankful. The last figure was recorded during the final leg of our return journey where empty roads (we arrived well after midnight) permitted an average speed of 121mph. Had we restricted ourselves to a steady 100mph, the 286bhp coupe would have bettered 21mpg.

Providing you can avoid wheelspin, first gear is good for 36mph, second will take you to 60mph, third reaches up to 118mph and fourth gives you 129mph. Although the torque curve peaks at a highish 4500rpm, this engine is surprisingly flexible. With 75percent of the maximum torque available at 2000rpm you can potter along at 50mph in fifth without changing down for slight gradients or even relaxed overtaking manoeuvres. But there are many sides to the character of this powerplant. Docile as it may be to a gentle driver, all hell will break loose once you push it past the 4000rpm mark. Accompanied by the most aggressive intake noise this side of a Ferrari, the 24valve six will storm to its 6900rpm redline. Whenever the right foot goes down, the volume goes up, but the noise that comes out of the large-diameter twin-pipe exhaust is sonorous and rorty rather than unpleasant.

We stop at a monstrous Pavesi service area. The filling station attendant is ideally suited for a supporting role in a third rate Mafia movie, and he acts like a family member, too, by squeezing 80litres of supercarburante into a 70litre tank. Any complaints, Sir? We climb back into the car and head for Milan which welcomes us with a congested ring road, an overturned fruit lorry and a bunch of grim looking Carabinieri who are out chasing tourists at the toll booth. Instead of checking your speed with radar traps, the Italian authorities prefer to check your toll ticket instead which tells them exactly when and where you entered the motorway network. 150 divided by 60 – thank God for the delay at the filling station...

As big Fiat billboards count down the final 100kilometres to Turin for us, Richard Davies quite rightly remarks that we shall arrive much too early for our lunchtime appointment. Arrivederci autostrada, and welcome to the ox-carts and speeding school buses of Piedmont. The BMW feels instantly at home on the narrow country lanes which meander through the endless apricot and apple groves. The engine responds to throttle inputs with a hungry growl that turns into a shirt-sleeved roar as the accelerator goes down.

The gearbox is ideally spaced for optimum performance in all speed ranges. Unlike the Getrag sport transmission which is an optional extra for the 218bhp 635CSi, the five-speeder of the 286bhp M635CSi features a conventional shift pattern. It works with ease and precision, and the gears drop in with the reassuring determination of a firm handshake. The power-assisted rack-and-pinion steering could be better damped, but maybe such damping would take away some of the sharpness and accuracy which make the M Coupe one of the best-handling sports cars on the market. The steering is comparatively heavy, and yet you would not want it any other way since this weight only adds to the overall balance.

The world of oversteer has always been a favourite playground for BMWs, and this coupe makes no exception to the rule. On dry tarmac, the car will remain neutral for a surprisingly long time. In the wet, however, it does not take exceptional g-force to kick out the tail. In first and second gear, any tight bend will offer an opportunity to play Niki Lauda, but in third and fourth one does appreciate that the rear end gives plenty of warning before it actually loses adhesion.

As we reach the first suburbs of Turin, the weather begins to improve. The haze that has been with us all morning is eventually lifting, and the air is hot and humid. The sunroof is in the tilt position, and the quiet fan is on full blast, but to cool down what looks like a real scorcher, a good air conditioning system is hard to beat. It was a tiring drive, but it was fun all the same. The M635CSi is one of the last remaining grand tourers. It is very fast and very thirsty, it is quick-handling and yet sure-footed, and it is refined and relatively quiet. Under that bonnet sits one of the most powerful volume production engines in the world, but this BMW has the suspension, steering and brakes to keep its power in check. Exploring its limits can be hard work. But it certainly beats flying – any time.

ERGONOMICALLY SOUND THE DASH IS LET DOWN BY PLASTIC FINISH. DRIVING POSITION'S GOOD BUT TIGHT ON HEADROOM

BMW M635CSi and 507 — ROAD TEST

Sisters under the skin?

Life is never fair. Despite the fact that, for many years, they aimed at distinctly different markets, there has always been a tendency (conscious or not) to measure BMW's products against the corresponding model in the Mercedes-Benz range. Of course, BMW should take that as the ultimate compliment, even though it often means judging a relatively inexpensive car unfairly against a more costly rival. What is more, this is not merely a recent trend. Thirty years ago people were already doing it!

Take the BMW 507, a rare bird indeed. We recently spent an enjoyable day prowling round the fringes of Salisbury Plain in one of these elegant two-seaters, in company with today's top-rung high performance coupé from Munich, the M635CSi. Thirty years ago the BMW company was barely alive. It had lost absolutely everything in the war and its very existence was threatened. At one point it seemed possible that the Eisenach factory would be crated up lock, stock and barrel to be transported to the USA.

Mercifully, BMW survived, but even by the mid-1950s it was touch-and-go whether it would ever re-establish a solid commercial bedrock on which to prosper anew. Yet the company was nothing if not audacious. All this gloom and depression did not deter it from producing the costly, exclusive 507, a quite beautifully balanced luxury machine whose lines look fresh and uncluttered even after the passing of a generation. It is hard to believe this was the same company that manufactured the Isetta bubble car!

The tubular chassis was, in effect, a shorter wheelbase version of the 502, using the saloon's running gear mated to an all-alloy, wet liner 3.2-litre V8. Developing 150 bhp at 5,000 rpm, it was a long-legged autobahn cruiser rather than an out-and-out sports car.

Styling was the work of Count Albrecht Goetz, who later went out to produce the distinctive profile of the Datsun 240Z coupé which sold like hot cakes in Europe during the early 1970s. When one looks at its specification and appearance in restrospect, one would have expected the 507 to sell like bratwürst at the Nürburgring. But remember, we are talking about the austere mid-1950s. And if those of you who remember life in Britain think things were hard, they had been a whole lot harder in Germany.

The 507's production run lasted three years, until 1959, but never averaged as much as two per week over that time. A mere 253 cars trickled out of the Munich factory, and only three found their way to England. Two were imported by AFN, who then had the BMW concession for this country during their pre-Porsche period, but since there were practical engineering problems in building them in RHD form (it would have meant re-designing the sump and the steering box) no more were brought in.

Fleeting interest in the 507 was expressed by Tommy Sopwith and Ken McAlpine, both of whom had wallets capable of sustaining the bruising inflicted by the sky-high tax-paid £4,201 price tag. They would have put their money on the table, but did not want LHD! As a useful comparison, at the time a Jaguar XK140 was less than £2,000, while even the Mercedes-Benz 300SL was priced at £4,651. Of course, with its cachet of recent racing success, the Mercedes coupé, with its distinctive "gull wing" doors, rather eclipsed the 507. And that was sad.

Although AFN only imported two of these LHD BMWs, one of which we tested, former

ROAD TEST

Engine bays compared: the 507 V8 (top) and the M635CSi six.

World Champion motorcyclist and F1 ace John Surtees purchased one new from Munich when he was riding for the MV Augusta team. At the time Surtees was criss-crossing Europe every week, hurrying to motorcycle races in just about every country on the map, and he needed a high speed cruiser which would be comfortable and relaxing. The BMW 507 fitted the bill perfectly, and John drove it regularly right through until 1963, when he signed for Ferrari. To this day, he retains his 507 as one of his most cherished personal possessions.

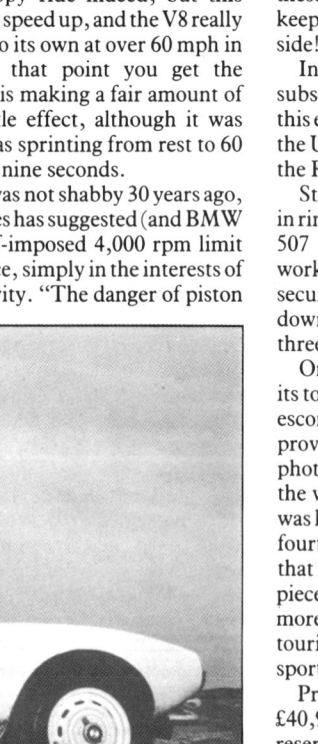

The stylish lines of the 507 have lasted well.

Carrying the distinctive registration number 5 BMW, the 507 we sampled is owned by BMW GB. We drove the M635 down to Westbury to meet its forebear, for the 507 is cared for by the specialist BMW restorers TT Workshops, run by John Giles and Tony Hutchings on the fringes of this pleasant Wiltshire town. TT has looked after the car for over ten years now, keeping it clean and in fine fettle for the handful of occasions when the precious beast is allowed out for promotional trips. It only completes a few hundred miles each year, so we were conscious of the privilege!

As you slide into the cockpit, first impressions are of far more room than the 507's external appearance suggests. The seats are quite comfortable and supportive, the painted fascia rather stark. The large steering wheel is close to one's chest and what looks like a Ford Zephyr-style horn half-ring is actually the control for dipping the headlights!

The footwells are comfortably wide, allowing plenty of room for the well-positioned pedals, but the four-speed gearbox takes a bit of getting used to. Although engagement of the gears is quite precise, the lateral movement across the gate is enormous. But the overall feel of the controls is one of lightness and precision.

At low speed, the torsion bar suspension gives a very choppy ride indeed, but this smooths out as you speed up, and the V8 really begins to come into its own at over 60 mph in top gear. Up to that point you get the impression that it is making a fair amount of noise for very little effect, although it was originally quoted as sprinting from rest to 60 mph in just under nine seconds.

That certainly was not shabby 30 years ago, although John Giles has suggested (and BMW GB agreed!) a self-imposed 4,000 rpm limit for this period piece, simply in the interests of mechanical longevity. "The danger of piston

Fifties cockpit: leather and painted metal.

Eighties cockpit: leather and plastic.

breakage is the ultimate rev limitation on these engines," explains John Giles, "so we keep it down to 4,000 just to be on the safe side!"

Interestingly, BMW's 3.2-litre V8 was subsequently sold to Buick, and derivatives of this engine eventually found their way back to the UK where they became the power unit for the Rover 3.5-litre V8!

Standing almost awkwardly high on its 16 in rims shod with thin Avon Turbospeeds, the 507 emits a shrill whine when the V8 is working hard. But it exudes a stability and security at high speed which has been handed down to its younger stablemates over the past three decades.

On a winding road across Salisbury Plain, its top-gear performance hardly extended our escorting M635, but it was sufficient to provoke a degree of surprise from our photographer, Steven Tee, who was behind the wheel of the latest coupé at the time. He was keeping his foot pretty close to the floor in fourth gear keeping pace with us. So put in that light, the 507 seemed a pretty impressive piece of machinery, although it is perhaps more honest to describe it as a high-speed touring car rather than as an out-and-out sports coupé.

Propelling ourselves forward 30 years, the £40,950 M635CSi hardly attracts the same reservations when compared to its rivals. The current range of big BMW coupés first appeared on the scene early in 1977, in the form of the 633, so has now been around for ten years as these words are written. The first 635 came our way towards the end of 1978,

BMW M635CSi and 507

The M635CSi has a classic elegance which has proved timeless over the big BMW coupé's ten year production run.

combining terrific performance with what I can only describe as worryingly inadequate braking performance, a trait which now, thankfully, has been buried in the far distant past. All the 6-series cars also have ABS anti-lock systems as standard.

I recall having a particularly nasty moment motoring in Sussex one November afternoon when, having braked very hard from a speed which I'd rather not record in print, I then had to call on the 635's brakes a second time soon afterwards. The pedal felt distinctly half-hearted, and the retardation a pale reflection of what it had been a few minutes earlier. Suffice to say, I was unimpressed.

As an interesting and not totally disconnected aside, I happened to mention this in conversation with DSJ and he rather surprised me by remarking: "well, things haven't changed that much, have they?" For a moment I thought he must have had recent experience of a BMW with marginal braking performance, but in fact he was referring back to the 507.

It turns out that, on a visit to Munich at about the time of its launch in 1956, Jenks was taken aback by the way in which its brakes seemed to fade in exactly the same way. What is more, he managed to get himself into a fairly trenchant debate on the matter with BMW

ROAD TEST — BMW M635CSi and 507

This BMW is docile and tractable, even in the snow!

technical chief Alex von Falkenhausen, who felt that such criticism was unjustified!

BMW's current coupé range on the UK market includes the 628CSi, 635CSi and the M635CSi, the last-mentioned being a really exclusive version of an already impressive machine. The heart of the 3.5-litre cars is a wonderfully smooth 286 bhp, 24-valve in-line six cylinder engine producing a hefty 246 ft lb of torque at 4,500 rpm. It combines silky smooth docility with what can only be described as mighty performance, this large two-door machine rocketing to 60 mph in fractionally over 6 sec.

Features distinguishing the M-designated machine from the straightforward "cooking" 635CSi include a sports suspension set-up incorporating gas filled shock absorbers all round, bigger ventilated disc brakes and forged light-alloy wheels with Michelin TRX 240/45 VR 415 rubber.

Air conditioning is fitted as a standard feature, along with heat insulating, green tinted glass, a deeper front spoiler and heated external door mirrors which are finished to match the bodywork. The M635CSi interior is also enhanced by the superb BMW sports seats trimmed in leather (although Highland cloth is offered as an alternative); these feature electric height adjustment, fore/aft movement, backrest inclination and headrest height adjustment. There is also an on-board computer, rear head restraints and a rear window blind, plus a larger battery installed in the boot.

For a car this size, its agility and sure-footedness is quite remarkable. Apart from instilling a sense of confidence and well-being into its driver, it seems taut and compact from behind the wheel. Directional stability is almost beyond criticism, even at very high speed, and the always-excellent BMW power-steering system provides an ideal combination of lightness and feel.

Sudden directional changes often prove to be the Achilles heel of big high performance machines, but the M635CSi comes out near the top of the class in this respect as well. It rolls perceptibly, but the firm suspension keeps it nicely in check and there is never even a hint of it getting into a worrying fishtail, even on a slippery surface. Wind noise is minimal, visibility and headroom more than adequate.

Its torque characteristics also make it a delightfully flexible car, equally at home inching along in the commuter jam as on the open road. The only area in which some obvious criticism comes to light is the need to depress the clutch pedal slightly further than one might expect on initial acquaintance, in order to avoid lightly crunching one's gearchanges. But the gearbox itself is fine, albeit with quite a long throw to the lever, particularly between first and second.

Like Mercedes, BMW instrumentation and interior layout makes little concession to passing trends and remains solidly unchanging over the years. You can slip behind the wheel of an M635CSi and notice precious little difference from the original 633s and 635s. But a great deal of work has gone on under the skin, and the current cars are infinitely more rounded in their character than they were ten years ago.

So are these two BMWs sisters under the skin? Certainly not! The 507 was not a sports coupé in the way we accept the role of the M635CSi. It was more of a boulevard cruiser, short on agility but long on gearing. The current high performance coupé is not only more of a sports car relative to its 1987 rivals, it is a shatteringly fast road machine which reminds us just how much we have come to expect from top-drawer machinery over the intervening three decades.

In the 1950s a sports car was a sports car, and a tourer just that. The twain seldom met in one bodyshell. Now the M635CSi underlines that a high performance car of the eighties can be a vehicle of many parts, blending the interior refinement of a limousine, the gait of a grand tourer and the acceleration and speed of a road racer. All things to all men, you might say. At a price, of course!

AH

156 mph tailpiece!

4 SEASONS TEST

BMW 635CSi

Twelve months and 24,500 miles in the premier Bavarian coupe: lots of delight and a little frustration.

BY JOHN STEIN

Ann Arbor—The calendar has run out on our four seasons with the BMW 635CSi. It was generally a year of good times, just as we had hoped it would be when we introduced this high-dollar coupe [July 1986]. But to understand what we have learned about the 635CSi after a year and 24,500 miles, we must consider both its assets and its liabilities.

Most important is the BMW's convincing performance as a real driver's machine. The 635 is immensely capable and likable, and it manages a wide variety of tasks equally well: the daily commuting grind, storming along a mountain road, and lazily touring. In all cases, the Bimmer's fundamental virtues—its superlative powertrain, its spot-on handling, and its excellent ergonomics—are above reproach.

The CSi struck a favorable chord during our initial experience with it in the spring of 1986, when we launched this Four Seasons review with a round trip from Ann Arbor to North Carolina. Since then, there have been trips to Virginia and Ontario and throughout the Midwest, plus a steady diet of general commuting and errand running.

A few telling notes in the 635's logbook explain our early enthusiasm more fully. Executive editor Kevin Smith wrote, "There is a combination of perfect ergonomics, fine control feel, good power, and proportions that are just right, and it makes the 635CSi sweet and rewarding.

PHOTOGRAPHY BY TOM DREW

Thanks to the hard but excellent seats, and the car's just-right size and shape, it feels like a close extension of my body. It makes me feel good to drive it."

Senior editor John Phillips III was similarly inspired by the BMW's competence, and particularly by its graceful shape. He noted, "After four months with the beast, and at least three 500-mile road trips, my impressions are obviously favorable. I also think it's about the most handsome coupe on the road today—perhaps because the lines are so smooth and utterly simple."

A major portion of the 635's easy charm comes from its flexible and refined powerplant. The 3.4-liter straight-six (developing 214 pounds-feet of torque at 4000 rpm and 182

BMW 635CSi

horsepower at 5400 rpm) is more than sufficient for the 3424-pound two-plus-two, especially in its mid-range punch. With the exception of its lumpy, 650-rpm idle, the mill is turbine-smooth, too, and the modest amount of noise that intrudes is positively an aural feast: a muted howl from up front, and a straight-six snarl out the chrome-tipped exhausts.

Our CSi's electronically controlled four-speed automatic is simply the finest and most versatile automatic we've tried. It is definitely one of the 635's strong points, and one of only a few automatics that tempt us to forgo our normal preference for a manual gearbox. To provide unusually smooth shifts, the automatic transmission is linked to the engine management electronics. In addition, there is a choice of three different operating modes: low- or high-rpm shift points, and complete semi-automatic control. In the semi-automatic mode, the transmission may be manually shifted through first, second, and third gears, or left in any of those positions. This, along with the optional limited-slip differential, helps stabilize the car on fresh snow or icy pavement, and it also maximizes the entertainment factor on a good set of switchbacks.

The 635CSi is most in its element as a high-speed tourer. With its eager engine, precise recirculating-ball steering, and high-speed stability, it can rocket across an entire state—

or a continent—with the consummate ease of a bullet train. The fine tracking, and the detailed feeling the BMW driver enjoys from the other controls, make a lasting impression.

If the splendid driving qualities sound too wonderful to be true, well, a car *should* be pretty good for $44,240. Yet our euphoria faded a bit when the time came to attend to the car's service needs. Numerous minor ills included continual stalling problems; sticking door hinges and a sticking driver's seat release; faulty rubber seals around the windshield, the sunroof, and the side windows; and various electrical component failures. Once, when a couple of relays in the injection system failed, the CSi required towing (the ultimate disgrace!) to our local BMW franchise.

Even though we religiously attended to the BMW's maintenance needs, our greatest sources of irritation were repeat difficulties with the rubber weather seals, the alignment of the doors and side windows, and the BMW's often unsteady idle. In studying the problems in retrospect, it is difficult to tell whether the recurring troubles were entirely the car's fault, or whether some of the blame may rest with ineffective servicing.

We should point out that BMW ranked sixth overall among manufacturers in the J.D. Power & Associates 1986 Car Customer Satisfaction Index. (BMW was surpassed by Honda, Mercedes-Benz, Toyota, Mazda, and Lincoln.) Among individual models, the 635CSi—which accounted for 3.1 percent of BMW's sales last year—rated slightly higher than any other BMW, and it finished twenty-eighth among 148 models in the study.

Following is a rundown of the 635's scheduled (and unscheduled) service visits for the year.

623 miles: 1200-mile inspection (warranty).

6248 miles: wheel alignment (nonwarranty).

7209 miles: corroded door hinges, squeaking door retarder, loose power window regulator, hot engine stalling (warranty); 7500-mile engine oil service (nonwarranty).

14,126 miles: leaking windshield seal, sticking seat release lever, engine hesitation (warranty); 15,000-mile inspection (nonwarranty).

16,391 miles: leaking windshield, sunroof, and door weather seals, irregular air conditioning operation, irregular windshield wiper operation, cold engine hesitation (warranty); worn windshield wiper blades (nonwarranty).

18,286 miles: replacement of discharged battery (warranty).

18,909 miles: replacement of second discharged battery because of a wiring short, misaligned door, squeaking instrument panel, loose seat trim (warranty).

22,378 miles: worn front and rear brake pads and wear indicators, 22,500-mile engine oil service (nonwarranty).

22,747 miles: leaking heater core (warranty).

23,815 miles: leaking door seal, misaligned window, squeaking instrument panel, center brake light failure (warranty).

23,846 miles: faulty fuel pump and fuel injection relays (warranty).

24,483 miles: misaligned windows, brake light switch recall (warranty); annual check (nonwarranty).

With the exception of the incident involving faulty fuel system relays, the rest of the BMW's complex electronics—including the engine management and automatic transmission systems, the Bosch ABS, and the simple-to-use cruise control—performed without so much as a hiccup. We also enjoyed using the eight-function trip computer and the eight-speaker AM/FM/cassette stereo.

Other good service news involves the BMW's two sets of Michelin tires—the original 220/55VR-390 TRXs, and the rugged TRX winter rubber. The standard Michelins were perfectly suitable for three out of four Michigan driving seasons, but when winter arrived, it quickly became apparent that the 635 could be a handful. Once fitted with its winter tires, the car's manners improved dramatically.

Proper traction helped us to keep the coupe out of the cornfields last winter, and when we returned the car to BMW of North America, little external damage was evident. Body injuries were limited to a few stone chips and parking lot blemishes (the low-jutting air dam escaped virtually unscathed), and interior wear and tear was held to the usual grubby floor mats and upholstery leather buffed by clothing. That's commendable, especially since the car has been used regularly to transport everything from grocery bags to dogs.

Mr. Phillips made a somewhat wistful final entry in the 635CSi log. He wrote: "I stood and stared at the car again last night. What a *pretty* automobile. We really should keep it, if only to keep the parking structure classy."

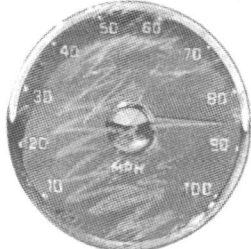

OVERALL RATING

1986 BMW 635CSi

We found BMW's highest-priced car to be an excellent driving machine. Over the months, the 635CSi's balanced nature was always a joy and a comfort. But if every coin has a flip side, then the BMW's is plainly the continual minor failures—and unscheduled repair shop visits—that we endured. In sum, we're both delighted and disappointed.

	EXCELLENT	GOOD	FAIR	POOR
ENGINE			●	
HANDLING				●
RIDE QUALITY			●	
COMFORT			●	
ISOLATION			●	
BODY INTEGRITY		●		
INTERIOR INTEGRITY			●	
RELIABILITY		●		

REVIEW PERIOD:
One year
24,500 miles

SCHEDULED MAINTENANCE:
Parts $91.87
Labor $593.00
Fluids $74.88

WARRANTY REPAIRS:
Parts $855.84
Labor $1277.50

NONWARRANTY REPAIRS:
Parts $121.62
Labor $97.20
Towing $18.00

FUEL ECONOMY:
20 mpg

RUNNING COSTS (gas, oil added between changes, and nonwarranty maintenance and repairs):
$181.18 per month
$0.089 per mile

PROBLEM AREAS:
Fuel injection
Electrical components
Door hinges
Windshield and window seals

ROAD TEST

BMW M6

The joy of six

by Phillip Bingham
PHOTOGRAPHY BY JIM BROWN

There were several things we took a deep disliking to during our week with the BMW M6. One was the 55/65-mph national speed limit. Another was the necessity to leave the car alone, at least for six or seven hours a day, to enter the other world of dreams. But by far the worst was handing the vehicle back. We had seriously wondered, at the start of the week, how a 10-year-old design with low-speed acceleration fairly similar to that of a Ford Mustang GT could possibly justify $59,000. We were left seriously wondering, at the end of the week, how we might *just possibly* raise the cash. The M6 is that kind of car. If you have any kind of business head, its price will, at first, raise serious questions. But if you have any kind of driver's heart, you'll soon find the answers.

The harder you press the throttle pedal, the harder it gets to criticize the cost of the M6. And you can press it all the way to 145 mph. But it's not the big numbers that impress most; it's the forceful rapidity with which the M6 stretches out and *grabs* them. You want 65 mph right now? Here, have it. You want 95 mph? No problem, take it before shifting up to 4th. The climbing speedometer needle knows nothing of gravity. It shows no signs of faltering until well past twice the legal limit. You've heard again and again how luxo-cruisers "eat up the miles." This one chomps them up in great big lumps and spits out other performance cars behind. Long hauls can be shrugged off with superior indifference. Only the short journeys annoy. They leave you craving more.

The heart and soul of this wild party is BMW's now-fabled S38 inline 6-cylinder engine, a refugee from the BMW M1 coupe that has since been profitably employed in the M635CSi and M5 sedans. The M635CSi has been on sale in Europe since mid-1985; it took nearly three years, a name change, and the sacrifice of 30 hp to get it across the Atlantic. But it was worth the wait *and* the sacrifice.

Not that the U.S.-spec M6 has lost much, other than an affordable 10% of its power, in the cause of federalization. We still get 256 hp at 6500 rpm and 243 lb-ft of torque at 4500 rpm—40% more power and 14% more torque than the 635CSi/L6. We still get an engine with a shorter stroke but a wider bore than that of the 635CSi/L6, increasing displacement from 3430 to 3453cc. We still get the addition of a second, chain-driven overhead cam and the 4-valve per cylinder format used to such good effect for the last 17 years by BMW in international motorsport. We still get the neat little touches, the machined intake and exhaust ports, the honed cylinders, the dual exhaust system, the (front spoiler-mounted) oil-cooler, the latest computerized Bosch Motronic electronic engine management system.

In other words, under the impressive prow of the BMW, U.S. buyers will still find enough latent energy to propel an 3600-lb curvaceous lump from 0-60 mph in 6.9 sec. Enough, at much higher speeds, to play in the same ballpark as the Porsche 928S. Enough, in short, to legitimately claim supercar performance.

But enough to justify $59,000, the price of a cozy country cottage someplace nice and quiet? This is where those who willingly inhabit a world of gray suits and calculators might well be advised to turn the page. What we have to say next will appeal rather more readily to the heart than to the head. The expense of the BMW M6 cannot be justified by reference to performance numbers alone. It has to be understood, right now, that there's more to this car than cold statistics. Much more.

Consider, first, that your $59,000 buys you engineering integrity—and not only the renowned BMW integrity, but something more, a very special pedigree. See the discreet "M" badges mounted on the trunklid and front grille, see the blue/mauve/red tricolor at the base of the bottom steering wheel spoke? With many other car makers, those logos would most likely denote M for Marketing, just another little sales gimmick, just another designer label to snare the Yuppies. Not here. M denotes Motorsport, the department that designed the M635CSi, the plain-Jane M5, and, most recently, the little M3 racer. Yes, that's the same department responsible for the engines that, for so long, dominated the nursery ground of Grand Prix stars, the European

Formula 2 championship. And, yes, that's the same M badge carried by Nelson Piquet's Formula One Brabham-BMW in 1983, the year he won his second World Championship. M is meaningful. M is for megalomania.

Or consider the build quality. This is a car hewn from rock. The two big doors slam shut solidly. The gutters between body panels are narrow, the tolerances constant. The metallic paint shimmers deeply. The interior leather is joined with arrow-straight seams and impossibly tidy stitches. The trunklid rises gently. The hood can be opened without enrolling Arnold Schwarzenegger's help. The handbrake pulls up through the shortest, firmest arc. The seatbelts slip through their rollers smoothly and quietly. The stalk controls all move crisply. Like a real sports car, you sit deep down, *in* rather than on the car. There are the clearest con-

The steering response isn't just diligently obedient, it's downright flattering

ceivable white-on-black instruments. The gearshift lever feels natural in your hand. Dammit, even the audible *clickety-click* accompanying the turn signals is just right.

Or consider the *standard* equipment. The long-sided seats offer superb comfort and support in all imaginable directions, largely because they're electrically adjustable in all imaginable directions. And they have memory settings for three different drivers. And there are dual electrically adjustable, heated outside mirrors. And a two-way electric sunroof (at the noticeable cost of some head room). And electric windows, with tinted glass and rear defroster. And central locking with a friction anti-theft system. And a heated driver's door lock with illuminating master key. And air conditioning. And cruise control. And a computer. And an AM/FM stereo radio/cassette player with power amplifier, graphic equalizer, and eight speakers. And a delightful leather sport steering wheel, one of the best in the business, easy to grip and perfectly sized, with an adjustable column. And that deep metallic paint. And a front spoiler incorporating foglamps (which a misty California coastal road revealed, for once, to be worthy of their name). And a tool kit in the trunk. And stunningly beautiful wire-spoke alloy wheels. And fat Michelin TRXs that lend the car a broad-shouldered stance. And anti-lock braking, with discs all around, the ventilated fronts increased from 11.2 to 11.8 in. diameter. And a limited-slip differential. And, last, but by no means least, an interior that literally reeks of opu-

lence, a sumptuous expanse of fine leather. The white seats of the test car were successfully balanced in tone with a black dash, gray door toppings and dash top, and white roof lining. (This tastefully simple color scheme created an unexpectedly light, airy atmosphere in what, quite honestly, is an unfashionably small passenger compartment for a car 16 ft 2 in. long.) But, oh, the aroma! You only have to inhale to be convinced that the M6 represents money well spent.

Or consider the exemplary road behavior. Like everything else about this car, the ride is reassuringly solid, the handling pin-sharp. Especially during the first 20° or so of movement at the wheel, the steering is so exquisitely sensitive it's almost neurotic. The wheel is loaded with feel. At 3.5 turns lock to lock, it's also instantly obedient. With a busy pitter-patter, the fat front tires tell all you need to know about the surface beneath. Assisted by an increase in front caster, the straight-line tracking is impeccable even at unmentionable three-figure speeds. Only a combination of the car's wide rubber and a sizeable undulation in the road will lead the M6 ever so gently astray. Even then, the steering is so communicative you'll immediately detect the tiniest deviation—and will delight in effecting an equally fine correction. This is a car you can place *exactly* where you want. The steering response isn't just diligently obedient, it's downright flattering.

The familiar 6-series suspension format has been retained, with front

Consider the build quality—this is a car hewn from rock

MacPherson struts and rear semi-trailing arms, but has generally been made more beefy. Firmer Bilstein gas-pressure shock absorbers have been fitted all around, with the twin-tube type at the front. There are shorter, stiffer coil springs, with progressive-rate components at the back. The front and rear anti-roll bars have accordingly been recalibrated. Although we judged the M6's slow-speed ride to be tolerable, it is possibly a little harsh for some tastes. If you've just stepped out of a BMW L6, it might seem that the M6 traverses potholes with a noticeable crash; that sharp ridges deliver a jolt.

But if you buy an M6 merely to squander it on suburban stop-and-go, frankly, you deserve to be beaten about. You can't expect a powerful tiger to be at its best if you keep it in a cage. Take it out into the open. Don't just jog along, have a run for your money. Then you'll find that the ride soon smoothes out, that the car

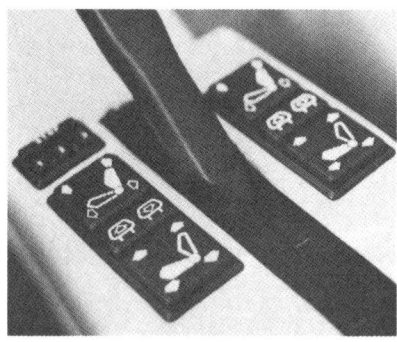 

corners at a brash gallop with admirable poise, that only brutality or stupidity or a determined mixture of both unsettles the M6's balance in sweeping, high-speed turns. Sure, it is fairly easy in the lowest two gears to fish up big gobs of power and send the car's tail flapping, and, sure, hooligan acceleration in dry conditions can also provoke some uncomfortable axle tramp. (In the wet, you'd better be more careful.) But the M6 more usually reprimands over-enthusiasm with easily controlled understeer. Besides which, in most circumstances, most of the time, the big Bimmer will be nowhere near its roadholding limits, even at the sort of speeds likely to invoke magisterial wrath.

At speed, you'll also be reminded how creamy smooth and free-revving the engine is, how it hammers with urgent power and emits an eery, jet-like whistle under hard acceleration, how its sound effects make the M6 seem like the next best thing to a front-engined Porsche 911. Turning the outside world into a blur, you'll forget the slight heaviness of the

clutch pedal, you'll revel in the superbly precise, light, short-throw gear shifts—and you'll be grateful that an automatic rev-limiter so rudely interrupts all the fun at 6800 rpm. It is sometimes needed. The S38 engine revs so eagerly it would otherwise be very easy to send pistons and conrods into orbit.

Which brings us to one of only two truly irksome weaknesses. The gap between the 2nd and 3rd gears seems too wide when hustling along a swift, curvy road. Too many twists and turns demand a 3rd-gear burble at about 3000 rpm, the engine turning too lazily for optimum throttle response when the road opens up again—or a frenetic, breathless *scream* near the redline in 2nd gear, all sound and fury, signifying nothing. (The compensation for this, though, is the usefulness of 3rd: It's a perfect chargin'-'n'-passin' gear, one of the M6's lovable little party pieces, a trick that'll pull 95 mph out of nowhere.)

Whether or not you're troubled by the 2nd- and 3rd-gear ratios, there's no escaping the other weakness. It stares you right in the face: The fuel needle descends like a free-faller with a tangled parachute. The cost of fun

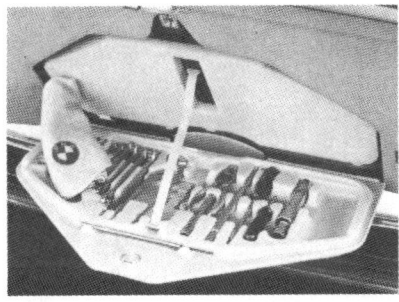

here, in our experience, is anything from 14 to 19 mpg. That's no great sin in itself, after all, the engine's only drinking money, and if you've got an M6, you've got plenty of that. But the gas tank only holds 16.6 gal, its reserve included. If you drive the M6 as its creators intended, you will regularly scrape something like 230 miles from the barrel, no more. It can take only three hours for the needle to dive into the red. Gas stations become familiar territory—and we all know what they say about familiarity.

Contempt, however, is an emotion very rarely stirred by the BMW M6. It is much more likely to raise a big, child-like smile. Why pretend otherwise? It makes you feel good, which means it is also likely to become addictive. And addictions have a way of seeming to be worth all the money in the world.

TECH DATA

BMW M6

GENERAL
- Vehicle mfr.: Bayerische Motoren Werke AG, Munich, West Germany
- Vehicle importer: BMW of North America, Inc., Montvale, N.J.
- Body type: 4-pass., 2-dr. sedan
- Drive system: Front engine, rear drive
- Base price: $55,950 plus $2250 gas-guzzler tax
- Major options on test car: None
- Price as tested: $58,200

ENGINE
- Type: L-6, cast iron block, aluminum head
- Displacement: 3453 cc (211 cu in.)
- Bore and stroke: 93.4 x 84.0 mm (3.7 x 3.3 in.)
- Compression ratio: 9.8:1
- Induction system: Bosch Motronic w/port fuel injection
- Valvetrain: DOHC, 4 valves/cylinder
- Max. engine speed: 6800 rpm
- Max. power (SAE net): 256 hp @ 6500 rpm
- Max. torque (SAE net): 243 lb-ft @ 4500 rpm
- Emissions control: Catalytic converter, closed-loop mixture control
- Recommended fuel: Super unleaded

DRIVETRAIN
- Transmission: 5-sp. man.
- Transmission ratios
  - (1st): 3.51:1
  - (2nd): 2.08:1
  - (3rd): 1.35:1
  - (4th): 1.00:1
  - (5th): 0.81:1
- Final drive ratio: 3.91:1

CAPACITIES
- Crankcase: 6.5 L (6.9 qt)
- Cooling system: 12.0 L (12.7 qt)
- Fuel tank: 62.8 L (16.6 gal)
- Luggage: 334.4 L (11.8 cu ft)

SUSPENSION
- Front: Independent, MacPherson struts, coil springs, gas-pressure telescopic shocks, double-pivot lower control arm, anti-roll bar
- Rear: Independent, semi-trailing arms, progressive-rate coil springs, gas-pressure shocks, anti-roll bar

STEERING
- Type: Recirculating ball, power assist
- Ratio: 16.2:1
- Turns (lock to lock): 3.5

BRAKES
- Front: 300 mm (11.8 in.), vented discs, power assist, ABS
- Rear: 285 mm (11.2 in.), discs, power assist, ABS

WHEELS AND TIRES
- Wheel size: 16.3 x 7.7 in.
- Wheel type: Cast alloy
- Tire size: 240/45VR415
- Tire mfr. & model: Michelin TRX
- Tire construction: Steel-belted radial

DIMENSIONS
- Curb weight: 1619 kg (3570 lb)
- Weight distribution, f/r: 52/48%
- Wheelbase: 2624 mm (103.3 in.)
- Overall length: 4922 mm (193.8 in.)
- Overall width: 1723 mm (67.9 in.)
- Overall height: 1353 mm (53.3 in.)
- Track, f/r: 1431/1463 mm (56.3/57.6 in.)
- Min. ground clearance: 133 mm (5.2 in.)

CALCULATED DATA
- Power-to-weight ratio: 13.9 lb/hp
- Top speed: 145 mph (est.)
- Drag coefficient: 0.41

SKIDPAD
- Lateral acceleration: 0.82 g

FUEL ECONOMY (mpg)
- EPA rating, city/hwy.: 10/19
- Test average: 17.2

ACCELERATION (sec)
- 0-30 mph: 2.66
- 0-40 mph: 4.16
- 0-50 mph: 5.32
- 0-60 mph: 6.89
- 0-70 mph: 9.27
- 0-80 mph: 11.52
- Standing quarter mile: 15.45 sec/94.0 mph

SPEEDOMETER
Indicated	30	40	50	60
Actual	26	36	46	56

BRAKING
- 30-0 mph: 30 ft
- 60-0 mph: 126 ft

BMW 635CSi

AUTOCAR TEST EXTRA

Far from being a dinosaur of a past decade, the 12-year-old 635CSi remains a superb package, and now BMW has improved it still further by fitting the 735 engine

Price £36,860 **Top speed** *134mph* **0-60** *8.4secs* **MPG** *17.2*
For *Drivetrain, ride, refinement* **Against** *Seats*

THE BMW 6-SERIES IS APPROACHING its 12th birthday. As revealed earlier in this issue, in 1989 it will be replaced by the 8-Series coupé. But far from being a dinosaur from an earlier decade, the latest version of the 635CSi is still an immensely satisfying car.

This year's 635CSi benefits from the 7-Series development programme. The 3430cc six-cylinder is lifted straight from the 735i, complete with its Bosch Motronic management system. It produces 220bhp at 5700rpm and 232lb ft of torque at 4000rpm, compared with the previous model's 218bhp at 5200rpm and 228lb ft torque at the same 4000rpm. A lower compression ratio of 9.2:1 allows the car to run on unleaded fuel without modification.

The four-speed ZF automatic is now equipped with electronic control and a choice of sport, economy or manual driving programmes. Combined with a final drive ratio of 3.64:1 and the low-profile Michelin TRX tyres also fitted to the 735i, this gives the 635 appreciably lower gearing than before.

Maximum speed is now achieved in fourth gear, the 635 managing a 134mph mean with the engine spinning over well below its power peak at 4900rpm. BMW claims 140mph for the automatic model, and with a few more miles on the clock this might well be feasible.

From a standing start the 635CSi performed as BMW predicted. The quickest launches were achieved simply by flooring the throttle in 'D', the car reaching 60mph in exactly the claimed 8.4secs. That's almost a second slower than the last 635CSi we tested, while the latest 3.6-litre automatic version of Jaguar's XJ-S coupé recorded exactly the same maximum and was half a second quicker to 60.

Selecting 'Sport' with the rotary switch to the right of the gearchange does two things — it locks out top gear, limiting the 635 to a top speed of 121mph, and moves the part throttle and kickdown points up the speed range. On full throttle the change points are the same in D or S, just nudging 6000rpm.

Using the 3-2-1 setting on the knob actually locks the gearbox into the gear selected, even from a standing start — useful on slippery surfaces or on very hilly ground. The rev counter has a two-stage red line beginning at 6100 and going solid at 6300 — this is supplemented by a cut-out at an indicated 6250rpm (6100rpm true).

The box actually changes gear very smoothly indeed, only a kickdown of two ratios causing any form of lurch. The selector handle itself seems to be the same as the 735, stiff and awkward to use — BMW would do well to examine the 'J' handle on Jaguar's XJ6. The programme selector switch isn't illuminated at night and there is no light on the facia to indicate when you are in sport mode.

Leave the car in sport and it goes very well indeed, with no hint of vibration or harshness even with the engine right on the redline. With such smooth power delivery you wonder

TEST EXTRA

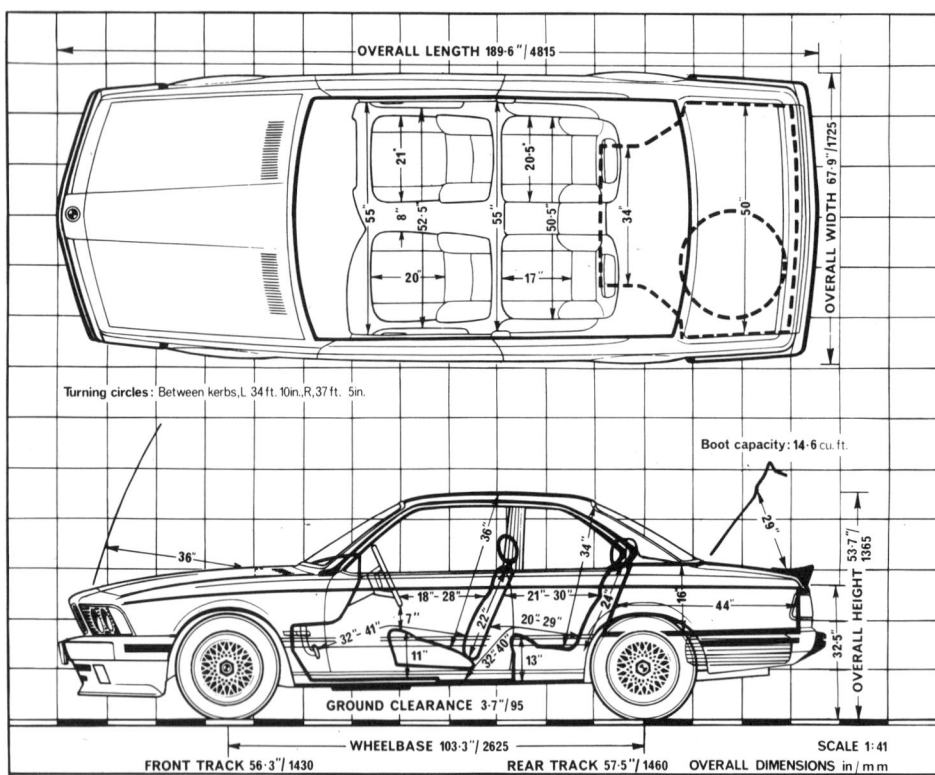

COSTS

Prices
Total in GB	£36,860.00
Road tax delivery, no plates	£395.95
Total on the Road	£37,252.95
EXTRAS (fitted to test car)	
M-Technic steering wheel	£30.00
Anti-theft device	£369.00
Total as tested	**£37,651.95**

SERVICE & PARTS

Change	*Oil service	*Insp 1	*Insp 2
Engine oil	Yes	Yes	Yes
Oil filter	Yes	Yes	Yes
Gearbox oil	No	No	Yes
Spark plugs	No	Yes	Yes
Air cleaner	No	Yes	Yes
Total cost	**£34.37**	**£167.71**	**£208.68**

(Labour cost £33.12 an hour inc VAT)

* As per service interval indicator

PARTS COST (inc VAT)

Brake pads (2 wheels) front	£60.32
Brake pads (2 wheels) rear	£33.00
Exhaust complete	£908.94
Tyre — each (typical)	£166.89
Windscreen	£248.41
Headlamp unit	£41.93
Front wing	£475.35
Rear bumper	£805.91

WARRANTY
12 months/unlimited mileage, 6-years anti-corrosion

EQUIPMENT

Cruise control	£324
Limited slip differential	£402
Power steering	●
Steering reach adjustment	●
Trip computer	●
Heated seats	£226
Height adjustment	●
Seat cushion tilt	●
Door mirror remote control	●
Electric windows F/R	●
Interior adjustable headlamps	●
Sunroof	●
Tinted glass	●
Headlamp wash/wipe	●
Central locking	●
Fog lamps	●
Metallic paint	●
Radio/cassette	●

● Standard ONC Optional at no extra cost

TEST CONDITIONS

Wind	8-10mph
Temperature	10deg C (50deg F)
Barometer	1004mbar (29.6ins Hg)
Humidity	60 per cent
Surface	dry asphalt/concrete
Test distance	760 miles

Figures taken at 1474 miles by our own staff at the Millbrook proving ground. All Autocar test results are subject to world copyright and may not be reproduced without the Editor's written permission.

Model: BMW 635 CSi

SOLD IN THE UK BY:
BMW (GB) Ltd
Ellesfield Avenue
Bracknell
Berkshire RG12 4TA

SPECIFICATION

ENGINE
Longways, front, rear-wheel drive. Head/block al. alloy/cast iron. 6 cylinders in line. **Bore** 92mm (3.62ins), **stroke** 86mm (3.39ins), **capacity** 3430cc (210.7 cu ins). **Valve gear** ohc, 2 valves per cylinder, chain camshaft drive. **Compression ratio** 9.2 to 1. Bosch Motronic ignition and fuel injection management system. **Max power** 220bhp (PS-DIN) (162kW ISO) at 5700rpm. **Max torque** 232lb ft (315 Nm) at 4000rpm.

TRANSMISSION
4-speed ZF automatic.

Gear	Ratio	mph/1000rpm
Top	0.73	27.3
3rd	1.00	19.9
2nd	1.48	13.4
1st	2.48	8.0

Final drive/ratio 3.64.

SUSPENSION
Front, independent, MacPherson struts, coil springs, telescopic dampers, anti-roll bar.
Rear, independent, semi-trailing arms, coil springs, telescopic dampers, anti-roll bar.

STEERING
Ball and nut, power assistance. Steering wheel diameter 15ins, 3.2 turns lock to lock.

BRAKES
Front 11.1ins (281.9mm) dia ventilated discs. **Rear** 11.1ins (281.9mm) dia discs. Antilock standard. Vacuum servo.

WHEELS
Al alloy, 210mm rims. Radial ply tyres (Michelin TRX on test car), size 240/45VR415.

PERFORMANCE

MAXIMUM SPEEDS

Gear	mph	km/h	rpm
Top (Mean)	134	216	4900
(Best)	135	217	4950
3rd	121	195	6100
2nd	82	132	6100
1st	49	79	6100

ACCELERATION FROM REST

True mph	Time (secs)	Speedo mph
30	3.4	33
40	4.7	43
50	6.3	54
60	8.4	64
70	10.7	74
80	13.5	84
90	17.2	95
100	21.5	106
110	27.1	116
120	34.8	127

Standing ¼-mile: 16.7secs, 89mph
Standing km: 29.4secs, 113mph

IN EACH GEAR

mph	Top	3rd	2nd	1st
10-30	—	—	4.0	2.4
20-40	—	—	4.7	2.6
30-50	—	5.7	4.1	—
40-60	—	6.4	4.1	—
50-70	—	6.9	4.5	—
60-80	—	7.0	5.1	—
70-90	—	7.1	—	—
80-100	—	8.1	—	—
90-110	—	9.8	—	—
100-120	—	13.5	—	—

FUEL CONSUMPTION
Overall mpg: 17.2 (16.4 litres/100km) 3.8 miles/litre
Grade and fuel: Premium, 4 star (97)
Fuel tank: 15.4 Imp galls (70 litres)
Mileage recorder: 0.5 per cent long
Oil: (SAE 15W/40) negligible

BRAKING
Fade (from 89mph in neutral)
Pedal load for 0.5g stops in lb

start/end		start/end	
1	30-32	6	32-34
2	30-34	7	38-34
3	30-34	8	40-38
4	34-30	9	42-42
5	34-36	10	42-40

Response (from 30mph in neutral)

Load	g	Distance
10lb	0.20	150.3ft
20lb	0.50	60ft
30lb	0.80	37.6ft
40lb	1.10	27.4ft
Parking brake	0.4	75ft

WEIGHT
Kerb 3467lb/1576kg
Distribution % F/R 55/45
Test 3839lb/1745kg
Max payload 722lb/328kg
Max towing weight 1800/918kg

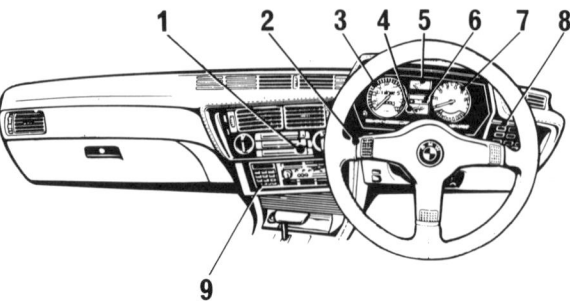

1 Heater controls, **2** Horn push, **3** Speedometer, **4** Fuel guage, **5** Temperature, **6** Service indicator, **7** Rev counter, **8** Check panel, **9** Trip computer

Balanced handling, excellent stability and ride comfort make the CSi an ideal cruiser. Equipment level is high, and leather upholstery adds style. The engine is taken straight from the 735

why there's any need for a 12-cylinder engine — the big BMW six is an object lesson to other manufacturers.

This effortless performance is not at the expense of economy. We managed an overall figure of 17.2mpg, giving the 635CSi a range of 260 miles from its 14.4-gallon tank.

With 11ins disc brakes all round, ventilated at the front, the BMW has the stopping power to match its performance. Although the pedal action was rather spongy, our tests showed them to suffer only slightly from fade, and a 40lb pedal load was able to generate 1.1g before the standard Bosch ABS came into action.

As well as a new nose incorporating the ellipsoidal headlamps from the 735i, the CSi inherits its wide alloy wheels and ultra-low profile 240/45VR415 Michelin TRX tyres. They provide impressive traction in the wet or dry, and for a big automatic the 635CSi can be flicked around with some verve. Strut front suspension and semi-trailing arms at the rear are almost a BMW trademark, and there is no sign of the tail-happiness sometimes attributed to this marque.

Steer the car into a corner at speed with the optional 15ins 'M Technic' wheel and it responds better than you might expect from a power-assisted ball and nut system. This is not the Servotronic speed steering system fitted to the 7 but nevertheless provides just the right balance between weight and feel — and at 3.2 turns lock to lock is higher geared than other BMW models.

Up the speed and the car responds with gently increasing understeer. With some lock wound on, the car can be unsettled by lifting off the throttle, but only then does flooring it again cause the car to drift into oversteer. Careful suspension development matched to good tyres makes handling surprisingly controllable — even on a damp road it would take a determined effort to spin the 635CSi.

This well-balanced handling is not achieved at the expense of ride comfort. While not quite up to Jaguar levels, the 635CSi is impressively supple at speed, with only a slight amount of bump thump at lower velocities. Despite the low-profile Michelins, tyre noise is generally well controlled, and in fact the 635 rides better and quieter than our long-term 735i.

Stability is excellent, even at maximum speed, displaying none of the nervousness around the straight ahead position from which the 735i suffers. At cruising speeds there is only slight wind noise from the frameless front windows, a murmur from the tyres and no sign at all of the engine noise.

The 635CSi justifies its hefty £36,860 price tag with an impressive range of standard equipment. As well as the usual BMW central locking and electric door mirrors, there are electric windows front and rear, electric seat adjustments with three position memory, trip computer, electric sliding sunroof, headlamp washers, heated door mirrors, locks and screen washer nozzles, air conditioning, internally-adjustable headlamps and leather upholstery.

Full leather means just that. Apparently BMW uses 27 square metres of hide for the interior of a 635 compared with the previous model's 11 square metres. Where vinyl *is* used, it is so well matched to the leather it's extremely hard to spot the difference. The stitching and fit of the panels is immaculate, and the whole interior has an air of class.

BMW has a few chinks in it's 'ultimate driving machine' armour when it comes to the driving position of the 635. Although the instruments and minor controls are excellent, the seats feel out of place in a car like this. Electric motors control fore and aft movement, backrest tilt, front and rear cushion height and even the headrest height, but to get any form of thigh support the front of the seat needs to be fully raised and the rear fully lowered. That in turn means the backrest has to be fully forward to provide any back support — and the top of the seat curves away from the spine above the base of the shoulder blades. There is very little side support which causes problems when cornering hard, neither is there enough support for the base of the spine.

A six foot plus driver will need the seat adjusted fully back for sufficient legroom, but even with the seat on its lowest setting the sunroof restricts headroom. The steering column is adjustable for reach but not rake.

The 635CSi has a capacious boot, the spacesaver spare wheel being mounted in a well under the floor. Because of the coupé design rear passengers are short of head and leg room, but less so than in an XJ-S. Apart from the glove box and a pocket in each door, the CSi is also woefully short of oddment stowage space.

As a mile-eater the 635CSi is a superb car. It handles well, cruises quietly and comfortably all day at autobahn speeds, and detail development over the years has kept it competitive. We would advise buyers to take up the no-cost option sports seats rather than the standard electric items, but otherwise the interior is hard to fault. £36,860 looks a lot of money compared with the £24,240 Jaguar demands for an automatic XJ-S, but the BMW is better built, better equipped and has the best engine in the business. ∎

COVER STORY

POWER TRIP!
BMW M6 vs. Porsche 928 S4

BMW M6 battles Porsche 928S 4

by Jack R. Nerad
PHOTOGRAPHY BY VIC HUBER

Some things are just meant to impress. Things like an Armani suit or a Dunhill lighter or a Waterford goblet. In classics like these, form follows function in the broadest sense because part of their function is to please the senses. They accomplish this by their look, their feel, and, yes, even their sound. Which brings us to the Porsche 928S 4 and the BMW M6, two landmark luxury cars that thrill all the senses in much the same way as these aforementioned design classics.

It was our solemn duty to take these classic examples of the luxury sports car and thrash them in all possible ways during the course of a trip from our posh Los Angeles digs to Sears Point Raceway in infamous Marin County, California, and back again. Before it was over, we'd logged more than 1000 highway miles on each of our subjects and put them through a veritable meat grinder of objective and subjective tests.

Of course, neither of these platforms is new. In fact, both have been soldiering on for a decade or so, but continual refinement has rendered them still world-class among the globe's luxury coupes. The 928S 4 is essentially a carryover from the '87 model year, as Porsche figures it holds a pat hand, but this time around the BMW 6-series has been dealt some new cards from the company's motorsports division to give it added firepower. Both offer enormous presence, style, comfort, and, lest we forget amidst the leather seating, premium sound systems, and automatic temperature controls, a bunch of performance. Owners of these cars often find themselves dressed in silk and serge at the latest gallery opening, but that doesn't mean the cars can't boogie when called upon, and probably boogie more than most of their owners will ever know. Or to paraphrase a pithy old saying, "It's a shame wealth is wasted on the rich."

Dime-store philosophy aside, here then is a play-by-play look at the individual examinations these

POWER TRIP!

cars were forced to endure before we came to our final verdict:

The Going-to-Dinner Test

After wending our separate ways up to Marin, we met at our hotel in San Rafael just in time to head to dinner. (Planning, planning.) The BMW won the first phase of this test when one of our number took a quick look at the Porsche's rear seats(?) and decided the Bimmer was a much more civilized alternative. When we reached our restaurant (Angelino's in Sausalito), the result of this test was a little less conclusive.

The parking lot attendent, no doubt jaded by many high-dollar cars, wasted nary a second glance on either of our contenders. Passers-by did express more enthusiasm for the Porsche, which had the advantage of its vivid red color. In our travels, both cars attracted a great deal of respect. First round: tie.

POWER TRIP!

The Photography Session Test

While our test subjects were under the lens at Sears Point, we had ample opportunity to cast the critical eye at both. We don't have to tell you which car looks better. You've got eyes, taste, and can certainly judge for yourself

The 928's 4-valve engine cranks out 316 hp compared to the M6's 256. Both cars take to the track with surprising aplomb.

from ace photog Vic Huber's excellent handiwork on these pages. Frankly, to our eyes, both are lovely cars that are beginning to show their age, and we think Porsche has done the more successful job of updating. The 928's winged rear quarters look lean and contemporary, while the Bimmer offers a rear spoiler that seems tacked on. The Porsche's swoopier looks, however, do come at the price of marginal rear passenger capacity and significantly less luggage space. You can judge for yourself how practical you want to be. Second round: slight edge, Porsche.

The Race Course Test

Looking at these two beauties is fun, but they really come into their own when you slip behind the wheel. In the interest of empirical testing, all that is holy, and just for fun, we next put the cars to the test around the challenging Sears Point track, home of the Bob Bondurant driver's school. Our test pilot for this session was none other than erstwhile engineering editor and bon vivant Ron Grable, who, coincidentally, held the overall track record at this venue for more than a decade. After zipping up hill, zapping down dale (who didn't object a bit), and getting kinky through the kinks, the Porsche bested the BMW by 2 sec, 2:04.88 to 2:06.88. (For your edification, Showroom Stock Corvettes in race trim circle the track in the low 1:50s.) Of course, two seconds on a racetrack is a millenium, but sheer speed doesn't tell the whole tale.

In the slalom test, the BMW was more precise and easier to place than the 928S 4

When the lapping was over, Grable opined that he could likely wring another second or two from the Porsche's clocking by making better use of the 928's power advantage (316 hp at 6000 rpm versus the BMW's 256 hp at 6500 rpm) if he had significantly more time to experiment. Relatively less stable than the BMW in these high-performance conditions, the 928S 4 proved much harder to precisely place on the course. With body roll and a significant amount of oversteer, the 928 was a handful, while the M6, though slower, was much more neutral and thus easier to drive fast. Third round: slight edge, Porsche.

The Dragstrip Test

With the test on the road course out of the way, we segued to the dragstrip to take a hard look at straightline acceleration. Here the Porsche seemed to have an obvious

advantage going in with 60-hp more on tap. When it was all over, though the Porsche did vanquish the M6, it wasn't all that easy. The problems: wheelspin and axle tramp. Simply put, it was hard to get all the 928's power to the ground. After experimenting with many techniques, we got a 0-60-mph time of 6.1 sec and plowed through the quarter mile in 14.43 sec at 102.4 mph. Those are stellar numbers, but they weren't accomplished without some drama. Porsche actually reports slightly better 0-60 numbers (5.7 sec), but then it has plenty of 928s, so if it blows one up, it doesn't much matter. Compared to the drama of the 928S 4, the M6 was easy as pie to get out of the hole, but not quite as fast. Sixty miles per hour arrived in a still rapid 6.96 sec (versus a claimed 6.8 sec) and the quarter mile vanished in 15.45 sec at 94.5 mph. Fourth round: clear edge, Porsche.

The Brake Test

It's great to have cars that go, but if they can't stop, you'll quickly find yourself in a bagful of trouble. Both of these behemoths are fitted with giant power-assisted disc brakes all around. The front discs are vented on the BMW, and all four are vented on the Porsche; both cars use sophisticated anti-lock hardware. The benefits from all this are simply incredible stopping numbers. The brake test was an instance where the BMW beat out the Porsche, if by just a little bit, with a 60-0-mph best of 126 ft to the Porsche's 130. From 30 mph, the 928

edged the Bimmer by a foot, 33 ft to 34. By any measure, these are terrific numbers, and, equally important, they can be easily attained by the average driver—just put your foot in it and hold it down. Oh, and don't forget to steer. Fifth round: slight edge, BMW.

The Skidpad Test

The skidpad tests the outer limits of steady-state cornering ability, but, as we'll see, it's not an infallible test of overall handling capability. When the tire smoke cleared, the Porsche was the winner, with an impressive 0.87g mark compared to the M6's still topnotch 0.83 g. Since tires are often the biggest factor on the pad, our guess would be that the 928's Dunlop 225/50VR16 fronts and 245/45VR16 rears were slightly stickier than the M6's Michelin TRX 240/45VR-415 metrics. Sixth round: clear edge, Porsche.

The Slalom Test

As we noted previously, skidpad alone doesn't measure handling. You also must look at braking prowess and the subject of this test, transient behavior. We flimflammed our two subjects through our standard 600-ft slalom (with gates 100 ft apart) several times and then picked out the best run turned in by each car. To our surprise, the BMW won by a margin of 0.07 sec, but again the numbers don't tell the whole story. Clearly, the

When the tire smoke cleared on the skidpad, the Porsche was the winner

BMW was more precise and easier to place. It made its transitions with little drama, just cool, calm confidence. The 928S 4, on the other hand, needed every bit of its considerable horsepower advantage to even come close to the BMW's best time. With significant amounts of power oversteer lurking under your right foot, a whole bunch of concentration and self-restraint was needed to keep from wasting a cone or two. From behind the wheel, the Porsche felt like a much heavier car than the BMW when in fact it isn't. Both cars weigh in at about 3500 lb, beefy but befitting their luxury status. Seventh round: clear edge, BMW.

The Mileage and Range Test

If you can pay for either of these cars, you can surely afford a tank of gas (even high-test), but we still think

fuel economy and, particularly, range are valid subjects for our comparison. In this test, the Porsche shone at least in relative terms. Its EPA mpg numbers of 15/23 will put smiles on the faces of Arab oil sheiks, but the M6's 10/19 numbers will have them laughing and scratching. Of greater concern to the bucks-up buyers of these cars is range, which is meager in the BMW. With its 16.6-gal fuel capacity, you'll find yourself filling up every 160 miles or so. In comparison, the Porsche's 22.7-gal gas tank gives you a theoretical city driving range of nearly 350 miles. Eighth round: clear edge, Porsche.

The Everyday Driving Test

Everyday driving might not be the most accurate term for this test because when you're at the wheel of either of these two beauties, nothing is really everyday. A trip to the corner market becomes a sensory experience; a cruise on the freeway becomes a magic carpet ride. In both cars, the driver is pampered by terrifically comfortable leather-covered seats, efficient and easy-to-operate automatic temp control systems, and multi-speaker stereos with concert-quality sound. Neither is lacking in standard features to enhance the luxury experience, and, as you've seen, both boast plenty of performance to turn even short trips into satisfying road forays.

To get specific, we'd judge the M6 more convenient and practical for

high-profile around-town cruising. It seats four adults in relative comfort, accommodates their luggage for trips to mountain lodges or country homes, and offers a tad more interior convenience. The two biggest pluses are the nine-function onboard computer and what BMW calls its Active Check Control system. The latter monitors things like headlights, brakelights, washer fluid, and coolant level to warn you of a possible problem. The onboard computer figures things like range, estimated time of arrival, and exterior temperature. The last is a particularly useful feature in mountain and cold-weather driving when

The Bimmer's engine is smooth all the way to the place where the rev-limiter says, "Hey, stop that!"

you're not sure if the puddle up ahead is frozen or not. The only gripe we have about the M6 is its abominable AM radio reception. We found it amazing that a car that costs this much has a radio that can't clearly hold a local 50,000-watt station.

The M6's around-town handling ability is good, and the all-independent Macstrut front/semi-trailing arm rear suspension damps potholes without giving passengers that otherworldly floating feeling of many luxury cars. The M-Squad motor is certainly powerful enough, though much of the power and torque lie fairly high up on the rev band. (Torque peak is 243 lb-ft at 4500 rpm.) This means you have a tendency to run the car up toward the redline as you zip into openings in traffic. That's all right, since the engine is smooth from the proverbial git-go all the way to the place where the rev-limiter says, "Hey, stop that!"

Although we think the BMW is slightly better for around-town duties, we find it hard to fault the 928S 4 in that capacity. It's hard to believe that a car with 160-mph capabilities in stock trim can be such a docile everyday performer. The 4-valve 5-liter V-8 engine produces so much power and torque (317 lb-ft at a low 3000 rpm) that it's easy to lug through traffic. In fact (forgive me, Porsche), around town it feels like nothing as much as a Chevy Camaro with a small-block V-8 underhood.

Actually, the Camaro reference al-

TECH DATA

BMW M6

GENERAL
Vehicle mfr. ....Bayerische Motoren Werke AG, Munich, Federal Republic of West Germany
Vehicle importer ....BMW of North America, Inc., Montvale, N.J.
Body type ....4-pass., 2-dr.
Drive system ....Front engine, rear drive
Base price......................$55,950
Price as tested................$58,705

ENGINE
Type ....Inline 6, liquid-cooled, iron block, alloy head
Displacement..................3453 cc (211 cu in.)
Compression ratio 9.8:1
Induction system............Electronic multi-point fuel injection
Valvetrain ....DOHC, 4 valves/cylinder
Max. power (SAE net) ....256 hp @ 6500 rpm
Max. torque (SAE net)....243 lb-ft @ 4500 rpm
Emissions control ....3-way catalytic converter, Lamba oxygen sensor, closed-loop mixture control
Recommended fuel ....Unleaded premium

DRIVETRAIN
Transmission ....5-sp. man., lockup torque converter
Transmission ratios
 (1st)............3.51:1
 (2nd)..........2.08:1
 (3rd)...........1.35:1
 (4th)............1.00:1
 (5th)............0.81:1
Axle ratio 3.91:1
Final drive ratio 3.17:1

CAPACITIES
Crankcase ....6.5 L (6.9 qt)
Fuel tank ....62.8 L (16.6 gal)
Luggage..........................334 L (11.8 cu ft)
Range (at EPA combined)......................267 km (166 mi)

SUSPENSION
Front...............................Independent, MacPherson struts, lower control arms, coil springs, anti-roll bar
Rear................................Independent, semi-trailing arms, coil springs, gas-filled shocks, anti-roll bar

STEERING
Type ....Recirculating ball
Ratio...............................16.2:1
Turns (lock to lock)........3.5
Turning circle.................11.8 m (38.7 ft)

BRAKES
Front...............................299 mm (11.8 in.), vented discs
Rear................................284 mm (11.2 in.), discs
Anti-lock ....Bosch/BMW

WHEELS AND TIRES
Wheel size......................16 x 7.0 in.
Wheel type.....................Cast aluminum
Tire size & construction.P240/45VR-415
Tire mfr. & model...........Michelin TRX

DIMENSIONS
Published curb weight ....1619 kg (3570 lb)
Weight distribution, f/r...52/48%
Wheelbase ....2624 mm (103.3 in.)
Overall length..................4923 mm (193.8 in.)
Overall width..................1725 mm (67.9 in.)
Overall height.................1354 mm (53.3 in.)
Track, f/r........................1430/1463 mm (56.3/57.6 in.)

SPECIFICATIONS
Power-to-weight ratio ....13.9 lb/hp
Drag coefficient..............0.41
EPA (combined)..............15 mpg

MEASURED PERFORMANCE
QUARTER MILE (TIME)....15.45
QUARTER MILE (SPEED) ....94.5 mph
BRAKING
 60-0 ....126 ft
 30-0 ....34 ft
SKIDPAD ....0.83 g
SLALOM (600 ft).............6.88 sec

SPEEDOMETER CALIBRATION
Indicated	30	40	50	60
Actual	30	40	49	59

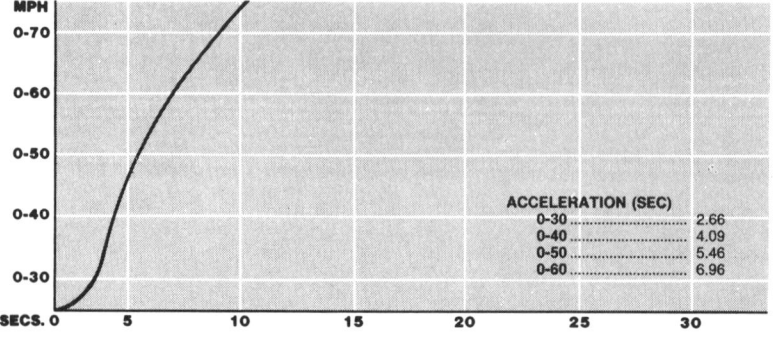

ACCELERATION (SEC)
0-30 ...... 2.66
0-40 ...... 4.09
0-50 ...... 5.46
0-60 ...... 6.96

so applies to the passenger and luggage accommodations. Room for two and their luggage is perfectly adequate; try to haul four and baggage and you'll have two unhappy people on your hands. In terms of specific likes and dislikes: We liked the adjustable steering column, since the dash pod moves with it to provide clear view of the excellent instrumentation, but we disliked the hidden power seat controls. It seemed like we were always hitting the wrong one. We also like the convenient tape storage area and the terrifically effective temp control, but we thought the stereo was rather confusing and, amazingly, didn't hold an AM station worth a damn, either. Ninth round: clear edge, BMW.

The You-Pays-Your-Money Test

Sure, we get to drive these cars for free, but even the wealthy have to check their current bank balance when it comes time to step up to the cashier's window for these battlewagons. As we said, both cars come fully loaded with all manner of niceties and both have their appeal. Both, too, have high pricetags. The M6 doesn't come quite as dear as the 928S 4. Its as-tested price was $58,705 including a rather stiff $2250 gas-guzzler tax. The Porsche was significantly more at $71,119 as tested with $2841 worth of options and a gas guzzler tariff of $650. Some might argue that at these prices it doesn't really matter *what* they cost, but we'd argue that even a millionaire wouldn't mind having an extra 10 grand or so in his jeans. Round 10: clear edge, BMW.

The Results

When all was said and done, we sincerely wished all wasn't said and done, since we didn't really want to return either of these cars to their rightful owners, much less other magazine wanks. But we had to, and here are the results. The Porsche 928S 4 had the clear edge in three categories and a slight edge in two; the BMW M6 had the clear edge in three categories and a slight edge in one. The winner of our power trip is the Porsche, on points with no knockdowns and by the narrowest of margins. Of course, as we speak, BMW is preparing its 6-series successor, so next time around, who knows which the winner will be? Stay tuned. MT

TECH DATA

Porsche 928S 4

GENERAL
Vehicle mfr. ....Porsche AG, Stuttgart, Federal Republic of Germany
Vehicle importer ....Porsche Cars North America, Inc., Reno, Nev.
Body type ....4-pass., 2-dr.
Drive system ....Front engine, rear drive
Base price $67,955
Price as tested $71,119

ENGINE
Type ....V-8, liquid-cooled, alloy block and heads
Displacement...................4957 cc (302.5 cu in.)
Compression ratio 10.0:1
Induction system.............Electronic multi-point fuel injection
Valvetrain ....DOHC, 4 valves/cylinder
Max. power (SAE net) .....316 hp @ 6000 rpm
Max. torque (SAE net) ....317 lb-ft @ 3000 rpm
Emissions control ....3-way catalyst with oxygen sensor, secondary air injection
Recommended fuel ....Unleaded premium

DRIVETRAIN
Transmission.....................5-sp. man.,
Transmission ratios
    (1st) 3.76:1
    (2nd) 2.51:1
    (3rd) 1.79:1
    (4th) 1.35:1
    (5th)............1.00:1
Axle ratio..........................2.64:1
Final drive ratio 2.64:1

CAPACITIES
Crankcase ....7.6 L (8 qt)
Fuel tank ....85.9 L (22.7 gal)
Luggage.............................178 L (6.3 cu ft)
Range (at EPA combined)......................544 km (340 mi)

SUSPENSION
Front..................................Independent, unequal-length A-arms, tube shocks, anti-roll bar
Rear....................................Independent, Weissach upper links, lower A-arms, tube shocks, coil springs, anti-roll bar,

STEERING
Type ....Rack and pinion
Ratio..................................17.8:1
Turns (lock to lock).........3.2
Turning circle ....11.5 m (37.7 ft)

BRAKES
Front..................................304 mm (11.9 in.), vented discs
Rear....................................299 mm (11.8 in.), vented discs
Anti-lock ....Bosch/Porsche

WHEELS AND TIRES
Wheel size.........................16 x 7.0 in. front, 16 x 8.0 in. rear
Wheel type........................Cast aluminum
Tire size & construction..225/50VR16 front, 245/45VR16 rear
Tire mfr. & model.............Dunlop

DIMENSIONS
Published curb weight....1589 kg (3505 lb)
Weight distribution, f/r...50/50%
Wheelbase.........................2499 mm (98.4 in.)
Overall length....................4524 mm (178.1 in.)
Overall width ....1836 mm (72.3 in.)
Overall height...................1275 mm (50.2 in.)
Track, f/r............................1551/1547 mm (61.1/60.9 in.)

SPECIFICATIONS
Power-to-weight ratio .....11.1 lb/hp
Drag coefficient................0.34
EPA (combined)................19 mpg

MEASURED PERFORMANCE
QUARTER MILE (TIME)....14.43
QUARTER MILE (SPEED) ....102.4 mph
BRAKING
    60-0 ....130 ft
    30-0 ....33 ft
SKIDPAD ....0.87 g
SLALOM (600 ft)...............6.95 sec

SPEEDOMETER CALIBRATION
Indicated	30	40	50	60
Actual	30	40	50	59

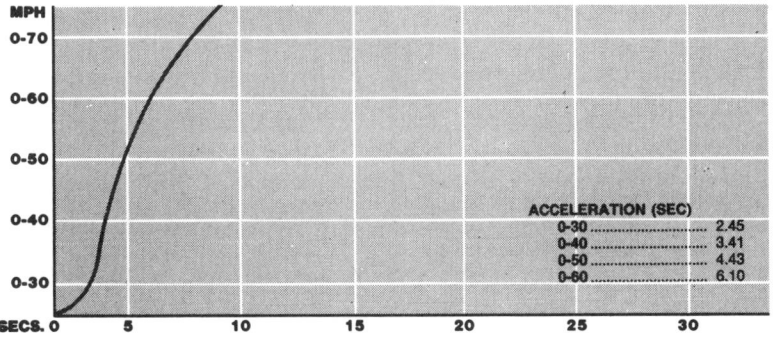

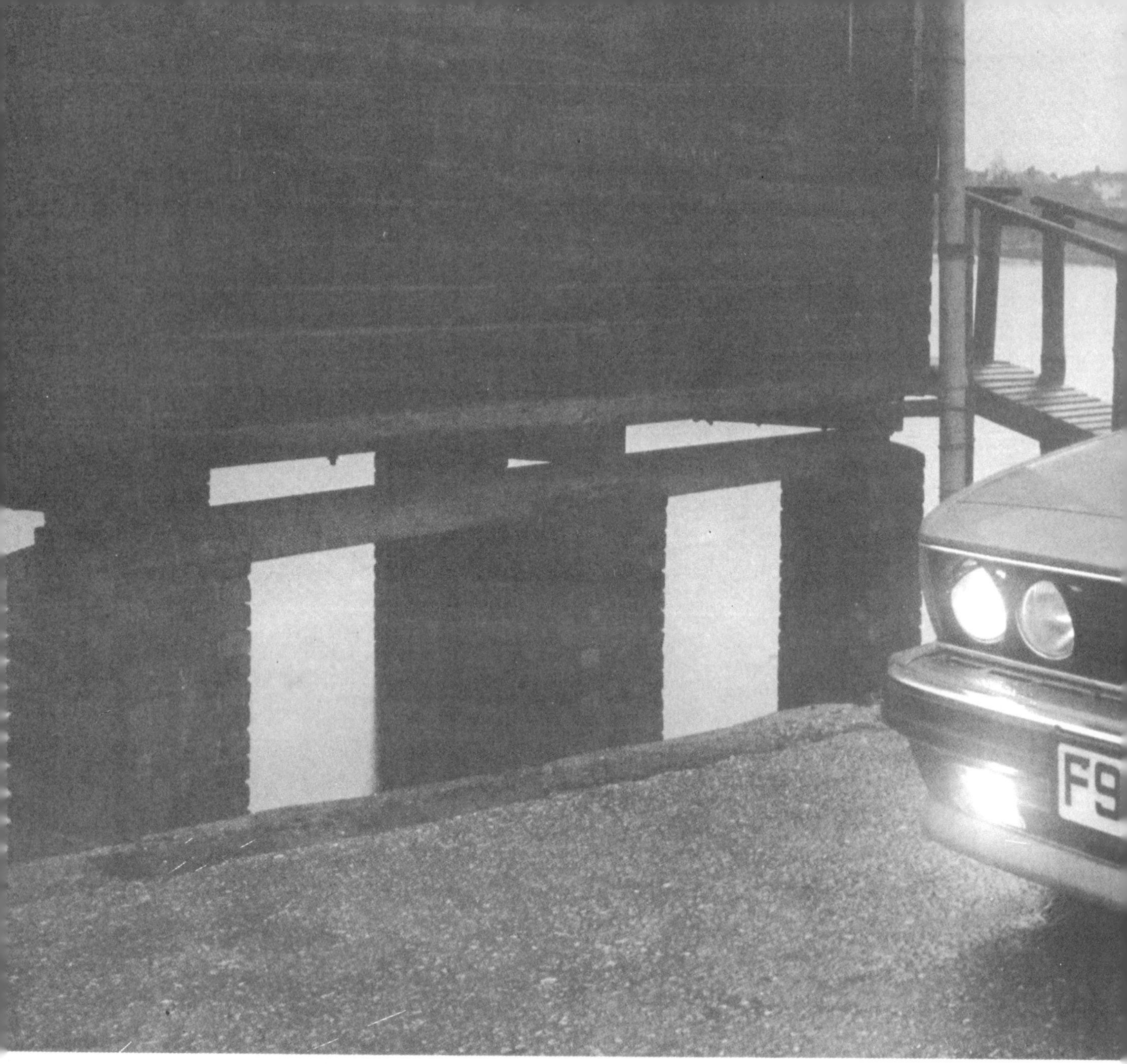

TEST EXTRA

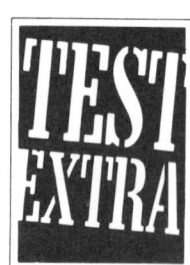

BMW M635 CSi

The M635 CSi has lost none of its driver appeal over the years. Its performance still stuns but refinement falls behind

Price *£45,780* **Top Speed** *150mph* **0-60** *6.0secs* **MPG** *20.6*
For *Performance, handling, brakes*
Against *Harsh ride, wind noise*

THE DISCREET *M* BADGING SAYS IT all. It marks this 635 CSi as one that has been massaged by BMW's Motorsport division to emerge a better, swifter car. It costs £46,000 but it will storm to 100mph in 15 seconds.

Unleash its towering 286bhp from a standstill and it lays two thick black lines of rubber on the way to a blistering 0-60mph time of just 6secs. The urge doesn't fade until 150mph. This car is *very* fast.

Yet the M6 launches its exceptional ability from a solid, practical base. Dated it may be but the 6-Series coupé still cuts it as a long-distance tourer. With the power of *M*, it's simply magnificent. And how.

It was back in 1984 that we last tested an M635 CSi. Little has changed mechanically. But the 1989 car gets a full leather interior, air conditioning and electrically adjustable seats as standard. Outside, though, the changes are less obvious. Ellipsoid headlights replace the round units, and the front spoiler and rear

A bootful of throttle in a tight bend unsticks the rear end

BMW M635 CSi

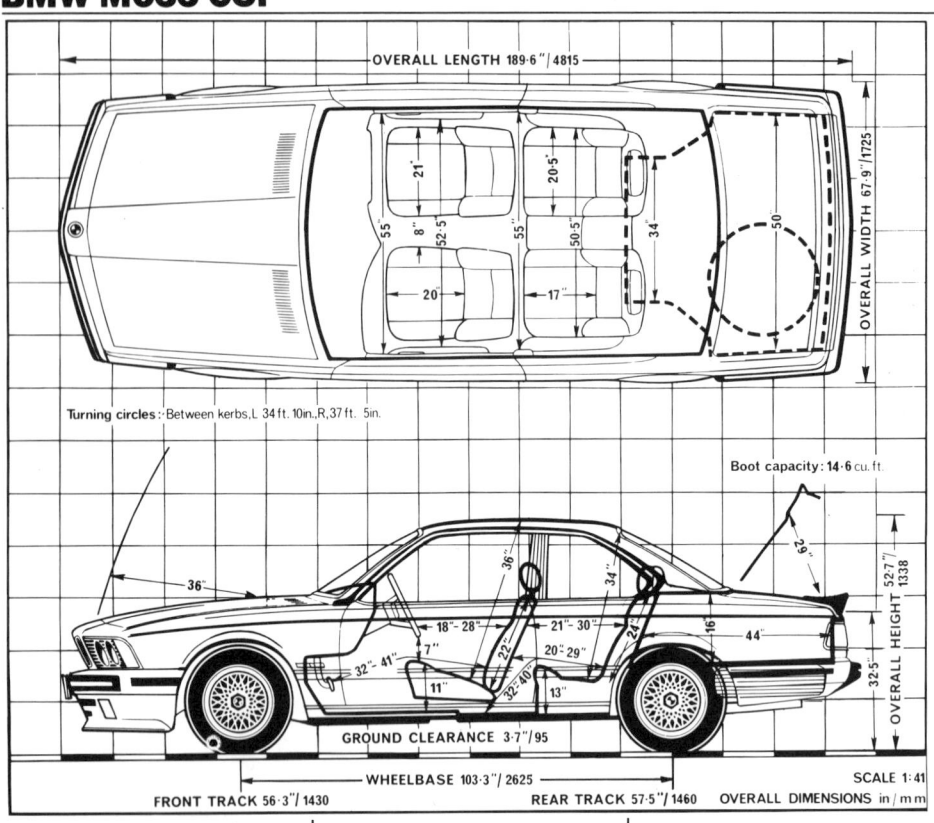

PERFORMANCE

MAXIMUM SPEEDS

Gear	mph	km/h	rpm
Top (Mean)	150	241	6200
(Best)	150	241	6200
4th	134	216	6900
3rd	99	159	6900
2nd	64	103	6900
1st	38	61	6900

Standing ¼-mile: 14.6secs, 99mph
Standing km: 26.3secs, 128mph

ACCELERATION FROM REST

True mph	Time (secs)	Speedo mph
30	2.4	33
40	3.5	44
50	4.6	54
60	6.0	65
70	7.9	75
80	9.9	85
90	12.1	95
100	15.1	106
110	18.4	116
120	22.5	126
130	27.5	137

ACCELERATION IN EACH GEAR

mph	Top	4th	3rd	2nd
10-30	—	7.7	5.2	3.2
20-40	9.0	6.8	4.9	2.8
30-50	9.2	6.7	4.5	2.5
40-60	9.4	6.6	4.2	2.5
50-70	9.3	6.2	4.0	—
60-80	9.0	6.1	3.8	—
70-90	9.4	5.9	4.1	—
80-100	9.9	6.0	—	—
90-110	10.1	6.5	—	—
100-120	10.8	7.6	—	—
110-130	12.8	9.3	—	—

FUEL CONSUMPTION
Overall mpg: 20.6 (13.7 litres/100km)
Touring mpg*: 22.4mpg (12.6 litres/100km)
Govt tests mpg: 17.1 (urban) 36.2 (56mph) 29.1 (75mph)
Fuel tank: 15.4 Imp galls (70 litres)
Max range*: 345 miles
Mileage recorder: 0.5 per cent long
* Based on Government fuel economy figures: 50 per cent of urban cycle, 25 per cent each of 56/75mpg consumptions.

BRAKING
Fade (from 99mph in neutral)
Pedal load for 0.5g stops in lb

start/end		start/end	
1	25-25	6	32-32
2	27-27	7	31-31
3	27-27	8	31-31
4	27-27	9	31-33
5	28-28	10	32-35

Response (from 30mph in neutral)

Load	g	Distance
10lb	0.12	251ft
20lb	0.34	89ft
30lb	0.55	55ft
40lb	0.89	34ft
50lb	1.10	27ft
Parking brake	0.35	86ft

WEIGHT
Kerb 3458lb/1570kg
Distribution % F/R 53/47
Test 3848lb/1747kg
Max payload 838lb/380kg

TEST CONDITIONS
Wind 12-15mph
Temperature 17deg C (63deg F)
Barometer 1012mbar
Humidity 60 per cent
Surface dry asphalt/concrete
Test distance 1301 miles

Figures taken at 3061 miles by our own staff at the Lotus Group proving ground at Millbrook.

All *Autocar & Motor* test results are subject to world copyright and may not be reproduced without the Editor's written permission.

SPECIFICATION

ENGINE
Longways, front, rear-wheel drive. Head/block al alloy/cast iron 6 cylinders in line.
Bore 93.4mm, **stroke** 84.0mm, **capacity** 3453cc.
Valve gear 2ohc, 4 valves per cylinder.
Compression ratio 10.5 to 1.
Ignition and fuel system Mapped electronic ignition and fuel injection.
Max power 286bhp (PS-DIN) (213kW ISO) at 6500rpm. **Max torque** 251lb ft (340 Nm) at 4500rpm.

TRANSMISSION
5-speed manual.

Gear	Ratio	mph/1000rpm
Top	0.81	24.0
4th	1.00	19.4
3rd	1.35	14.4
2nd	2.08	9.3
1st	3.51	5.5

Final drive ratio 3.73:1; limited slip differential.

SUSPENSION
Front, independent, MacPherson struts, coil springs, telescopic dampers, anti-roll bar.
Rear, independent, semi-trailing arms, coil springs, telescopic dampers, anti-roll bar.

STEERING
Ball and nut, power assistance. Steering wheel diameter 15ins, 3.1 turns lock to lock.

BRAKES
Front 11.8ins (300mm) dia ventilated discs. **Rear** 11.1ins (282mm) dia discs. Vacuum servo.

WHEELS/TYRES
Al. alloy, 8ins rims. 240/45VR415 Michelin TRX tyres.

Sold in UK by
BMW (GB) Ltd
Ellesfield Avenue
Bracknell
Berks RG12 4TA

COSTS

Prices
Total (in GB)	£45,780.00
Road tax, delivery, no. plates	£395.95
Total on the road	£46,175.95
Insurance group	OA

EXTRAS (fitted to test car)
Anti-theft lock	£369.00
Blaupunkt Toronto radio/cassette	£558.57
Total as tested	£47,103.52

SERVICE & PARTS
*Oil

Change	service	*Insp. 1	*Insp. 2
Engine oil	Yes	Yes	Yes
Oil filter	Yes	Yes	Yes
Gearbox oil	No	No	Yes
Spark plugs	No	Yes	Yes
Air cleaner	No	Yes	Yes
Total cost	£28.46	£196.00	£230.64

(Assuming labour at £33.12 per hour inc VAT)
*As per service interval indicator

PARTS COST (inc VAT)
Brake pads (2 wheels) front	£60.66
Brake pads (2 wheels) rear	£38.36
Exhaust complete	£454.79
Tyre — each (typical)	£166.89
Windscreen	£248.41
Headlamp unit	£41.93
Front wing	£475.35
Rear bumper	£264.78

EQUIPMENT
Five speed	●
Limited slip differential	●
Power steering	●
Steering reach adjustment	●
Head restraints F/R	●
Heated seats	£226
Height adjustment	●
Seat cushion tilt	●
Electric windows F/R	●
Heated rear window	●
Interior adjustable headlamps	●
Sunroof	●
Headlamp wash/wipe	●
Central locking	●
Metallic paint	●
Radio/cassette	DO
Aerial	DO
Speakers	DO

● Standard, DO Dealer option

WARRANTY
12 months/unlimited mileage, 6 years anti-corrosion

1 Heater controls, **2** Horn, **3** Speedometer, **4** Fuel gauge, **5** Temperature gauge, **6** Service indicator, **7** Revcounter, **8** Check panel, **9** Trip computer

TEST EXTRA

apron now incorporate 2.5mph, impact-absorbing bumpers.

The M6 carries a price premium of almost £9000 over the standard 635 CSi, placing it in the same price league as the Ferrari Mondial, Lamborghini Jalpa, Mercedes 420 SEC and, of course, right in the middle of 911 territory. The M6 is roomier than all but the Mercedes and potent enough to embarrass many of its more overtly sporting rivals.

Open the bonnet and the reason is obvious. BMW M Power is the legend on the crackle black cam cover. Beneath are twin camshafts and 24 valves to boost power from the standard 635 CSi's 220bhp at 5700rpm to 286bhp at 6500rpm. Torque is increased too, from 232lb ft at 4000rpm to 251lb ft at 4500rpm.

But M Power means more than just a multi-valve head. There's a stouter crank with 2mm shorter stroke, longer con rods, and stronger, high-compression pistons in a 1.4mm larger bore block. This gives a capacity of 3453cc, a slight increase over the single-cam engine. Manifolding is changed too, but both engines share Bosch Motronic mapped fuel injection and ignition. A beefier clutch feeds an increased output to a ZF close-ratio, five-speed manual gearbox.

The changes from standard specification don't end there. A limited slip differential and massive 240/45 Michelins keep each wheel firmly clamped to the tarmac, while firmer suspension settings and a lower stance add further to cornering capabilities. Braking is taken care of by huge discs all round with huge four-pot racing calipers.

At small throttle openings there is little indication of the huge performance available. The big BMW glides along on just a whiff of throttle in top gear. Change down and floor the throttle, though, and the M6's character is transformed. There's a squawk from the hefty Michelins as they grip the tarmac, the nose rises with a sense of purpose and the BMW's great bulk shoots forward.

There's no peakiness in the power delivery, just an overwhelming gain in momentum backed up by a hard-edged mechanical howl from the engine, the whine of hard-worked transmission gears meshing together, and an urgent exhaust wail.

Performance figures back up the subjective impressions. From rest, 60mph comes up in just 6secs — an exceptional time for a front-engined car. Change into third and, at 100mph, only 15.1secs have elapsed.

In-gear figures are equally impressive. Third gear is the best overtaking ratio in most conditions, each 20mph increment from 40 to 90mph taking around 4secs. Fourth carries the big BMW all the way to 130mph before a change into top is necessary.

The car feels rock steady at maximum speed with little more than the roar of the wind for company. The maximum of 150mph has the big engine turning over at 6200rpm, just below peak power in fact.

The BMW's brakes are reassuringly powerful and free from fade. The weighting is just right with no sudden response in gentle use and a firm, progressive action.

While never quiet, the 24-valver always feels smooth and potent, though power comes on much stronger beyond 3500rpm. Fine mechanical smoothness is enhanced by a beautifully progressive throttle linkage.

Considering the terrific performance of the M6, fuel economy is a pleasant surprise. The

Cam cover says it all. The power of M is 286bhp at 6500rpm and 251lb ft torque at 4500rpm. BMW shoots to 60mph in just 6secs and 100mph in 15.1secs. Full leather interior incorporates 27 metres of hide. Driving position is good but wheel set too high. M635 is loaded with equipment

overall figure of 20.6mpg gives a range of about 300 miles on one tankful. Motorway driving hiked the overall figure to 23.5mpg while, remarkably, even during performance testing the consumption never dropped below 17mpg.

The M635 CSi has impeccable road manners. Outright grip and traction are simply outstanding. Thankfully, BMW's engineers spare us the Servotronic steering fitted to the 7-Series, opting instead for the weightier recirculating ball arrangement.

A bootful of throttle in a tight bend will unstick the back end. On a wet or greasy road, restraint is essential. But the M6 is so well balanced that applying the right amount of corrective lock seems to come naturally. High-speed sweepers, for example, induce nothing more than mild understeer — you really do need to push hard to provoke anything beyond on-rails handling.

By contrast, ride refinement is not a strong point. Compared with the supple ride of the regular 635 CSi, the M-badged car lacks subtlety. It feels as though everything has been stiffened up for the sake of good handling with scant regard for ride quality or road noise suppression. This is quite deliberate.

When worked hard the engine gets noisy, too, though at cruising speeds it settles down to a subdued thrum. Then it is the wind roar that intrudes, mostly from around the sunroof and frameless door windows. But on abrasive and concrete surfaces, tyre rumble competes with wind roar as the worst offender.

Inside, the M6 has a full leather interior, stitched from enough hides to clothe a small herd — 27 square metres in all. The quality, style and finish are tasteful and classy.

The driving position is basically fine with nicely set pedals inviting heel and toe gearchanges, a conveniently positioned gear lever and a range of powered seat adjustments that should suit most drivers. However, the steering wheel is over-large and set just too high. And taller drivers could well find leg and headroom at a premium.

The rear seats are beautifully shaped for two and would be comfortable were it not for a lack of knee room. On the passenger side this is no problem as the front seat can be moved forward enough to make life easier. Taller people will find the lack of rear headroom an annoyance.

The amount of luggage space gives no such cause for complaint. The commodious boot is marred only by having a high rear lip which makes loading heavy items rather a chore. Beneath the floor is a 'space-saver' spare.

To match the luxury image — and the price tag — the M6 comes loaded with goodies. All the usual electrical gadgets are in evidence, including air conditioning. Thoughtful touches include a rechargeable torch in the glovebox, a first-aid kit, a fire extinguisher, spare bulbs and a comprehensive tool kit. Not so thoughtful is the omission of a radio-cassette.

So BMW's muscle car lingers on towards the time when its replacement — the 8-Series — is due. It has a few flaws but none which affects its desirability as a performance car. Some might even argue that this gives it character. Certainly the M6 is not short of charisma.

Occasionally, a model is built that you just *know* will one day become a classic. The M6 is such a car. For its prodigious performance, many owners will be quick to forgive the harsh ride and loud engine. ∎

PRACTICAL SUPERCAR

Deprived of a company car, Simon Firullo had to choose between a grand but high mileage Classic or cheap new car. He picked a 73,000-mile BMW635CSi and the gamble has paid off

Above, what shall I buy, a new Lada, secondhand Escort or well-kept BMW? No contest. Right, tidy interior, far right, sparkling engine bay

SEVEN months ago, I was faced with a decision which many people find difficult, myself included: buying a car after having had a company car. With company cars one has no worries about servicing costs, economy, insurance costs and the like. This problem arose when I changed jobs and a company car was not included in the deal.

I was looking to spend in the region of £5,000. What could I buy? How about a brand new Lada Samara? I didn't think so. Or maybe a Ford Escort, a couple of years old? Not that either. After much deliberation my choice ended in the shape of an eight-year-old BMW 635CSi. Why, you may ask? Mainly because the BMW 635CSi is big, comfortable and luxurious, oozing quality and refinement; it is a real thoroughbred, downright aggressively sexy in its sleek looks. For around £5,000 it is also relatively low priced, or at least it is when compared to the £36,000 needed to drive a new one away from the showroom.

Now that that was sorted came another difficult job. Finding one worth having for that price. I made this job even trickier as I wanted certain things on my car. It had to have a five-speed manual gearbox which is rare enough, but I also wanted mine to have the BBS Multispoke alloy wheels and not BMW's own alloys. As for the interior, I wasn't too bothered whether it was leather or velour. Good job really, because as it happens my car has beige velour trim which looks great wrapped in dark metallic blue coachwork.

After scanning the classified advertisements for a few weeks, I spotted one which sounded about right. On arrival, my first impression of the car was that it looked immaculate. While inspecting it closer, I was told that it had, in the not too distant past, had a total respray. A good one from what I could see. A common problem on the six and seven series BMWs is rust on the top edge of the front wings. This car's wings had already been sorted out but the nearside wing was showing small signs of bubbling beneath the paintwork. The whole car was very clean, including the interior which had neither cuts, cigarette burns nor worn patches and all the electrics worked, the sun-roof, windows – front and rear, mirrors and central locking. The odometer read 73,000 miles.

Other than needing four new tyres because three were bald and the offside rear was a remould – handy on a 140 mph car! – the exterior seemed fine. All that was left now was to see and hear the fuel-injected in line straight six engine which resides beneath the great expanse of metal bonnet. Raising this revealed an amazingly clean engine compartment. I checked it with anxious eyes, as the condition of the engine was the decider in whether or not to buy. Even the underside was clean and free from oil and dirt, there seemed to be no evidence of leaks of any fluid. Time to bring this big six into life to hear what the internals had to say. It started well, idled smoothly and had no uninvited noises, no chattering camshafts, which is not an uncommon fault. Running round to the rear and looking down at the centrally protruding exhaust tailpipes revealed no blue smoke or unburnt petrol. Everything pointed towards healthy mechanicals. The test drive revealed no problems either, other than slightly too much play in the gear lever which was at times a touch stiff going into its dogleg first cog.

Time to discuss finances. The advert said £5,250 and after a little haggling we agreed on £5,000. All this happened on Saturday and I became the proud owner of a blue 635CSi on the Monday.

Familiarising myself with all the switchgear was easy because, as is normal BMW practice, the instrumentation was very good, clear and placed behind a single reflection-free pane of glass, and all the major and minor controls are well arranged; nothing on the facia rattled or squeaked. The large glass area and low waistline allow for a bright interior and an excellent view over that huge bonnet which slopes off slightly to end in that wonderfully curvacious, shark-like nose and double kidney-shaped grille, which is synonymous with BMW.

"It's thoroughbred and downright sexy in its good looks. I love it"

They have made light work of obtaining the perfect driving position for one's build with the introduction of a reach-adjustable steering column and the firmly padded seats that can be adjusted for height and tilt. There's ample leg and headroom for the front occupants but despite its massive overall dimensions of 187in in length, the 635 is very much a 2+2. The rear seats are severely cramped and, even if a compromise is reached with the front passengers, those in the back will find their heads touching the roof. Pity, as the seats themselves are very comfortable and well shaped. On the other hand, it is a sports saloon.

Before indulging in any serious driving, I had to buy a new set of rubber for KGX 396V to run on. I chose Avon Turbospeeds, having been very pleased with their performance on another of my cars. I opted for a 205/60 section tyre instead of the 195/70 with which it left the factory, hoping that a larger footprint in each corner would improve the grip, especially in the rain.

The level of grip has been transformed, and the car is now highly impressive in the dry and very reassuring in the wet. This is just as well as the camber and rear suspension set up means this big BMW can easily be provoked into oversteer, especially when powering out of a tight bend. The resulting tail slides are progressive and controllable but require quick thinking at the helm. This is particularly true in the wet although it is not mandatory to travel sideways round each corner even when pressing on! It is a large and heavy car but the 3.5-litre 218bhp engine packs good low down punch resulting in an extremely quick motor car, revving freely and sweetly all the way to its 6,000rpm red line, near which the well-bred hum hardens and deepens into a sophisticated, spine-tingling howl. High speed cruising is a real pleasure, though maybe it's not one of the quietest cars around. Yet I do not find the engine noise intrusive and wind noise does not become irritating until very high cruising speeds are reached, and even then it only seems to happen around the driver's frameless window's leading edge. There is no road roar on motorways but there does tend to be a fair bit of suspension thump when driving on broken roads.

With Bosch L-Jetronic fuel injection system the engine fires up from cold at the first turn of the key, idling evenly and pulling cleanly from the word go. Its tractability is quite extraordinary, full throttle being accepted from a mere 600rpm in fifth without a tremor of protest.

KGX 396V is wonderfully satisfying to drive hard across country with handling and roadholding of a very high order. On motorways the BMW just purrs along and even in town the car behaves impeccably, never showing signs of discontent at sitting in traffic queues for long periods of time and is immediately eager and responsive when the road clears. All this performance is impressively countered by the huge 11in and 10.7in ventilated disc brakes front and rear respectively which are squeezed upon by four-pot and two-pot calipers respectively.

The ZF worm and roller steering is power assisted and well weighted giving as vivid a feel of the road as a good unassisted set up. Parking is made easy despite the fat tyres and heavy front end.

Recently, the first gear stiffness I sometimes found when I bought the car progressed into a big problem with the lever feeling like a spoon in a bowl of porridge, hitting the sides of the centre console and making it absolutely impossible to engage first without a prior visit to one of the other ratios. Even then it was possible that first seemed to have been selected until a little power was summoned to the scene at which time the lever would jump violently back to the neutral position. There was definitely something wrong. You could also say I was worried; gearbox rebuild perhaps? After a friendly chat with the BMW Car Club, of which I am a member, it became apparent that things were probably not so serious. The Getrag five-speed gearbox has a plate above it holding all the linkage in place and it was this which had come loose, causing the problem. It was soon rectified by myself and my brother-in-law. Gear changes are now smooth and swift.

Since last August, I have never been left stranded, although I have spent money on two oil changes, an air filter and a set of spark plugs after covering close to 10,000 miles.

I had to replace the front brake discs and pads and the offside front oil-filled shock absorber after it began to weep. Apart from the tyres and copious amounts of four star fuel that's been it.

Basically, it's all been money well spent on maintaining the car to its high standard, which I feel has been very moderate considering the miles I have covered.

I LOVE IT!!

BMW 635CSi

BMW's stylish big coupé combines comfort with supercar performance, at prices which are now well in the affordable bracket

History

Produced as a successor to BMW's sporting 3.0CSi coupé, the 6-series was launched at the 1976 Geneva show.

The design philosophy of the new coupés continued the gradual move away from the earlier unashamedly sporting BMW models. The 6-series was altogether more upmarket, comprising a luxurious and refined executive package.

The well-proportioned coupé bodywork — styled in-house by BMW but built by Karmann — gave the cars distinctive, purposeful looks.

Three versions of the 6-cylinder engines were available, in 2985cc, 3210cc and 3453cc capacities. Most desirable is the 3.5-litre version (sharing the same bottom-end as the twin-cam M1) which was fitted to the 635CSi in 1978. Producing 218bhp and matched to a close-ratio 5-speed 'box (three-speed auto was an option), this engine gives the model supercar performance, with a top speed touching 140mph.

Changes in 1982 included a revised 3430cc engine, 4-speed 'switchable' auto, new bumpers, spoiler and side mouldings, anti-lock brakes and a trip computer.

A breathtaking high performance version — the M635CSi — was launched in 1985. With uprated suspension and brakes, and 286bhp, this model is capable of nearly 160mph.

PARTS	
Front wing	£400
Exhaust	£230-£400
Radiator	£110-£150 (exchange)
Windscreen	£345
Tyre	£70+

Practicality

The 635CSi excels as a long-distance grand tourer. It lopes along at a mile-eating pace, but combines its relaxed *autobahn* composure with the ability to show a more sporting side to its character when road conditions permit.

Comfort levels are

Smaller-engined 633CSi is a less powerful alternative

Prices

	A	B	C
Pre-'82	£9,000	£7000	£4000
'82-'87	£18,000	£12,000	£7000

Fuel-injected 3.5-litre looks intimidatingly complex

and aerial, and service warning lights.

Regular maintenance is vital, and because of the car's complexity, a recognised BMW agent or specialist is probably the best person to do it, but will charge accordingly.

BMW's high standards of build quality mean that if these cars are properly looked after they can keep their youthful looks and vitality for years. But the drawback of owning this sort of older luxury high performance car is, of course, that if things start to go badly wrong, they can be very expensive to put right.

Buying

Enormous bills lie in wait for the unwary buyer who gets palmed off with a dodgy car. The best advice is to insist on a car with a full service history, and then get it checked over by someone who has expert knowledge of the model.

Despite BMW's stringent rust-prevention measures, early cars are starting to rot. Check the inner flitches, inner and outer wings and sills, bearing in mind that prices of replacement panels from BMW are in the Rolls-Royce league. The front bumper assembly, for instance, costs around £1000. Interior trim is similarly expensive.

Engines are basically reliable, though cylinder heads are known to crack. Traces of water emulsion on the dipstick or filler cap will give this away, as will an unusually noisy top end.

Start the car from stone cold and listen for rattles from the bottom end. Check that it idles smoothly, and produces no blue smoke on acceleration or overrun.

Best gearbox to go for is the close-ratio Getrag five-speed, but make sure it works properly as replacements are very costly. The switchable four-speed auto fitted to later cars is noted for its unreliability, and is best avoided.

When used hard these BMWs eat brake discs, so make sure that the car pulls up in a straight line, and check the discs for signs of scoring.

Useful contacts

Corry, Harrow, Middx (081 863 9944)

Linwar Motors, Southport (0704 40047)

BMW Car Club (0865 741229)

BMW Drivers Club (0362 694459)

increased by the vast number of extras fitted: as well as variable-speed power steering (which gives progressively less assistance, and so more feel, as road speed increases), there are supportive Recaro seats, headlamp wash/wipe, electric windows

Rear accommodation is reasonable for a coupé

BMW 635CSi

Year	1978-89
Engine	In-line 6-cyl, ohc, alloy-head
Capacity	3453cc
Bore/stroke	89x86mm
Max power	218bhp @ 5200rpm
Max torque	210lb ft @ 4000rpm
Transmission	five-speed manual/ auto
Suspension	Front: ind wishbones, MacPherson struts, coil springs Rear: ind semi-trailling wishbones, coil springs
Brakes	Discs all round, servo-assisted
Steering	Power-assisted worm and roller
Length	15ft 7in
Width	5ft 9in
Max speed	140mph
0-60mph	8.5sec
Fuel cons	20mpg

Hot Performer
BMW Alpina 635CSi

Waving the wand
Those magicians at Alpina have done it again.

photography by Ashley Westerman

by Ashley Westerman

COULD there be a model in the BMW line-up more suited to a wand-waving session with an aftermarket tuning specialist than the big 6-Series coupé? I doubt it.

The body shape shouts performance — low waisted with beautifully slim, elegant pillars and that aggressive, thrusting nose. Unfortunately (in standard form at least) the engine hasn't always been able to deliver a spread of power in keeping with the exterior's masculine lines. Sure, 136 kW and low 16s for the standing quarter mile are more than adequate, but what the stock engine really lacks is full low-end flexibility and enough urge to hustle 1500 kg off the mark with real urgency.

In standard trim, the 635's engine delivers a healthy 290 Nm of torque, but it's developed high up in the rev range at 4000 rpm, meaning that frequent downchanges are needed to maintain a brisk touring pace if the terrain is anything other than dead flat.

So go-fast merchants Alpina have come up with an engine and suspension package that transforms this big Barvarian into one of the most mouth-watering luxury performance coupés on the market.

Thanks to a total engine transplant, power jumps from 136 kW to 187 kW, and torque climbs to 325 Nm. The major change is Alpina's alloy cylinder head with its oversize valves and hemispherical combustion chambers, along with a more aggressive chain-driven single overhead camshaft. There are also cast Mahle pistons, balanced rods and an upgraded lubrication system that includes an oil cooler mounted neatly below the radiator. The Bosch fuel injection system is reprogrammed to meet the additional fuel demands, while the exhaust system gets an efficient set of headers and a freer breathing passage past the catalytic converter.

While this is nothing unconventional, it is extremely effective, and thankfully the engine loses none of its formidable smoothness. There is just a hint of lumpiness at idle, but it's a characteristic shared by the unmodified car and it disappears the moment you pull away from the kerb. Gone is the lethargy below 3500 rpm — the big Alpina hums sweetly in this part of the rev range and pulls with real vigour. In fact, there's no need to use more than around 2500 revs when loafing around in traffic, and the engine will pull fifth (admittedly a little lazily) from below 1500 rpm.

Sight an opening in the traffic, and it's yours. Any speed below 100 km/h means that second gear can be used for the downshift, and it's the car's cue to do what it does best. With a loud yelp from the fat rear tyres, it squats down slightly as it bolts forward, showing just what some top-end breathing work can do for its sprinting ability. Whipping it through the gears is a real joy, thanks to the wonderfully light, precise clutch action and the slick, positive Getrag gearbox.

Third gear at the 6500 rpm redline translates to 160 km/h, and the change to fourth sees the substantial shove continue to 210 km/h. Around this point the 12-year-old body shape — with its non-flush fitting glass and protruding rain gutters — starts to lose its battle with the airstream. Progress to the claimed 240 km/h top speed is noticeably less spirited; we ran out of road at an indicated 220 km/h in fifth.

The lack of aerodynamics is one of the 635's few shortcomings, and one that Alpina can obviously do very little about. Wind noise around the A-pillars and exterior mirrors starts to become noticeable between 90 and 100 km/h, and above 140 km/h I found myself pressing the power window buttons, thinking the glass must be down a centimetre or two. It's a little disappointing, because otherwise the car is beautifully relaxed. The engine noise is never more than an efficient, muted hum, while tyre and suspension noise is also heavily suppressed.

The MacPherson strut front end receives firmer Bilstein inserts from Alpine, teamed with heavier and lower coil springs. At the rear, the semi-trailing arms are also fitted with uprated coils and heavier gas Bilsteins. The anti-roll bar at the rear is enlarged to 18 mm; the front bar is left unchanged.

Michelin tyres are retained, but both front and rears are now specially imported MXX 235/45 ZR17s.

Inside, the changes are limited to a wooden gearknob and an Alpina emblem set into the steering wheel, along with recalibrated Alpina gauges and the air-vent digital display of the car's vital (and not so vital) functions. Also on board are a superbly effective cruise control, a comprehensive trip computer and great front seats fitted as original equipment — electronically adjustable and three-way programmable buffalo hide Recaros. The rear seats are a pair of well-shaped buckets split by a central fold-down armrest. After sitting in the passenger's seat for a short time listening to another tester unfurl a string of favourable adjectives, I was trying hard to remain objective. "Any criticisms?" I asked. "Yeah, one major one", he shot back.

"I don't own it . . ." □